Experimental Marine Biology

CONTRIBUTORS

ARNOLD F. BRODIE

JOSEPH H. CONNELL

MILTON FINGERMAN

WILLIAM F. HERNKIND

JAMES S. KITTREDGE

M. S. LAVERACK

RALPH S. QUATRANO

FINDLAY E. RUSSELL

STEPHEN SPOTTE

EXPERIMENTAL MARINE BIOLOGY

Edited by Richard N. Mariscal

DEPARTMENT OF BIOLOGICAL SCIENCE
FLORIDA STATE UNIVERSITY
TALLAHASSEE, FLORIDA

ACADEMIC PRESS New York and London 1974

A Subsidiary of Harcourt Brace Jovanovich, Publishers

TO MY PARENTS

ACADEMIC PRESS, INC.
111 Fifth Avenue, New York, New York 10003

United Kingdom Edition published by
ACADEMIC PRESS, INC. (LONDON) LTD.
24/28 Oval Road, London NW1

Library of Congress Cataloging in Publication Data

Mariscal, Richard N Date
Experimental marine biology.

Includes bibliographies.
1. Marine biology. 2. Marine biology–Technique. I. Title.
QH91.M26 574.92 73-9438
ISBN 0-12-472450-7

PRINTED IN THE UNITED STATES OF AMERICA

Contents

Chapter 7. **Toxicology: Venomous and Poisonous Marine Animals**

Findlay E. Russell and Arnold F. Brodie

Chapter 8. **Developmental Biology: Development in Marine Organisms**

Ralph S. Quatrano

List of Contributors

Numbers in parentheses indicate the pages on which the authors' contributions begin.

ARNOLD F. BRODIE, Department of Biochemistry, University of Southern California, Los Angeles, California (269)

JOSEPH H. CONNELL, Department of Biological Sciences, University of California, Santa Barbara, California (21)

MILTON FINGERMAN, Department of Biology, Tulane University, New Orleans, Louisiana (165)

WILLIAM F. HERRNKIND, Department of Biological Science, Florida State University, Tallahassee, Florida (55)

JAMES S. KITTREDGE, Division of Neurosciences, City of Hope National Medical Center, Duarte, California (225)*

M. S. LAVERACK, Gatty Marine Laboratory and Department of Natural History, University of St. Andrews, Fife, Scotland (99)

RALPH S. QUATRANO, Department of Botany, Oregon State University, Corvallis, Oregon (303)

FINDLAY E. RUSSELL, Laboratory of Neurological Research, School of Medicine, University of Southern California, Los Angeles, California (269)

STEPHEN SPOTTE, Aquarium Systems, Inc., Eastlake, Ohio (1)

*Present address: The Marine Biomedical Institute, 200 University Blvd., Galveston, Texas.

Preface

There has been a tremendous increase of interest in recent years in the general area of "marine biology," a term which few would care to define, yet one which possesses meaning for nearly everyone. The general public, legislators, undergraduate and graduate students, and researchers from a variety of areas have become increasingly intrigued by the great phylogenetic diversity, the variety of structural and physiological adaptations, and the potential of marine organisms to aid in the solution of many human problems.

This interest is evidenced by the many new courses in marine biology and oceanography, many of them at inland universities, which have sprung up in the last few years. In addition, many new programs have been initiated at seaside laboratories and coastal colleges and universities. Furthermore, both scientists and the general public are expressing concern for the future of the oceans as a viable and functioning ecosystem. The necessity of such concern may be easily seen when one remembers that the oceans cover nearly three-fourths of the earth's surface. Thus, any changes that tend to disrupt the stability of the oceanic environment will have far-reaching and perhaps disastrous consequences for all of us.

Many people today feel that the oceans are, or contain, a panacea capable of solving all our present and future problems. Without debating the *pro* or *con* of such an argument, it is possible to find many recent examples of startling new research discoveries with marine organisms which do indeed have direct human applications. One example which might be mentioned is the current biomedical interest in natural products research using marine organisms. Although the phrase "drugs from the sea" has been considered by some to be little more than a catchy cliche, many in the medical research community are now realizing that marine organisms may be the key to unraveling a number of perplexing clinical problems which have defied solution by the more traditional approach using warm-blooded vertebrates as experimental materials.

Because much of the current research interest involving marine organisms has an experimental basis and because of the far-flung nature of the literature in the many pertinent disciplines, it was felt that there was a need for a volume or a series of volumes which would attempt to bring together some of the recent work in these areas, both for the benefit of the interested student and the experienced researcher. The rationale behind each contribution was partly to review the pertinent literature, partly to present some of the more useful research techniques in the various disciplines, and partly to integrate the above with the author's own work. In addition, because of the general lack of reliable information concerning the setting up and maintenance of a closed-system marine aquarium, I felt it important to include a chapter dealing with this basic topic.

Hopefully such an approach will not only allow the interested reader insights into the problems and perspectives of each discipline, but will also point out new directions which future research endeavors might most profitably follow.

I would like to thank my many colleagues and friends who, either knowingly or unknowingly, have provided me with the many ideas and insights which have led to the evolution of this volume. I would especially like to acknowledge my intellectual debt to Donald Abbott and Rolf Bolin of Hopkins Marine Station of Stanford University, Cadet Hand and Ralph Smith of the University of California at Berkeley, Howard Lenhoff of the University of California at Irvine, Peter Marler of The Rockefeller University, Arthur Giese and Paul Ehrlich of Stanford University, and Joel Hedgpeth of Oregon State University.

Richard N. Mariscal

Chapter 1

Aquarium Techniques: Closed-System Marine Aquariums*

Stephen Spotte

*The term *aquaria* is considered archaic and the more accepted plural "aquariums" is used throughout this chapter.

I. Introduction

A marine aquarium is not, as many imagine, a little slice of the ocean brought indoors. It is instead an aberration, a marginal environment in which success often depends upon how well the animals can adjust to adverse conditions. When our animals manage to survive, we probably should congratulate them and not ourselves; chances are they have lived in spite of us.

Are the above statements true? Yes, In large measure. The marine aquarium is, indeed, an abnormal environment. It is isolated from the diluting and rejuvenating effects of the biosphere and certain of its biotic components differ from those in the sea by several orders of magnitude. But despite these handicaps, it is possible to consistently maintain many species of marine animals in captivity. The important thing is to understand how the environment deteriorates, then take corrective action before such deterioration has been allowed to progress too far.

Aquarium tanks with cloudy water, diseased animals, and rampant growths of algae often seem as commonplace in marine laboratories as do beakers and test tubes. Many competent scientists actively engaged in experimental marine biology still regard the workings of a successful aquarium as something akin to magic, unexplainable and certainly not reproducible. They live in fear of a promising experiment ending because they could not keep their animals alive. This is unfortunate because the aquarium is a basic tool in marine research. With a general understanding of how it works and a little practice, the risks of half-completed experiments and unreproducible results can be greatly alleviated.

Besides failing to understand how aquariums function, there are two other reasons why many people cannot maintain marine animals. First, experimental biologists are often interested in only a single aspect of an animal's biology rather than its whole needs. Second, and perhaps more surprising, is the failure of many marine biologists to identify with the basic physiologic problems of aquatic animals. Unlike land creatures, the external surfaces of a fish or aquatic invertebrate must be bathed continuously in its watery medium. Their surfaces are composed of living and often delicate tissue, in intimate contact with the environment. A fish, it has been said, can be defined as a physiologic extension of the water in which it lives.

II. Advantages of Closed Systems

There was a time when every marine biologist was forced to do his research near the sea. But recent progress in the closed-system approach have made it possible to successfully maintain marine organisms anywhere, even hundreds of miles inland. Three significant advances have made the difference. The first has been the development of stable synthetic sea salts, eliminating natural seawater as a requisite for most routine maintenance or culturing. Second, there has been a rapid improvement in culturing techniques, based on sound management practices of captive water systems. Third, there has been an increased availability of marine organisms to inland researchers. It is now possible to telephone an order to a seaside dealer and receive the animals by air freight the same day.

In closed-system aquariums the water is filtered and recycled rather than pumped in from a natural source and then discarded. Closed systems offer one distinct advantage over open systems: stable and reproducible water conditions. Many researchers near the coast are finding open systems risky because of widespread pollution and because raw natural water is subject to fluctuating temperature and may introduce predators and infectious agents into the culture systems.

Good water quality, more than any other factor, is the key to success. Seawater decomposes in captivity. Without adequate management its properties can change from life-sustaining to lethal in a few hours. In fact, maintenance of the water transcends everything else in importance, even nutrition. The proper methods for setting up small, closed-system marine aquariums, along with discussions of how each piece of equipment functions to maintain water quality, is the subject of this chapter. A detailed discussion of the biologic and chemical functions taking place in both large and small closed systems has been given elsewhere by Spotte (1970).

III. Setting Up

A checklist of equipment needed to set up a small aquarium is given in Table I. A 15-gallon tank will be used for illustrative purposes, although the procedure for setting up larger culture systems is much the same (Fig. 1).

A marine tank should be "all glass" to eliminate problems of rusting. All new tanks should be filled with tap water and left standing for 2–3 days. Put the subgravel filter plate inside the tank and soak it too. Soaking dissolves any water-soluble residues the tank might have picked up during manufacturing or shipping. It also exposes leaks and gives you time to patch them before adding seawater and animals. After the soaking period is up,

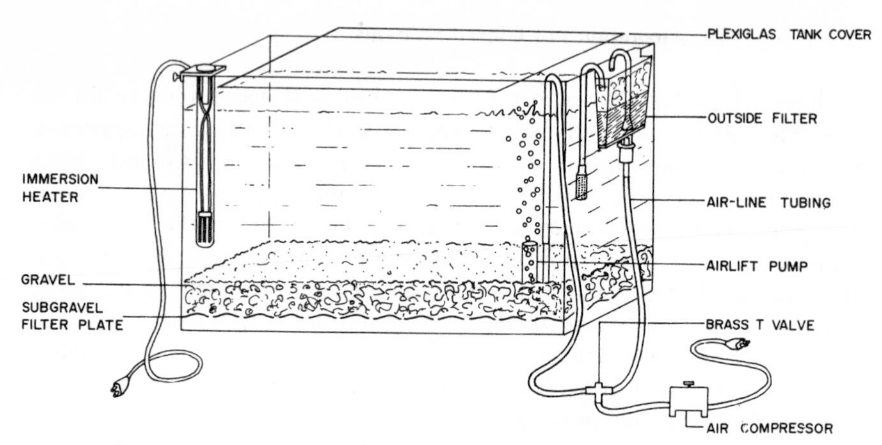

Fig. 1. A 15-gallon marine aquarium showing equipment.

TABLE I

Checklist of Equipment for a Typical 15-Gallon Marine Aquarium

15-gallon "all-glass" tank
Gravel
Subgravel filter plate
Air compressor
Brass T valve
Air-line tubing
Airlift pump
Outside filter
Filter fiber
Activated carbon
Immersion heater
Plexiglas tank cover
Seawater

flush the tank and filter plate with liberal amounts of tap water. Never use soaps or detergents to clean aquarium equipment.

Next, wash the gravel in running tap water until the water turns clear, then spread it evenly on top of the filter plate. The gravel should be at least 3 inches deep. The gravel and filter plate together are referred to as the *filter bed.*

Hook the air-line tubing to the lift pipe of the subgravel filter and connect the other end of the tubing to the T valve. From the center of the T connect another length of tubing and terminate it at the lift pipe of the outside filter. Then connect the other end of the T to the air compressor.

The outside filter should be filled half-full with granular activated carbon (charcoal). Place the polyester filter fiber on top of the carbon so that influent water will pass through it first. Once the outside filter has been installed and connected, add the seawater and hook up the immersion heater. Follow the instructions that come with the heater and do not add animals to the tank until the temperature has stabilized at the desired level for at least 24 hr.

Put the tank cover in place. Synthetic seawater will be a little cloudy at first, but clears up in a few hours. Allow the system to recycle for 48 hr before adding animals. Adjust the air flow so that a greater volume passes through the subgravel filter than through the outside filter. The rate at which water moves through a system is known as the *turnover rate*. A finished unit, as depicted in Fig. 1, can be referred to as an aquarium, a culture system, or simply a system.

Now that the system has been set up, it would be a good idea to examine its components in detail and see how each functions. A summary of these functions is given in Table II.

TABLE II

The Function of Aquarium Equipment

Equipment	Component parts	Functions	Mechanisms
Filter bed	Gravel	Biological filtration	Mineralization Nitrification Denitrification
		Mechanical filtration	Straining Electrokinesis Electrostatic attraction Van der Waals forces
	Filter plate	Support/circulation	—
Air-flow system	Air compressor Air-line tubing	Circulation	Aeration of filter bed
	Airlift pump	Surface agitation	Gas exchange Detritus formation
Outside filter	Filter fiber	Mechanical filtration	Straining
	Activated carbon	Chemical filtration	Adsorption of organics
	Siphon tube Airlift tube Plate	Circulation	—
Heater	—	Temperature regulation	—

IV. The Filter Bed

A. Biological Filtration

Biological filtration has been defined by Spotte (1970) as "... the mineralization of organic nitrogenous compounds, nitrification, and denitrification by bacteria suspended in the water and attached to the gravel in the filter bed." Biological filtration is the most important single aspect of aquarium keeping. An excellent summary has been given by Miklosz (1970).

A newly set up aquarium is no more hospitable than the surface of the moon. In addition, if new gravel and synthetic seawater are used, it may be nearly as sterile. An aquarium is not fit to sustain life until it is "conditioned."

A *conditioned system* is one in which the bacterial population is in equilibrium with the routine input from various energy sources (Spotte, 1970). The bacterial population in a conditioned system is shown in Fig. 2. Conditioning implies that as soon as energy sources are produced in the aquarium, the resident population of bacteria removes them.

The filter bed provides attachment sites for bacteria. In conditioned systems, the two major groups of bacteria are the heterotrophic and the

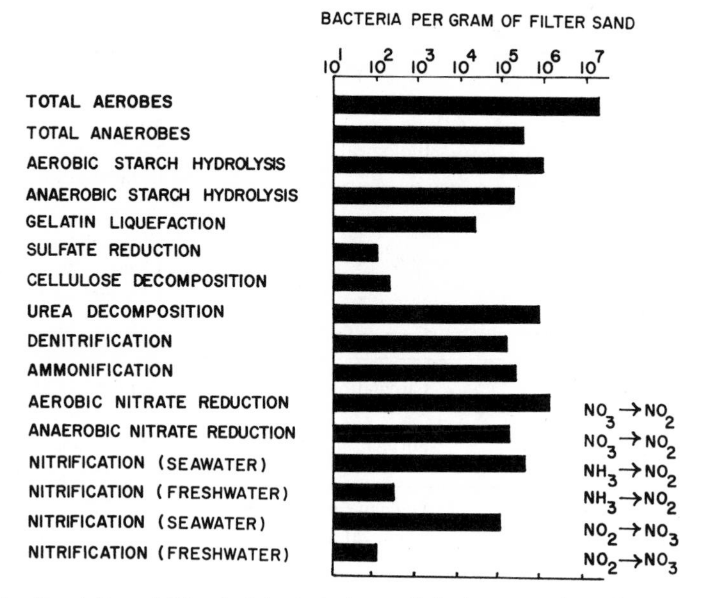

Fig. 2. Population of filter bed bacteria in small freshwater and marine systems after 134 days. (From Spotte, 1970; after Kawai *et al.*, 1964. Reprinted by permission of Wiley, New York.)

autotrophic species. Heterotrophic bacteria utilize organic compounds as energy sources and convert them to inorganic compounds such as ammonia. Autotrophic bacteria use inorganic substrates for energy. Both groups, as we shall now see, provide vital services.

The accumulation of toxic metabolites in a closed system is the major limiting factor in aquatic animal culture (Spotte, 1970). Without bacteria, these substances eventually reach lethal levels. Ammonia is the toxic metabolite of primary concern, although in conditioned systems not subjected to overcrowding it seldom exceeds 0.1 ppm, measured as total NH_4^+ (Spotte, 1970). Ammonia enters the water from two sources: (a) direct excretion by the animals, and (b) from the breakdown of organic compounds by heterotrophic bacteria. Ammonia is well known as one of the main nitrogenous wastes excreted by marine animals (Wood, 1958; Prosser and Brown, 1961; Binns and Peterson, 1969).

Organic materials gradually accumulate in old culture water. Excretory products, spent cells discarded during growth and maintenance, and uneaten food are all important sources. Part of the ammonia produced by the breakdown of organic material is used by the heterotrophic bacteria, but the greatest portion of it provides an energy source for the autotrophs. *Nitrosomonas* and *Nitrobacter* are the two principal nitrifying bacteria. The function they perform is nitrification which, in turn, is part of the aquarium nitrogen cycle, shown in Fig. 3. As seen in the figure, *Nitrosomonas* oxidizes

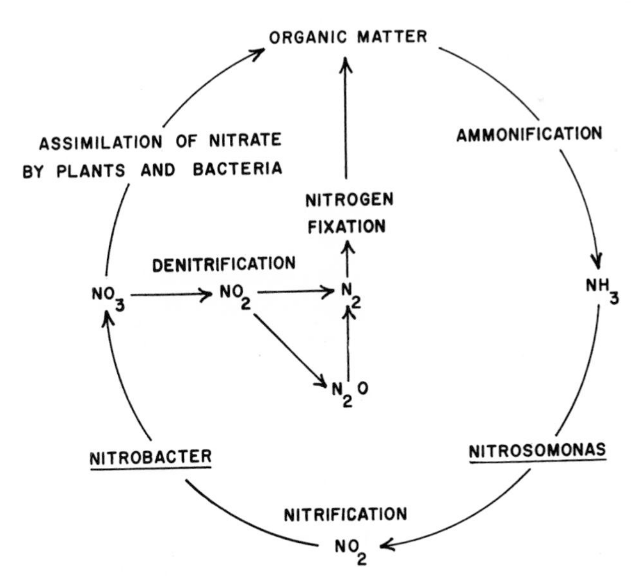

Fig. 3. The nitrogen cycle in aquariums. (From Spotte, 1970; redrawn from Stanier *et al.*, 1963. Reprinted by permission of Wiley, New York.)

ammonia to nitrite. *Nitrobacter* oxidizes nitrite to nitrate. The simplified reactions are shown below.

$$NH_4^+ + OH^- + 1.5\,O_2 \longrightarrow H^+ + NO_2^- + 2\,H_2O$$
$$NO_2^- + 0.5\,O_2 \longrightarrow NO_3^-$$

In poorly managed culture systems, nitrate accumulates. When biological filtration is functioning at optimum level, however, much of the nitrate is reduced to molecular nitrogen through the process of *denitrification*, as shown in Fig. 3 and by the equation below.

$$4\,NO_3^- + 3\,CH_4 \rightleftharpoons 2\,N_2 + 3\,CO_2 + 6\,H_2O$$

Denitrification can be defined as the reduction of nitrite or nitrate to either nitrous oxide or free nitrogen (Vaccaro, 1965).

B. MECHANICAL FILTRATION

Mechanical filtration is the physical removal of particulate matter from circulating culture water. The result is reduced turbidity. Detritus particles are trapped in the interstices between gravel grains as water passes through the filter bed. There are several mechanisms by which particles are filtered out, but the most prevalent is *straining*. Straining is the simple removal of particles by physical entrappment against the gravel grains and against other detritus particles, as illustrated in Fig. 4.

Substances are also removed through the action of *physical adsorption*, which includes electrostatic attraction, electrokinetic attraction, and van der Waals forces (Tchobanoglous, 1970).

The value of mechanical filtration is the consolidation of unwanted particulate matter in the upper layer of the gravel where it can be more easily removed than if it were in suspension.

C. BUFFERING

The average pH of the open sea is 8.3. In closed-system marine aquariums, the pH often drops below this level. This is because the oxidative processes by bacteria in the filter bed result in the production of hydrogen ions. Respiration by the animals in the tank also contributes to the general acidity of the water.

A suitable pH range for culturing marine animals is 7.5–8.3 (Spotte, 1970). Only gravels containing calcium or magnesium carbonate will sustain a pH value within this range. Such gravels are referred to as being *calcareous*.

Suitable calcareous materials are crushed oyster shells, crushed coral rock, and crushed dolomite. Dolomite is probably the best of the three so

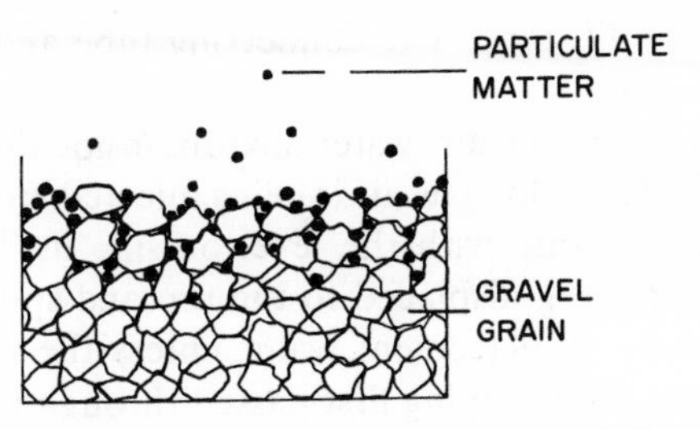

Fig. 4. Mechanical filtration through gravel showing particulate removal by staining.

far as buffering is concerned because of its high percentage of magnesium (Spotte, 1970). Several workers (e.g., Schmalz and Chave, 1963; Chave and Suess, 1967) have demonstrated that calcites with traces of magnesium were more soluble than pure calcite, or limestone. Crushed dolomite, bagged and graded specifically for marine animal culture, is marketed by Aquarium Systems, Inc.*

V. The Air-Flow System

A. MECHANISMS OF AIR FLOW

The air-flow systems of small aquariums consist of the airlift pump, air compressor, and air-line tubing. Their combined functions are to agitate the surface of the water, and to circulate it.

A typical air-flow system for a 15-gallon aquarium is shown in Fig. 5. Air from the compressor is forced, under pressure, through the air-line tubing and into a small connecting tube extending underneath the filter bed. From

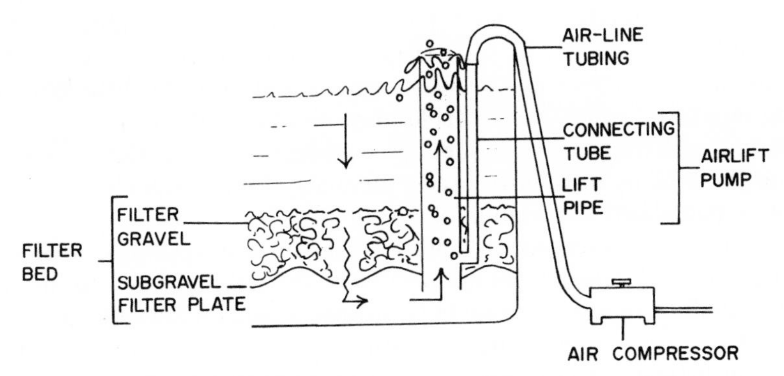

Fig. 5. The air-flow system for a 15-gallon marine aquarium.

* 33208 Lakeland Blvd., Eastlake, Ohio 44094.

there it goes into the lift pipe. The connecting tube and lift pipe together make up the *airlift pump*.

Since air rises in water, the air–water mixture inside the lift pipe is lighter than the water outside. So long as air is being injected, the water level inside the pipe cannot equilibrate with the level outside it. The result is water from inside the pipe being displaced to the top and spilled across the surface of the aquarium. Replacement water enters the lift pipe at a point underneath the plate, after having first passed through the filter bed.

B. CIRCULATION

When water comes in contact with the gills and various cell membranes of organisms in a culture system, certain things are removed from it, notably oxygen. All the animals under culture and the vast majority of bacteria utilize oxygen. Nitrifying bacteria, for example, are aerobic. Hirayama (1965) measured the BOD (Biological Oxygen Demand) of nitrifiers in filter sand from a conditioned aquarium and found it to be considerable. Water entering a column containing 48 cm of the sand had an oxygen content of 6.48 mg/liter; after passing through the sand, it measured 5.26 mg/liter.

Circulation through the filter bed is essential if the filter bed bacteria are to receive sufficient oxygen. Adequate circulation also prevents stagnant areas from forming in the bed and reduces the possibility of methane and hydrogen sulfide formation as a result of anaerobic processes (Spotte, 1970).

C. SURFACE AGITATION

The air entering an airlift pump gives up little of its oxygen to the water. The bubbles seen rising in the lift pipe are commonly thought to be oxygenating the water directly, but this is not so. They are doing it indirectly. The air bubbles serve mostly to move the culture water from the bottom of the aquarium back to the top. Gas exchange takes place at the disturbed air–water interface caused by water splashing out of the lift pipe.

Surface agitation accomplishes three things: (a) it facilitates oxygenation, (b) it helps to remove excess amounts of other volatile gases, and (c) it perpetuates the formation of detritus.

The first two functions are interrelated. Surface agitation, combined with circulation, brings a greater portion of the respired water to the air–water interface in a given period of time than would otherwise be possible. This allows oxygen to enter from the atmosphere. At the same time, surface agitation helps to remove excess free CO_2 produced from the combined respiration of the animals and the filter bed. Molecular nitrogen, resulting from denitrification, is also driven back into the atmosphere at the air–water interface.

Detritus formation is largely a function of surface agitation. The mechanisms of detritus formation in seawater have been given by several workers, notably Baylor and Sutcliffe (1963), Riley (1963), and Sutcliffe *et al.* (1963). Spotte (1970) has summarized the process of detritus formation as it occurs in culture systems. Essentially, surface-active organic compounds in solution become aggregated by air bubbles at the surface of the water. The particles adhere to each other, growing in size as they clump together. Eventually, they become visible as the loose, brown material seen on the surfaces of old filter beds.

VI. The Outside Filter

A. INTRODUCTION

The outside filter removes substances both mechanically and chemically. It consolidates these filtrates into a small, convenient space which can be serviced without disturbing the aquarium itself. Several types of outside filters are commercially available, including the "power filter," which moves water with a mechanical pump. The most practical model for small aquariums, however, is the type that is powered by an airlift pump, such as the one shown in Figs. 1 and 6.

The turnover rate through an outside filter can be regulated in airlifted models; those powered by mechanical pumps operate at only one speed. The turnover rate through power filters is often too high when tiny larvae are cultured and many of them are sucked into the filter.

A Bubble-Up Filter is shown in Fig. 6.* The filter consists of a case that holds the two filter media. Water enters the filter through a siphon tube and passes down through the filter media, then through a porous plate at the bottom. From there it is airlifted through the lift tube back into the aquarium. An airstone is used to disperse the air, making the airlift pump more efficient. A continuous siphon can be maintained from the tank into the filter because the water level in the filter is always lower.

B. MECHANICAL FILTRATION

The outside filter contains two filtering media: the *filter fiber* composed of spun, inert polyester fibers, and granular activated carbon (Fig. 6). The filter fiber filters the water mechanically by straining. Detritus particles are trapped within the fibers. When the fiber becomes soiled, it can be removed and either washed in clean tap water, or discarded. Filter fiber is available at most pet shops.

* Manufactured by Metaframe Corporation, Maywood, New Jersey 07607.

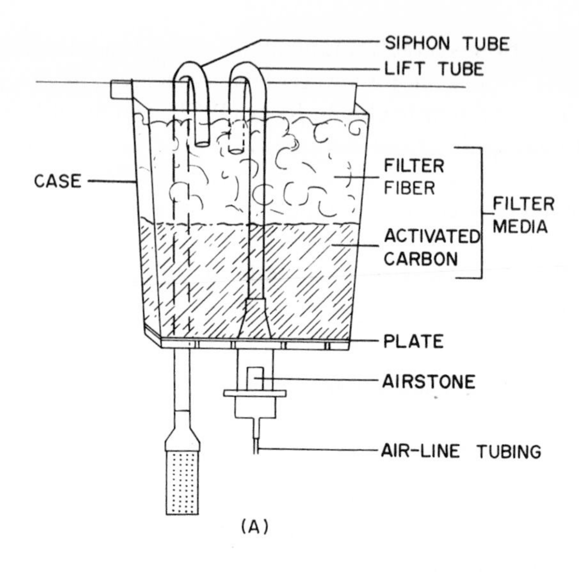

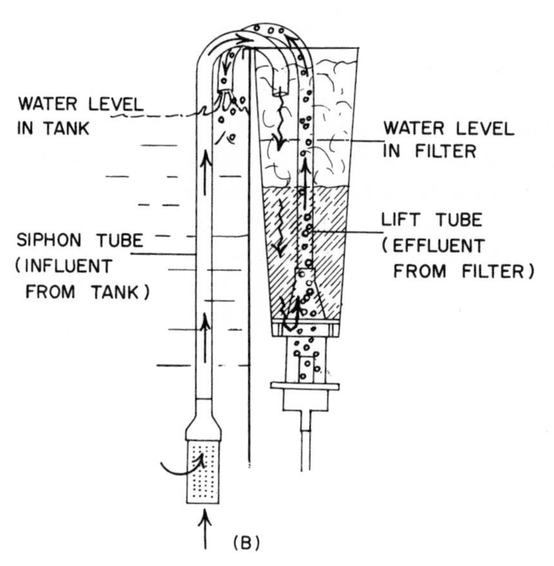

Fig. 6. Bubble-Up® outside filter for a small aquarium. (A) Component parts. (B) Side view of water flow through the filter.

C. CHEMICAL FILTRATION

The activated carbon filters the culture water chemically. *Chemical filtration* has been defined by Spotte (1970) as "... the removal of substances (primarily dissolved organics, but also nitrogen and phosphorus compounds) from solution on a molecular level by adsorption on a porous substrate, or by direct chemical fractionation or oxidation." When activated carbon is used, the removal mechanism is adsorption.

Activated carbon augments the mineralization activities of heterotrophic bacteria by removing dissolved organic compounds from solution. Parkhurst *et al.* (1967) demonstrated that activated carbon, as used in waste water processing, could remove twice its own weight in dissolved COD (Chemical Oxygen Demand). Morris and Weber (1964) investigated the mechanisms by which substances were removed using activated carbon. They found that a decrease in the pH of water resulted in a reduction of negatively charged organics being adsorbed. Also, temperature was an important determining factor and the efficiency of the carbon increased with increasing temperature. Spotte (1970), however, pointed out that temperature is unlikely to be a determining factor in culture water because it seldom fluctuates significantly.

As activated carbon becomes saturated, its efficiency diminishes. No regenerating procedure except the use of steam pressure is truly effective in restoring spent carbon to its original adsorptive capacity (Parkhurst *et al.*, 1967). Baking in an oven, the traditional method used by hobbyists, is inadequate. On a small scale, replacing the carbon is the only practical technique.

A portion of the activated carbon in the outside filter can be replaced biweekly during the partial water changes. Remove the filter fiber and take out about a third of the spent carbon. Replace it with new carbon which has been thoroughly washed in tap water, then soaked for 3 days in clean seawater. Mix in the new carbon so that it blends with the old granules to form a homogeneous mixture. Replace the filter fiber and start up the outside filter. Changing a third of the carbon by this method prevents "shocking" the system with an entirely new bed and its potent adsorptive capacity.

It is important to always store carbon in tightly sealed plastic bags or containers. Activated carbon will readily adsorb substances from the air, such as insecticides and tobacco smoke, concentrate them, and later release them into the culture system.

Dissolved organics can be removed by other methods, such as foam fractionation (airstripping) and oxidation (ozone and ultraviolet light). These methods have been discussed in detail by Spotte (1970). It has yet to be established, however, whether such techniques have a deleterious effect on the culture water. Also, any adverse effects may be magnified in small systems in which turnover rates through the equipment are too rapid. On the whole, adsorption by activated carbon appears to have no harmful effects on culture water and actually increases the carrying capacity of a system by reducing organic substrates for heterotrophic oxidation. Much more work is obviously needed in the area of chemical filtration before exact parameters can be recommended.

VII. The Immersion Heater

Temperature may be the most underrated factor in aquarium keeping. It directly affects the respiration of cold-blooded animals, and also the quantity of dissolved oxygen and most of the chemical reactions occurring in the water. All aquatic animals—without exception—are affected in one way or another by changes in temperature. Minor shifts having little effect on animals in the wild are often fatal when experienced by those same animals in captivity. Besides turnover rate, temperature is the one other factor that merits daily checks. An accurate thermometer should be kept in the aquarium at all times.

From a culturing standpoint, there are two temperatures to consider. The one is *natural ambient temperature*, which is the temperature of an animal's natural environment at any point in time. The other is *captive ambient temperature*, which is the final temperature to which a captive animal is acclimated. The main difference between the two is this: while the natural ambient temperature fluctuates at the whims of nature, the captive ambient temperature is a constant factor.

To place newly captured animals in water either warmer or colder than the natural ambient temperature subjects them to unnecessary thermal stress. Instead, put them in culture water preadjusted to the natural ambient temperature, then raise or lower it at a tolerable rate until the captive ambient temperature is reached. Saunders (1962) showed that even hardy freshwater fishes, such as carp, acclimated slowly to different temperatures. When raising the temperature, he found a rate of 1°C/24 hr to be sufficient. Tyler (1966) used a rate of 0.5°C/24 hr in acclimating freshwater minnows to lower temperatures.

VIII. The Cover

The cover prevents excessive surface evaporation. It may be constructed of glass or plastic. Some covers (generally metal) may include a built-in light source. However, steel or metal covers are subject to rusting and are generally unsatisfactory for marine systems.

IX. The Seawater

Either natural or synthetic sea water can be used for closed-system culturing. Each has certain advantages, depending on the situation.

A. NATURAL SEAWATER

Natural seawater is a living substance, frail and easily killed. Its chemical and biotic forms are never static, but forever shifting as dictated by pressures from within and from the atmosphere. One thing is certain: seawater in captivity loses some of its ability to support life. When you scoop up water from the surface of the open sea and confine it in a bottle, inperceptible, irreversible changes start taking place. The most startling, and perhaps most important from the standpoint of culturing, are those changes involving bacteria. Atz (1964) has given an excellent summary.

Zobell and Anderson (1936) studied the bacterial population in samples of stored seawater and found three things. First, the total number of bacteria increased inside the container. Second, the rate of proliferation was inversely proportional to the size of the container. And third, while the total population increased, the number of species decreased.

The samples collected by Zobell and Anderson had an initial population of 932 organisms per milliliter. After storage, a 10 ml sample had 1,050,000 organisms. Further breakdowns were as follows: 680,000/ml in a 100 ml sample; 251,000/ml in a 1000 ml sample; and 164,000/ml in a 10,000 ml sample.

When seawater is stored, some species of bacteria disappear altogether. Zobell and Anderson (1936) found that a newly collected sample of seawater yielded 25–35 species of bacteria. After storing, only 4–5 species remained.

Small volumes of natural seawater should be stored in the dark before use. Atz (1964) also recommended filtration or the use of ultraviolet light to reduce the number of bacteria. But storage alone is often enough to kill most parasitic protozoans, arthropods, or helminths. Storage for 30 days also serves to "condition" the water in the same sense that a population of filter bed bacteria condition aquarium water. Harvey (1941) found that bacteria in stored seawater liberated ammonia, phosphate, and free CO_2. Dying plankton provided substrates for heterotrophic bacteria. Autotrophic species initiated nitrification, converting ammonia to nitrite, and nitrite to nitrate. Storage obviously does *not* result in bacteria-free water. Zobell and Anderson (1936), in fact, found a bacterial count of 209,000/ml in water which had been stored for 4 years at 2°–6°C!

Always collect water from the open sea if delicate organisms are being cultured. Coastal waters are often polluted. Storage in covered glass containers for 30 days at room temperature is normally adequate to start the conditioning process.

B. SYNTHETIC SEAWATER

Unlike newly collected natural seawater, which may have too many bacteria, freshly mixed synthetic seawater seldom has enough. Most mixes rely on packaged reagent grade salts to be hydrated by the purchaser with chlorinated tap water. The result is a starkly barren medium, devoid of the tempering effects provided by bacteria. But here the disadvantage ends, because packaged synthetic sea salts are easily transported, easily stored, inexpensive, and readily available. Several brands, notably Instant Ocean® Synthetic Sea Salts, have excellent credentials in the culture of a wide selection of difficult marine organisms.*

Synthetic sea salts should be mixed in clean, inert containers. The new solution should be mildly aerated with an airstone for 48 hr prior to use to assure complete dissolution of all components. This practice also stabilizes the temperature and pH and brings the oxygen level to near saturation. Spotte (1970) has given instructions for mixing large volumes of synthetic seawater.

C. HOW TO ACCELERATE CONDITIONING

Animal losses in new culture systems can be drastically reduced simply by accelerating the conditioning process. The techniques are discussed in detail by Spotte (1970). The best method is to take a quantity of gravel from an already conditioned filter bed and spread it on the surface of the new bed. This inoculates the aquarium with the proper species of bacteria. The number of bacteria, of course, will fluctuate in the ensuing weeks until equilibrium is finally reached.

Another technique is to condition the system first with hardy animals, such as turtles or groupers, whose water quality requirements are not so stringent. They will provide continuous sources of organics and ammonia to initiate the processes of mineralization and nitrification and bring the system through its dangerous, formative stages. Afterwards, these animals can be removed and the more delicate specimens added.

X. Maintenance

A marine aquarium needs a certain amount of care. But moderation and common sense are the watchwords. Perhaps as many animals are killed by

* Manufactured by Aquarium Systems, Inc.

misdirected attempts at disinfection as by neglect. A conditioned system cannot function under sterile conditions. Sterility, therefore, is not the objective in aquarium maintenance. The following are a few important rules and practices.

A. OVERFEEDING

Every bit of uneaten food left in an aquarium contributes directly to the ammonia level, and serves as a major cause of undesirable bacterial "blooms."

B. OVERCROWDING

At its best, overcrowding causes stunted growth, loss of fecundity, and decreased resistance to disease in captive marine animals. At its worst, the animals die.

The number of animals that a system can support is known as its *carrying capacity*. The carrying capacity has been exceeded when the animal population produces more toxic metabolites than the filter bed bacteria can efficiently remove (Spotte, 1970). A formula for calculating carrying capacity has been given by Hirayama (1965).

When territorial animals are crowded, there may not be enough space to go around. Those animals without territories will be harassed from one territorial boundary to the next. If these animals are not killed, their wounds will invite invasion by bacteria and other disease organisms.

C. PARTIAL WATER CHANGES

Partial water changes, at a rate of 10% every 2 weeks, reduce the quantity of dissolved organics and keep the nitrate level at less than 20 ppm (Spotte, 1970). Partial changes also replenish ionic species which may have been lost to the detritus, utilized in growth and maintenance of the animals, or complexed with organic substances in solution.

When chemical filtration devices relying on oxidation are used, certain compounds are removed on a continuous basis. For example, ozone removes manganese from water (Bean, 1959). Continued use of ozone on a marine system soon results in a manganese deficiency.

Replacement water should be of the same temperature, specific gravity, and pH as the water in the system.

D. REMOVAL OF DETRITUS

No evidence exists that detritus is harmful. It is, nevertheless, unsightly and may dangerously increase the BOD in small, overcrowded aquariums with low turnover rates.

Detritus can be removed from small aquariums with vacuum tubes sold in most pet shops. An easier way is to siphon out some of it along with the 10% of the water that is changed biweekly. If this practice is followed regularly, detritus will not accumulate. After siphoning, turn up the air going to the outside filter to trap suspended detritus in the filter fiber.

E. LEVELING THE GRAVEL

The depth of the filter bed should be uniform all around. Some animals may dig in the gravel, eventually piling it up in one corner of the aquarium and exposing the filter plate. This allows the water to follow a path of less resistance, bypassing the thick piles of gravel containing most of the filter bacteria. The result is often insufficient oxygenation of the filter bed and the subsequent formation of anoxic regions where anaerobes proliferate.

References

Atz, J. W. (1964). Some principles and practices of water management for marine aquariums. *In* "Sea-Water Systems for Experimental Aquariums: a Collection of Papers" (J. R. Clark and R. L. Clark, eds.), U.S. Dept. of the Interior, Bur. of Sport Fish. and Wildlife. *Res. Rep.* **63**, pp. 1–192.

Baylor, E. R., and Sutcliffe, W. H. (1963). Dissolved organic matter in seawater as a source of particulate food. *Limnol. Oceanogr.* **8**, 369–371.

Bean, E. L. (1959). Ozone production and costs. *Advan. Chem. Ser.* **21**, pp. 1–465.

Binns, R., and Peterson, A. J. (1969). Nitrogen excretion by the spiny lobster *Jasus edwardsi* (Hutton): the antennal gland. *Biol. Bull.* **136**, 147–153.

Chave, K. E., and Suess, E. (1967). Suspended minerals in seawater. *Trans. N. Y. Acad. Sci.* (*Ser. II*) **29**, 991–1000.

Harvey, H. W. (1941). On changes taking place in sea water during storage. *J. Mar. Biol. Ass.* (*U. K.*) **25**, 225—233.

Hirayama, K. (1965). Studies on water control by filtration through sand bed in a marine aquarium with closed circulating system. I. Oxygen consumption during filtration as an index in evaluating the degree of purification of breeding water. *Bull. Jap. Soc. Sci. Fish.* **31**, 977–982.

Kawai, A., Yoshida, Y., and Kinata, M. (1964). Biochemical studies on the bacteria in aquarium with circulating system. I. Changes of the qualities of breeding water and bacterial population of the aquarium during fish cultivation. *Bull. Jap. Soc. Sci. Fish.* **30**, 55–62.

Miklosz, J. C. (1970). Biological filtration. *Mar. Aquarist* **1**, 22–29.

Morris, J. C., and Weber, W. J., Jr. (1964). Adsorption of biochemically resistant material from solution. I. PHS Publ. No. 999-WF-11, pp. 1–74.

Parkhurst, J. D., Dryden, F. D., McDermott, G. N., and English, J. (1967). Pomona 0.3 MGD activated carbon pilot plant. *J. Water Pollut. Contr. Fed.* **39**, R70-R81.

Prosser, C. L., and Brown, F. A. (1961). "Comparative Animal Physiology," 2nd ed. Saunders, Philadelphia, Pennsylvania.

Riley, G. A. (1963). Organic aggregates in seawater and the dynamics of their formation and utilization. *Limnol. Oceanogr.* **8**, 372–381.

Saunders, R. L. (1962). The irrigation of the gills in fishes. II. Efficiency of oxygen uptake in relation to respiratory flow activity and concentrations of oxygen and carbon dioxide. *Can. J. Zool.* **40**, 817–862.

Schmalz, R. F., and Chave, K. E. (1963). Calcium carbonate: factors affecting saturation in ocean waters off Bermuda. *Science* **139**, 1206–1207.

Spotte, S. H. (1970). "Fish and Invertebrate Culture: Water Management in Closed Systems." Wiley, New York.

Stanier, R. Y., Doudoroff, M., and Adelberg, E. A. (1963). "The Microbial World," 2nd ed. Prentice-Hall, Englewood Cliffs, New Jersey.

Sutcliffe, W. H., Baylor, E. R., and Menzel, D. W. (1963). Sea surface chemistry and Langmuir circulation. *Deep-Sea Res.* **10**, 233–243.

Tchobanoglous, G. (1970). Filtration techniques in tertiary treatment. *J. Water Pollut. Contr. Fed.* **42**, 604–623.

Tyler, A. V. (1966). Some lethal temperature relations of two minnows of the genus *Chrosomus*. *Can. J. Zool.* **44**, 349–361.

Vaccaro, R. F. (1965). Inorganic nitrogen in sea water. *In* "Chemical Oceanography" (J. P. Riley and G. Skirrow, eds.), Vol. 1. Academic Press, New York.

Wood, J. D. (1958). Nitrogen excretion in some marine teleosts. *Can. J. Biochem. Physiol.* **36**, 1237–1242.

Zobell, C. E., and Anderson, D. Q. (1936). Observations on the multiplication of bacteria in different volumes of stored sea water. *Biol. Bull.* **71**, 324–342.

Chapter 2

Ecology: Field Experiments in Marine Ecology

Joseph H. Connell

I. Introduction

In this chapter the term "ecology" is used to mean the study of organisms in relation to all aspects of their environment, with emphasis on the population level of organization. This definition includes as "ecology" most aspects of population biology except population genetics and systematics. Ecology is thus the study of natural systems of organisms interacting with their abiotic and biotic environment; this is essentially the same as Barnes' (1967) use of the term.

Barnes (1967) has given a definition of the term "experimental" as applied to ecology, which I will adopt here:

> I suggest that for our purposes we may define an experimental procedure as the activity in which a given system is observed when it is subject to a set of conditions whose values, numerical or otherwise, are selected and, ideally, controlled by a series of transformations, passing from one parameter to another, on the variables of the system under investigation. In the case of a fully natural ecosystem—existing, that is, only under natural conditions—rarely can an experiment, as so defined be carried out. But only if this is done and the interactions of the biotic components allowed full rein can the procedure be termed—if both the above definitions are accepted—experimental ecology.

This definition seems very restrictive, but is recognizes the fact that organisms do not live alone, and that organisms are constantly interacting with other organisms of their own or different species and that they are being subjected to regular and irregular changes in the abiotic environment. They have evolved in this complex environment, and to understand the reasons for their distribution and abundance we must study them in this system.

I will use the term "field experiments" for the manipulation by the investigator of environmental factors in natural ecosystems. Field experiments are thus synonymous with "experimental ecology" in the above quotation by Barnes. Ideally, all factors but the one under study are allowed to vary naturally, while the one of interest is controlled, if necessary at a series of different levels. This is, of course, the reverse of the classic laboratory experiment, where all factors but one are kept rigidly controlled. Such laboratory experiments are termed "experimental biology" by Barnes (1967). The reason why such laboratory experiments have limitations in ecological investigations is that they require every aspect of the environment of an organism to be altered by being held constant, and every biological interaction to be excluded. If organisms are taken from natural open systems which vary in materials, energy, and all other aspects of the environment, and they are placed in a closed and constant laboratory vessel, there is every reason to believe that they will not behave normally. Friedrich (1961) summed it up as follows (pp. 257–258):

> To some extent, these disadvantages can be overcome by laboratory experiment. In tests it is possible to vary one or more factors in the environment while keeping others constant either under optimal conditions or under any other controlled situation. Nevertheless, in determining ecological behaviour, the value of the results obtained in this manner is limited for the following reasons: (1) biological factors, especially those due to the presence of other members in a biocenosis, are lacking in an artificial environ-

ment; (2) it is impossible to reproduce the physical and chemical environments completely; (3) in nature, environmental factors represent an indivisible system. Alternation of any one factor, whether of biological origin or not, modifies all other factors.

It is difficult to gain evidence of just how abnormal such laboratory behavior is, but some data exist. Mullin (1969, p. 310) states that "Petipa (1966) . . . found that the respiratory rate of active *Calanus* during vertical migration might be as much as 35 times the standard respiratory rate as measured in the laboratory." The respiratory rate in the field was calculated from the difference in volume of the lipid storage organ between the beginning and end of a downward migration near dawn. The method may have been imprecise but the difference in the data was great enough to raise doubts about the wisdom of extrapolating from laboratory to field.

Laboratory experiments have obviously been very useful at the lower levels of organization where one is dealing with tissues, cells, and subcellular components. This is partly because the entities at these levels are buffered and protected from much environmental variation, so that laboratory conditions are not as foreign to them as to whole organisms.

The sorts of laboratory experiments which have proved most useful to ecologists are those in which the ideal pattern of complete control has been modified to introduce variation and to study interaction as such. Examples are the studies of populations enclosed in large outdoor tanks through which natural seawater is pumped (Trevallion *et al.*, 1970; Blake and Jeffries, 1971), observations of populations which live and breed for long periods in the laboratory (Connell, 1963; McIntyre *et al.*, 1970), and experiments in which biological interactions themselves are studied, for example, feeding behavior (Landenberger, 1968; Murdoch, 1969; Lasker, 1970).

The ideal field experiment is one in which the investigator manipulates one particular factor, while all others vary naturally. This usually means that only small systems can be studied, since it is difficult to manipulate a system on a larger scale. It is tempting to try to increase the scale by observing the effect of natural changes in some environmental factor over a large area, as for example, the effects of a cold winter or a flood. However, there are at least two reasons why such a procedure is less useful than it may seem.

First, the standard against which to compare the effect is usually taken to be the situation before the event; there is usually no "control" area where contemporaneous observations can be made in conditions which have not changed. The only study which I know of where such a control area was available was that of Sandison and Hill (1966), to be discussed later. Second, widespread environmental variations usually involve changes in several factors: floods which lower salinities also cause changes in nutrients and increase turbidity and colder winters usually differ in their patterns of wind and precipitation.

Useful ecological experiments cannot be done without a great deal of preliminary work. The ecosystem must first be accurately described, using proper methods of sampling and measuring the relevant environmental variables. Once this is done, the ecologist looks for patterns of correlation between the distribution and abundance of the different organisms or between the organisms and their abiotic environment. By asking questions about the causes of the patterns, answers suggest themselves which can be formulated as testable hypotheses. Then experiments can be designed to test these hypotheses.

Unless the original description is accurate and comprehensive, the hypotheses and experiments will be of no use. In this chapter the examples are taken mainly from intertidal habitats which, because of their accessibility, have been more adequately described than other marine ecosystems.

II. Experimental Manipulation of the Physical Environment

A. TEMPERATURE

Temperature has direct and indirect effects on marine organisms. Indirectly it affects the stability of the water column, determining the amount of time planktonic organisms are in the euphotic layers, the supply of nutrients to these layers, and many other aspects. In more direct ways it affects the rate of metabolism and, in the view of many marine biologists, may determine the limits of distribution of many species as well as their changes in abundance in different seasons.

The belief that temperature directly determines distribution and abundance is based upon correlations between environmental measurements and laboratory tolerance experiments. However, because of the existence of contrary instances and of the uncertainty of extrapolating laboratory results to the field, evidence from field experiments would be welcome. In most instances deliberate modification of sea temperatures is difficult if not impossible to do, except on a small scale such as in tidepools. But some workers have taken advantage of the heated effluent from power stations to examine the direct effects of temperature on a larger scale. Naylor (1965) has summarized much of this work. For example, a British native species of barnacle withstood increases in seawater temperature up to 7–10°C higher than normal, then disappeared, reappearing in years when the effluent was less than 7°C higher. It is not known whether the elimination was due to direct effects or because an artificially introduced subtropical species of barnacle displaced the native species in competition during the warm period. In the heated regions near the outfall the reproduction of some species was prevented, but in others it was extended over longer periods. Some changes in abundance have also been found.

In summary, the studies from heated effluents show that marine organisms are tolerant of large changes in temperature. The organisms studied by Naylor (1965) tolerated sustained increases of up to 7°C which correspond to temperature regimes of locations far south of their normal range. In addition, they show remarkable powers of acclimatization to changes (Kinne, 1963). This being the case, it is difficult to believe that the absolute boundary of most species ranges is determined directly by temperature. Transplanting organisms beyond the boundary of their range would test this point, particularly transplanting warm-water species to these heated areas. No experiments have been performed in which temperature alone was varied with controls for interactive effects such as influences of heat on dissolved gases, pH, etc., and of the presence of competitors or natural enemies.

Although these physical factors are being considered in separate sections, it is important to realize that they may interact; Curl and McLeod (1961) and Kinne (1964) give examples of this sort of interaction in marine plants and animals.

B. LIGHT

The amount of light obviously affects the rate of photosynthesis of marine plants. Gaarder and Gran (1927) and Marshall and Orr (1928) were among the first to demonstrate this for phytoplankton in the field by lowering light and dark bottles to various depths. These experiments have been repeated many times since then (Raymont, 1966). Variations in both quantity and quality of light with depth are correlated with variations in the abundance and species composition of benthic algae. Klugh and Martin (1927) suspended different species of algae at various depths and found variations in growth. Neushul and Powell (1964) repeated this, but used the more natural treatment of attaching the plants to the bottom. Shading by dense growths of both planktonic and benthic plants undoubtedly occurs. This has been demonstrated in the field by Dayton (1970) who removed large benthic algae and showed that many of the understory "shade" species quickly died in the open while other short-lived algae were able to colonize the exposed surfaces.

Reef building corals require light, growing only in shallow water. To study the effect of reduced light on corals, I erected small opaque plastic domes which cast shadows on parts of a reef crest. But since the domes also changed the water movement, I installed clear plastic domes as controls which had the same effect on water movement, but transmitted light (see Fig. 1). Since water flowed rapidly beneath the domes, the temperature and chemical characteristics were the same in both cases. The corals in the shade first turned white indicating that they had expelled their zooxanthellae and then died. Under the clear control domes the corals were not affected.

When incident light is experimentally varied, either by removing overstory plants or by providing shading structures or dark bottles, the manipulations usually result in changes in other factors, such as water movements, presence of grazers or competitors, increase of organisms growing on the new surfaces, etc. Therefore, it is essential to provide controls, such as the transparent light bottles or the plastic domes.

C. THE CHEMICAL ENVIRONMENT

1. *Salinity*

Besides its indirect effect on the density of seawater, it is possible that salinity may directly affect the distribution and abundance of inshore organ-

Fig. 1. Shading of corals on the reef crest at Heron Island, Great Barrier Reef; view at low tide. The plastic domes, about 75 cm in diameter, were supported on blocks so that the seawater could circulate over the corals. A small hole at the top allowed air to escape during rising tides. The two opaque domes of black plastic were painted white to prevent their being heated at low tide; they excluded over 90% of the incident light. The clear control domes were colonized by algae; in this photograph they had not been cleaned for 2 months and the growth excluded about 15% of the incident light.

isms. In regions with pronounced wet and dry seasons, the salinity of the near-shore water may drop precipitously during the first floods. In such a situation in Nigeria, Sandison and Hill (1966) found that four species of marine invertebrates died during each season of low salinity, whereas in a nearby bay where the salinity did not drop as much, they survived. During the next dry season they repopulated the area which had been flooded.

This is one of the few such studies in which an adjacent area where salinity did not change greatly, was available to serve as a control. However, since such major changes in salinity are usually brought about by runoff from the land there are accompanying changes in turbidity and nutrient concentrations which make it difficult to know just which factors produced the observed changes. To evaluate the effect of salinity alone would require a field experiment, such as lowering the salinity during the dry season by adding fresh water to a small enclosed area of sea shore.

The only such field experiment which I have found is that of Gersbacher and Dennison (1930). By removing the seawater and adding fresh water to tide pools at low tide, they found that certain invertebrate animals reduced their activity while others increased it. However, it is not clear whether other factors, such as temperature, were kept the same, since no pools were bailed out and then refilled with seawater again as controls.

As with temperature, it is difficult to do field experiments on a large scale in which salinity alone is varied. Perhaps the advent of large desalination plants will provide this opportunity as did power plants with their heated effluent.

2. *Dissolved Gases*

Certain deeper layers of the sea as well as certain basins have very low oxygen concentrations, but little is known of the ecology of the organisms living there. Certainly no field experiments have been done. However there have been studies of the biota of mudflats without oxygen and of rockpools where oxygen varies considerably. Some species obviously can tolerate large changes in concentrations of dissolved gases and the accompanying changes in pH (Newell, 1970). However the degree to which their distribution and abundance is determined by these factors remains to be demonstrated by appropriate experiments. Intertidal rockpools are ideal places in which to perform the pertinent field experiments, but I am not aware of any such experiments in the literature.

3. *Nutrients*

The effects of variations in the supply of mineral nutrients essential to plant growth, particularly phosphates and nitrates, have been studied for many years in the sea and in laboratory cultures. Rates of phytoplankton production are closely linked to rates of supply of these nutrients, and productivity is high in regions of the sea where the nutrients are renewed by upward water movements. Conversely, where stratification results from surface heating or low salinities, the rate of nutrient renewal is low, as is productivity.

Some field experiments have been done which demonstrate the direct effect of these nutrients. Gross *et al.* (1947, 1950) have fertilized small natural bodies of seawater (sea lochs in the west of Scotland) with phosphates and nitrates and observed the changes in the plankton, benthos, and fish. Nutrients initiated phytoplankton blooms and grazing by zooplankton probably caused them to diminish. The planktonic productivity was about doubled by the addition of fertilizers, compared to the productivity of a nearby sea loch which served as a control (Gauld, 1950).

Besides these major nutrients, other substances are required, both inorganic and organic, for phytoplankton growth, as demonstrated in laboratory culture experiments. Some of the organic matter may supply energy to the plants; this heterotrophic nutrition may be important when illumination is low in winter or when the cells are carried below the upper lighted waters. No field enrichment experiments have as yet been carried out. These could be done in an apparatus such as the large clear plastic sphere suspended in the sea which McAllister *et al.* (1961) have developed.

Finally, the introduction of large amounts of chemical substances such as pollutants into the sea provides an inadvertent source of experimental material. The effects of these substances on marine organisms must be determined. Present government regulations requiring ecological surveys provide opportunities to ascertain their effects, if proper experimental procedures are followed.

D. WATER MOVEMENTS

1. *Currents*

Horizontal and vertical water movements are clearly among the most important environmental variables. They determine much of the dispersal of the plankton, the supply of nutrients to the surface waters, the grain size of sediments, etc. The local and geographic distribution of larvae of benthic and nektonic species must largely be determined by currents.

Hatton (1938) transplanted rocks bearing the alga *Fucus vesiculosus* to regions where the currents were much faster than in their normal habitat. They grew much more slowly and were gradually detached and washed away.

Other fields experiments have been done on the effects of current speed on settlement of fouling organisms. Smith (1946) set up experiments in which discs were rotated in the sea. The distribution of newly settled barnacles at varying distances from the axis showed the maximum speed of current allowing attachment. In other experiments seawater was pumped through glass tubes of different diameters. Other field experiments on currents have been done in a stream which drains a nearly enclosed bay, Lough Ine in Ireland. The tidal drainage creates currents of up to 3 m/sec. One snail, *Gibbula cineraria*, which lives on the fronds of large algae, does not occur in the regions of rapid current. When transferred on fronds to both quiet water and into the fast current, they maintained their hold only in quiet water (Kitching and Ebling, 1967). These and other observations indicate that the direct action of rapid currents does influence the distribution of some benthic species.

2. *Wave Action*

It is well known that the distribution and growth form of plants and animals from rocky shores varies with the degree of wave action (see review by Lewis, 1964). Objective methods of measuring wave action have only recently been developed (Jones and Demetropoulos, 1968; Harger, 1970a; Dayton, 1971). Field experiments on the effect of variation in wave action have consisted in transplanting organisms. Hatton (1938) transplanted the alga *Pelvetia canaliculata* to a wave-beaten point where it did not normally occur. When placed in containers which protected them from wave shock they survived and grew well. Kitching *et al*. (1966) transferred *Thais lapillus* from a sheltered to a wave-beaten area and also moved them within the wave-beaten site as a control. More of the *Thais* from the wave-beaten site maintained their position. Reciprocal transfers to sheltered areas showed that *Thais* from the wave-beaten areas could survive the quiet conditions, provided that they were protected in cages from predation by crabs. Transplanting experiments have also been done by Harger (1970b) with *Mytilus*. Large *Mytilus edulis* are seldom found on wave-beaten shores in California; *Mytilus californianus*, which does occur there, has much stronger byssal threads and a much greater force is required to detach it. *Mytilus edulis* were transferred to wave-beaten localities and allowed to become attached in cages. When the cages were removed, the larger individuals were rapidly torn from the surface; the smaller ones survived longer, presumably because they fitted closer to the rock surface.

Growth is also affected by wave action. Harger (1970a) enclosed mussels in 2 sorts of containers, one with large openings to admit the waves, the others with smaller openings. Both sorts of cages were hung from a pier, either intertidally where they were exposed to wave action or subtidally with little wave action, as a control. *Mytilus edulis*, which normally does not live in wave-beaten areas, grew slower in the less protected cage.

Mytilus californianus, which lives in wave-beaten places, showed no difference in growth.

E. DESICCATION

The effect of drying in air has often been used to explain the upper limits of intertidal species (Lewis, 1964). To separate the effects of desiccation at high shore levels from those caused by higher air temperatures or greater insolation or more fresh water, field experiments are required. One of the first field experiments of any sort to be performed in the area of marine ecology dealt with this problem. At the very top of the intertidal zone in Brittany, Hatton (1938) fixed a basin of seawater with a tiny hole so that water dripped slowly out and ran down the rock surface. Young barnacles (*Balanus balanoides*) transplanted to this high level on rocks survived and grew for 3 months where the rock was wetted, but not beyond it. Frank (1965) did a similar experiment in Oregon. Water was trickled down a steep rock from a large bottle placed above the intertidal zone. More algae grew there, but since there was much wave splash and the weather was not dry for any long period, the added moisture was evidently not enough to cause upward migration of the limpets which he was studying. Castenholz (1963) noticed that diatoms extended higher on the shore where water leaked out of a tide pool and ran down the rock surface. Since these species tolerated high temperatures in pools and were exposed to strong insolation in the trickle of water from the pool, he reasoned that desiccation set their upper limit. Dayton (1970) built a tide pool at high shore level and found algae living higher than usual in the stream leaking out of it.

Hatton (1938) transplanted barnacles on pieces of rock to levels above their normal limit of distribution. Newly settled *Balanus* died quickly but older ones survived much longer, indicating that the factors determining the upper limit probably operated soon after attachment of larvae. When Foster (1971) transplanted barnacles to high shore levels he also found that the larger barnacles survived better. He interpreted this to mean that desiccation caused the mortality since smaller barnacles have proportionately more surface area in relation to their volume than do larger ones. High temperatures would probably have killed both large and small barnacles.

Animals near their upper intertidal limit of distribution may be directly affected by other aspects of the harsh physical conditions there. Hatton found that barnacles on south-facing surfaces had lower densities of settlement and lower growth rates than those on other orientations. However, when he provided shade the settlement was the same in all directions, even though the rock dried out quickly in the shade. This seems to indicate that direct sunlight and/or higher temperatures, either directly or through increased rates of desiccation, cause early mortality of barnacles at high shore levels.

The larger algae may produce a region of higher humidity underneath them at low tide. When Hatton (1938) removed the larger individuals of *Fucus vesiculosus*, newly attached ones underneath disappeared; they survived under any large plants left in place. Whether the young plants survived better under the canopy of large algae because they were protected from desiccation or from greater light intensity cannot be decided. The effect on other species may be similar; Dayton (1970) also removed the larger algae and found that other species, which live only beneath this canopy, decreased.

The effects of harsh weather may be felt only intermittently. During 2½ years of measuring survivorship of barnacles in Scotland, I found that at the upper boundary of distribution, the highest rates of mortality of barnacles over 6 months of age occurred during one period of a few days when neap tides, calm weather, and sunny weather coincided (Connell, 1961a). During 2½ days, the populations were wet only 6 times by waves from steamers passing by. Throughout the remainder of the study the mortality rate of these older barnacles was very much lower.

Organisms on the upper flats of coral reefs are sometimes exposed in the air during extremely low tides. Mass mortalities sometimes occur in either very hot or cold weather. To test the effect of direct exposure to air and sun, Mayer (1918) placed corals in the air for varying periods; if their bases were in seawater they survived better. Glynn (1968) exposed echinoids in the air to direct sun, others in the shade of rocks or covered with debris or sediment. They survived longer if protected from direct sun.

F. TURBIDITY AND SILTATION

Marine organisms must continually clean themselves of particles which settle out on them from the water mass. Although they are usually quite efficient at doing this, occasional heavy floods may bring loads of silt from the land which may overwhelm the cleansing abilities of some species. Corals seem to be particularly sensitive to excessive siltation, and Mayer (1918) did some field experiments to test the relative ability of different species to withstand siltation. He buried corals of several species under 5 cm of limestone mud on the reef flat for different lengths of time. Species which lived in the quieter water of the inner reef flat, where siltation should be heavier, withstood burial for longer.

Water movements not only affect organisms directly, but also determine the rate of sedimentation. Round *et al.* (1961) found that when rocks bearing a species of hydroid which lived in the rapid current at Lough Ine were transferred to quiet water, the hydroids became choked with sediment and died. A species of anemone similarly transferred on rocks survived in the quiet conditions if placed facing downward, but died if facing upward, with or without being shaded.

G. SUBSTRATE

The effects of the physical, chemical and biological characteristics of the substrate on marine organisms has been studied with field experiments for many years, in contrast to most of the variables already discussed. Obviously it is easier to manipulate the substrate under field conditions than to change the characteristics of water masses.

1. *Hard Substrates*

Antifouling research has stimulated the production of a body of useful data based upon both field and laboratory experiments. Panels treated in various ways have been exposed intertidally on the shore or subtidally on piers or below rafts. The distribution and abundance of settling stages of marine plants and animals has been measured on treated vs control panels exposed close together.

The physical structure of the surface is easiest to modify in controlled ways and this was the first aspect studied. Surface contour and roughness, orientation to light, gravity and currents, etc., have been studied during antifouling research (see Woods Hole Oceanographic Inst., 1952; Pomerat and Reiner, 1942; Pomerat and Weiss, 1946; Barnes *et al.*, 1951; Crisp and Barnes, 1954). A great deal of progress has also been made in manipulating the chemical nature of surfaces (see review by Crisp, 1965).

Chemical substances produced by previously settled individuals affect settling larvae of the same species. On the intertidal shore, Knight-Jones (1953) exposed slates from which barnacles has been removed, leaving only the bases. Control slates which had never borne barnacles were exposed with them. After 3 days, the slates with bases had significantly more newly settled barnacles than the controls. Although this phenomenon of gregariousness has been well studied in the laboratory (Crisp, 1965), this and another similar experiment by Knight-Jones and Stevenson (1950) remain (to my knowledge) the only field experiments designed with proper controls specifically to test it.

2. *Soft Substrates*

Many laboratory experiments have shown that larvae of marine animals living in soft substrates can distinguish between sediments having different particle sizes or different organic material on the surface of the grains. However the only field experiment which I know of is that of Boaden (1962). On the beach he buried open tubes of sand which had been depopulated and sorted into classes of different ranges of particle sizes. An unsorted but depopulated sample and a freshly dug sample were also buried, as controls. After 20 days the interstitial organisms were counted. The unmodified sample was similar to another taken at the start, indicating that enclosure in a tube did not affect the existing population. The unsorted treated sample also had regained an assemblage of organisms very similar to the untreated one, indicating that the organisms were not discouraged from entering the tube. The colonists in the graded series differed from these, having fewer species and individuals. One species of the archiannelid *Protodrilus* colonized only the large grain sizes while another colonized only the smaller grain size. This difference could have been due either to different preferences or to competitive exclusion between the species.

H. SHELTER

Another important aspect of the physical environment is the degree of shelter provided; some field experiments have demonstrated this fact.

Fig. 2. A "synthetic holdfast" after immersion for 7 months at about 14 m depth off La Jolla, California. The upper part is a plastic bucket with holes punched in it, filled with a jumble of plastic-covered rope to simulate a holdfast of the giant kelp, *Macrocystis pyrifera*. The base is a concrete block (Ghelardi, 1960).

Rocks, discarded trams and automobiles, etc., have been placed on sandy bottoms with the result that the local marine plants and animals of rocky reefs have colonized these experimental reefs (Carlisle *et al.*, 1964). Randall (1965) built reefs of concrete blocks which were soon colonized by fish from nearby coral reefs. E. W. Fager (1971) has built small artificial "rocks" on open sandy bottoms, demonstrating the variability in colonization between nearby sites.

In the intertidal zone, Frank (1965) attached shelters of rubber inner tube. Kensler and Crisp (1965), studying the fauna of crevices, followed the colonization of artificial crevices formed of two slates placed at a slight angle.

Ghelardi (1960, 1971) has studied the animal community living in holdfasts of kelp, *Macrocystis pyrifera*. To see whether the organisms lived there because of the shelter offered, or because of some characteristic of the living algae, he prepared artificial holdfasts. These were plastic buckets with holes punched in them, filled with a jumble of plastic-coated rope (Fig. 2). Many of the same species as in natural holdfasts colonized these artificial ones, indicating that shelter was important in determining the composition of the community.

III. Experimental Manipulation of the Biological Environment

A. WITHIN SINGLE-SPECIES POPULATIONS

1. *Population Density*

The influence of population density on recruitment, growth and mortality of a species has been studied in a few field experiments. These have all been done in the intertidal region or very shallow water where population densities could easily be manipulated.

Some of the first attempts to manipulate density consisted of scraping all organisms off rock surfaces and recording recolonization (Wilson, 1925; Hatton, 1938). Unfortunately, there were no controls in which colonization could be followed on undisturbed areas nearby, so the effect of population density on the rate of colonization could not be estimated. In more recent experiments using controls, high population densities have been shown to prevent or reduce recruitment. In natural populations of barnacles, I found that settlement stopped after high densities (up to 80/cm) had been reached, whereas on adjacent areas where barnacles were removed daily, settlement continued until the supply of larvae in the plankton ceased Connell (1961a). Older individuals also prevent or reduce larval settlement. Stimson (1968) changed the population density of larger limpets on pier pilings and found greater rates of settlement with lower numbers of adults.

Growth may be faster at lower population densities; this has been demonstrated in a few field experiments. For example, Frank (1965) built a series of vertical fences of wire mesh separating a steep intertidal rock slope into 6 strips, each about 65 cm wide. He then increased the population density of limpets by a factor of about 2 and 4 in two strips, keeping the other four strips as controls. After a year there was a slight but statistically significant lowering of growth rate in the populations at higher densities.

Behrens (1971) placed groups of *Littorina* on the shore in plastic mesh cages varying the densities. A concrete slab in each cage provided a standard surface for grazing. During the summer when the snails were most active, there were no algae visible on the slabs in the high density cages. Growth was slower at higher densities than at lower densities; in the latter case algae were visible on the slabs, which may have indicated that food was superabundant. In the winter, activity and growth were reduced, algae grew thickly on the slabs and yet the animals still grew more rapidly at lower densities. Evidently growth is limited both by food supply and by some form of interference between the snails.

Mortality probably is increased in many instances at higher population densities. When barnacles are growing rapidly in the first 6 months after attachment, the mortality is greater at higher population densities (Connell, 1961a). At high density when many of the barnacles are in contact with each other, growth results in a large number of them becoming crushed or buried; this accounted for about 65% of the mortality in one such population (Connell, 1961a). Yet Behrens (1971) has found that whereas *Littorina sitkana* had a higher mortality rates at higher densities, the reverse was the case for *L. scutulata*.

In Frank's (1965) experiments with limpets, he could detect no differences in mortality between the populations at normal vs increased density. However the limpets did adjust their population densities by movements. When the counts within each strip were tabulated into different horizontal zones, Frank found that at normal densities 81% were in the midshore zones 2 and 3, there being 6 zones between high and low tide level. But in a strip where the density had been increased 2½ times, the limpets had moved upward and downward so that only 53% were in zones 2 and 3.

Another field experiment illustrating this sort of adjustment of density by movement was done by Holme (1950) with the bivalve *Tellina tenuis*. He dug *Tellina* from an area of sandy shore and found that they were more evenly spaced than random. He then buried about 100 in a small patch within the edge of a circle 22 inches in diameter bounded by a steel strip extending 6 inches into the sand. After 8 weeks he found that they had moved out from the dense patch until they were again spaced out rather uniformly;

the density of this artificial population was about twice that of the natural population.

Species which occupy individual territories and defend them may limit their population density by this means. In a natural population of a pomacentrid fish, *Hypsypops rubicunda*, Clark (1970) found that the breeding males defended a territory all year round. When he removed some of them, their territories were immediately occupied by younger males who, up until then, had had no territory. Thus the population density of breeding males within suitable habitats was determined by this territorial behavior. Stephens *et al.* (1970) studied small fish (*Hypsoblennius*), one species of which occupied crevices in pier pilings. The population density was evidently determined by the number of these refuges, because when 42 fish were added to a piling supporting only 17, the population returned to 17 in 50 days. A much greater proportion of the introduced individuals were missing than the residents. When fish were put on a piling from which all residents had been removed, many more became residents than when a similar group was put on a piling having a normal resident population, similar to the above example.

2. *Food*

Although the distribution and abundance of animals is probably often correlated with their food, few field experiments have been done to test this association. One such was performed accidently when curious children removed a cage which I had attached a year and a half earlier to protect barnacles from their predators. Inside the cage the barnacles were much older and larger than those in the normal populations, and I had found that the predatory snails preferred to eat the larger barnacles. When the cage was removed, the predators were offered an increased supply of preferred food in an unusual place, and when I discovered it, I found the area almost completely covered with feeding predators, at a considerably greater density that usual (Connell, 1961a). A deliberate experiment in manipulating food was done by Landenberger (1967). Starfish (*Pisaster giganteus*) occurred in great numbers under a pier where they fed upon mussels which had fallen from the intertidal portion of the pilings. They were almost never seen on the open sand beside the pier. To see whether they stayed under the pier because of the presence of food, he placed piles of live mussels out in the open beside the pier. After some weeks large numbers of starfish were feeding in the open, indicating that food in great part determined their local distribution.

Mobile predators or scavengers quickly discover a new source of food. This was demonstrated by Isaacs (1969) who lowered a bucket of dead fish to the sea bottom at 1400 m depth. Photographs taken by a camera mounted above the bucket showed that large fish arrived to feed very quickly after the arrival of the food.

Growth rates may be limited by food supply. Ebert (1968) tested this idea by placing sea urchins in 2 tide pools, one of which had a greater supply of algae because the urchins had been removed from it 3 years earlier. The growth rate over the next year was significantly greater in the pool with more food.

Experiments in large outdoor tanks having a high rate of exchange of natural seawater have been used to estimate the relationship between food supply and growth rate. Edwards *et al.* (1970) set up experiments with different ratios of fish and their bivalve prey. The growth of fish was faster with higher ratios of prey to predator, independent of the population density of either. The bivalve also grew faster in tanks in which nutrients were added which increased the growth rate of the planktonic food of the bivalve (Trevallion and Ansell, 1967).

The relationship between food and growth is complicated by the fact that energy may be allocated to producing offspring instead of new tissue. An ingenious experiment demonstrating this has been done by Crisp and Patel (1961). Some species of barnacles cannot develop larvae without cross-fertilization, so that if they are far enough apart, they cannot reproduce. Crisp and Patel arranged barnacles on panels so that some were close enough to cross-fertilize and some were not. To control for the possibility of greater competition for food between the close pair, some pairs were made up of 2 different species which would not cross-fertilize. Nonbreeding individuals grew faster than breeding ones.

3. *Movements*

A few field experiments have indicated the degree of mobility of natural populations. For example, Stephens *et al.* (1970) removed populations of a territorial species of fish, *Hypsoblennius*, from pier pilings to measure the rate of repopulation. Other undisturbed pilings served as controls. Although there were pilings nearby with populations on them, the recolonization of the depopulated pilings was very slow, indicating that this territorial species moves very little.

Several field experiments have been done by displacing marine animals (fish, limpets, etc.) to study homing behavior, but these are beyond the scope of this chapter.

4. *Variations Between Populations*

In different locations, populations may vary in their morphologic or life history characteristics. Whether these variations are the results of genetic or local environmental differences can sometimes be determined by transplanting groups from one locality to the other. Crisp (1964, 1968) has transplanted the barnacle *Balanus balanoides* in both directions across the North Atlantic and compared the life cycle of the transplants with the residents. The original times of fertilization, development, and liberation of larvae, the size of eggs, and the susceptibility to parasitism are retained by the transplanted individuals, indicating genetic differences.

The mussel, *Mytilus edulis*, is very widespread, occurring in both the Atlantic and Pacific, and showing great morphologic variations. Seed (1968) found that individuals transplanted to localities several kilometers apart in Britain grew to resemble the residents of the new locality. Evidently the genotype in this species is capable of wide phenotypic expression in Europe. However, when individuals from the Atlantic east coast of North America were transplanted to the Pacific west coast (Harger, 1971) they retained their original morphologic characteristics during a month's growth in cage suspended from a pier. Small east-coast mussels grew at the same rate as west-coast ones, but remained narrower; larger east-coast mussels grew slower than west-coast ones.

These experiments show that populations of barnacles and mussels separated by great distances have evolved genetic differences, whereas those closer to each other have not, presumably because of the exchange of genetic material by planktonic larvae. In species with nonplanktonic larvae, such as the snail *Thais lapillus*, differences in chromosome numbers have been detected between populations from different sorts of habitats even when they are very close together (Staiger, 1957). However, no transplanting experiments have been done to determine the relative importance of these chromosomal variations as compared to local environmental conditions in determining phenotypic variation; such experiments should produce interesting results.

Berry (1961) has transplanted groups of snails (*Littorina saxatilis*) from high to low shore levels and vice versa, with controls transplanted to the same level nearby. This species does not have a planktonic larval stage, so one might assume that it would have evolved genetic differences similar to those of *Thais lapillus*. However, the transplanted populations assumed the characteristics of the local residents (in size of ovary and number of young in the brood pouch). Thus these traits are more influenced by environmental conditions than by genetic differences, if any. Genetic investigation and further field experiments would seem to be called for.

B. BETWEEN TWO OR MORE SPECIES

1. *Interactions at the Same Trophic Level: Competition*

If two species require the same resource, and if the populations of one or both increase to the point where the resource is in short supply, competition may be expected to occur. Such competition can be demonstrated by removing or reducing one species; if competition had been occurring the other species should show faster rates of recruitment or growth, or lower mortality rates, or should invade the territory from which the other was removed. For example, I removed all individuals of the barnacle *Balanus balanoides* which were touching those of a second species *Chthamalus stellatus*, leaving adjacent populations undisturbed as controls. In the controls the faster-growing species (*Balanus*) eliminated *Chthamalus* by crowding, but where the faster-growing species was removed, *Chthamalus* survived well. Competition for space was occurring, and in great part determined the distribution of these two species (Connell, 1961b). Another field experiment was performed by Stimson (1970), in which one species of large intertidal limpet, *Lottia gigantea*, held territories from which it excluded, by shoving, other species of grazing gastropods, as well as competitors for space such as mussels and anemones. When a *Lottia* was removed from its territory, the other species quickly invaded. This did not happen on adjacent control territories where *Lottia* remained. If the *Lottia* was transplanted to a new place, it soon shoved these competing species off an area the size of its territory.

Competition for space between anemones and barnacles has been demonstrated in controlled field experiments by Dayton (1971). When he removed large barnacles from an area, adjacent anemones spread into the cleared space. This happened in moderately quiet waters, but not where wave action was severe. Attached algae also compete for space in the light; as described earlier, when Dayton removed the large "canopy-forming" species, understory species decreased abruptly, while other short-lived species colonized the area. In all these experiments, no such changes occurred on adjacent control areas where the original species were left undisturbed.

Competition between two predatory starfish has been demonstrated by Menge (1972). He reduced the numbers of a large species, *Pisaster ochraceus*, on one small isolated reef and increased them on another, keeping a third reef undisturbed as a control. The growth rate of a smaller species, *Leptasterias hexactis*, was fastest on the reef with few *Pisaster* and slowest with increased numbers of *Pisaster*, indicating that competition was occurring.

Intertidal mussels (*Mytilus*) sometimes grow in thick beds, and Harger (1968) has found that the individuals living on the inside of the bed, beneath others, showed signs of stunted growth and of being partially crushed by their neighbors. He arranged various mixtures of two species in cages suspended from a pier and discovered that, regardless of the original arrangement, *Mytilus edulis* always ended up on the outside of the clump, with *M.californianus* inside. *M.edulis* gained this favored position by actively crawling toward the surface when it was inside the clump, whereas *M.californianus* did not do so. In quiet water where sediment accumulates, the mussels inside the clumps may be smothered; *M.californianus* may thus be eliminated in quiet bays.

Another effect of the competitive interaction between these two species of *Mytilus* is associated with differences in the strength of attachment to the rock. As described earlier, *M.edulis* is much more weakly attached than *M.californianus*. Harger (1970b) set up clumps of single species and of mixed species, using either large or small mussels. The clumps were first enclosed in cages on the rocky shore or in cheesecloth bags under a pier until they had attached themselves to the support. After the enclosures were removed the mortality of both species was greater in direct proportion to the proportion of *M.edulis* in the mixture. Thus the effect of wave action in tearing mussels off the shore varies with the proportion of the two species present.

In addition to the single-species populations of *Littorina* described earlier, Behrens (1971) also set up populations of two species in cages on the shore. Growth, egg production, and survival were affected in certain seasons by the presence of another species.

Sometimes one species will interfere with another, even though they may not be competing for a scarce resource. For example, the starfish *Pisaster giganteus* climbs up pier pilings to feed on mussels in the intertidal region. Landenberger (1967) found that there were fewer starfish on the mussel clumps if the lower part of the piling had large populations of the anemone, *Corynactis*. In laboratory aquaria, the starfish would not cross a barrier of mussels covered with the anemones. When he removed the anemones from two pilings, many starfish had climbed up to feed on the mussels within a very short time. There was no change on control pilings on which the anemones were left. Thus the anemone was interfering with the access of the starfish to its source of food.

Many other observations have been published from which one may infer that competition or interference is happening in natural circumstances. However, I know of no other publications in which field experiments, with controls, have demonstrated unequivocally that competition was occurring.

2. *Interactions between Trophic Levels*

In this category I include all attacks by natural enemies, such as herbivores grazing on plants as well as predator–prey, parasite–host, and pathogen–host interactions. All field experiments so far have been done with benthic species in shallow water. Grazing or predation of plankton is clearly important, but the technical difficulties in doing experiments in the field are immense.

a. *Grazing on Benthic Marine Plants.* Jones (1948) removed all limpets and larger algae from a strip of intertidal rocky shore and found that the distribution and abundance of algae was completely altered (Fig. 3). Southward (1964) and Haven (1973) have repeated this experiment with similar results (Fig. 4). Stephenson and Searles (1960) excluded grazing fish and mollusks from beach rock on a coral cay and found that algae grew more densely. Randall (1961) found a similar effect by excluding fish from sublittoral reefs. Randall (1965) extended the habitat of fish on coral reefs by building new shelters of concrete blocks; the fish then completely grazed the turtle grass within range of the shelter. Castenholz (1961) found that benthic diatoms on a rocky shore were abundant in winter when their growth was slow, and absent in summer. When he excluded grazing snails (*Littorina*) in the summer, the diatoms grew much faster than in the winter and quickly covered the surface. Kitching and Ebling (1961), Jones and Kain (1967), Paine and Vadas (1969), and Dayton (1970) removed sea urchins from intertidal and

Fig. 3. Effect after about 6 months of removal of all larger algae and about 15,000 limpets (*Patella vulgata*) from a strip 10 m wide and 115 m long across the intertidal zone of the Isle of Man in 1946 (Jones, 1948). The darker color of the strip is due to a complete cover of green algae.

Fig. 4. Effects of excluding small limpets (*Acmaea digitalis* and *A. scabra*) with stainless-steel mesh fences at high shore levels in central California. The photograph shows the growth of algae after 4 months in the autumn of 1964. From left to right: no limpets, *A. digitalis* removed, *A. scabra* removed, unaltered control. Pale squares are sites from which algae were removed for analysis (Haven, 1973).

sublittoral rocks; the algae grew quickly and the species composition changed. Stimson (1970) found that the amount of algae growing within the territory of *Lottia gigantea* was greater than outside it. When he removed the *Lottia*, other species of grazers moved in and removed the algae. When the *Lottia* was placed in a new position, the algae within its territory increased as before. These field experiments show clearly the important role of herbivores in determining the distribution and abundance of benthic marine plants.

b. *Predator–Prey Interactions*. Experimental manipulations of marine predators and prey in natural populations have mainly been done in the rocky intertidal region, although some recent work has been done sublittorally. Many observations and calculations have indicated that predatory fish and smaller predators in the plankton must undoubtedly have important influences on their prey, but field experiments will clearly be difficult to carry out. It may be possible to do them in enclosed waters, as Hall *et al*. (1970) have done in freshwater ponds.

The usual experiment has been to exclude predators from some prey populations while leaving others undisturbed. Exclusion of predators has been done using cages (Connell, 1961a,b, 1970; Luckens, 1970; Dayton, 1971; Scanland, 1971) or by picking them off by hand at frequent intervals (Paine, 1966, 1971; Dayton, 1970). Sometimes predators have been placed inside cages to measure feeding rates in the field, or to see which of two species of prey is preferred by the predator (see Fig. 5). These experiments have shown that predators often determine the age structure of a population and also its distribution. The lower limits of distribution of several intertidal species, and the horizontal limits of others, has been shown to be determined by intense predation. For example, the barnacle *Balanus glandula* survives better during the vulnerable period immediately after settlement, and also grows faster, at lower intertidal levels. But on most shores on San Juan Island, Washington, predators ate all individuals in the lower three-quarters of the intertidal zone before they reached sexual maturity, so the breeding population was limited to the upper part of the zone where growth and early survival are very poor (Connell, 1970). In Scotland, the predation was less intense so that some barnacles (*Balanus balanoides*) at lower levels reached

Fig. 5. View of a large concrete piling under a small pier at Friday Harbor, Washington. The stainless-steel wire mesh cages excluded predators from barnacles in the intertidal zone. "Roofs" of the mesh extended between some cages acted as controls for the effect of the mesh itself. They also excluded large predators such as the starfish *Pisaster ochraceus* while allowing smaller predators to enter (Connell 1970).

sexual maturity, but only when they were protected from predation did they survive to reach sizes as large as those living high on the shore, above their predators (Connell, 1961a). Paine (1966) working on the outer coast of Washington, has found that when the predatory starfish *Pisaster* was excluded, barnacles and mussels survived much lower on the intertidal zone. Field experiments by Luckens (1970) and Paine (1971) in New Zealand and Dayton (1971) in Washington have confirmed that predators determine the lower limit of intertidal animals. Scanland (1971) used cages to exclude predators from panels placed at 100 ft depth. Large predators, mainly fish, ate certain species completely so that the assemblage of species colonizing the panels was quite different in the absence of predation.

As an example of the sort of complex experimental design required to evaluate the effects of natural enemies on their prey, I have summarized in Table I the experiments which have been carried out to date using barnacles as the prey species. Grazing limpets, predatory snails, and predatory starfish are three of the most important natural enemies of barnacles. As shown in the table, of the 8 possible experimental treatments, 5 have been applied, 3 have not. One of the latter, the removal of grazers without removing predators, has been done by workers investigating algae, but unfortunately the effect on animals was not recorded. Fig. 6 shows the effect of removing all three types.

An interesting experiment on the effect of benthic suspension feeders on the rate of recruitment of other benthic animals was reported by Goodbody (1961). Bare panels hung from rafts in the harbor at Port Royal, Jamaica, normally become colonized with a dense growth of barnacles, worms, ascidians, etc., in 2 months. However, when bare panels were placed near others which had been in the sea for over 2 years, there were almost no colonists after 2 months. Controls hung at the same time from a nearby raft without the older panels were colonized as usual. The older panels were covered with anemones, sponges, and ophiuroids. A reasonable interpretation of this result is that the animals on the older panels ate most of the planktonic larvae before they could reach the adjacent bare panels. Further experiments with panels at varying distances from older aggregations would test this hypothesis.

c. *Parasites and Pathogens.* The experimental manipulation of parasites and pathogens has scarcely been attempted as yet. Crisp (1968) transplanted the barnacle *Balanus balanoides* across the Atlantic from Nova Scotia to Wales, and found that the Canadian form was more susceptible than the Welsh one to parasitism by the isopod *Hemioniscus balani.* However the effect of the parasite has not yet been estimated.

TABLE I

Experimental Methods used to Estimate the Effect of Grazing Limpets and Predators on Natural Populations of Barnacles on Rocky Shores

	Experimental treatments				
	Grazing limpets	Predatory snails	Large Predatory starfish	Methods	References
1. None removed	+	+	+	Observations of natural situations	All references below
2. One type removed	–	+	+	Limpets alone removed	Not done
	+	–	+	Predatory snails alone removed	Not done
	+	+	–	Predatory starfish alone removed, either by hand or by low roofs of wire mesh or by plastic rings	(Roof or rings): Connell (1961a, 1970), Dayton (1971) (Hand): Paine (1966, 1971), Dayton (1971)
3. Two types removed	+	–	–	Predatory snails and starfish excluded by wire mesh cages	Connell (1961a, 1970), Luckens (1970), Dayton (1971)
	–	+	–	Limpets and predatory starfish excluded by plastic rings	Dayton (1971)
	–	–	+	Limpets and predatory snails removed	Not done
4. Three types removed	–	–	–	Limpets and all predators excluded by wire mesh cages	Connell (1961a), Dayton (1971)

Rodella (1971) has performed the only field experiment known to me on a marine parasite. In the shore fish *Hypsoblennius gilberti* two populations about 24 km apart differed in the incidence of parasitism by the cloacal trematode *Diphtherostomum macrosaccum*. He took fish from the area of high incidences, removed the trematodes from the cloaca, and transplanted them to the area of low incidence. He also did the reciprocal transplant, but since these fish had few parasites, did not try to remove them. After several months, the fish which had been moved to the area of high incidence had gained many parasites, whereas the ones moved to the area of low incidence (after having their parasites removed) did not gain many new ones. The incidence of parasitism was evidently a property of the local en-

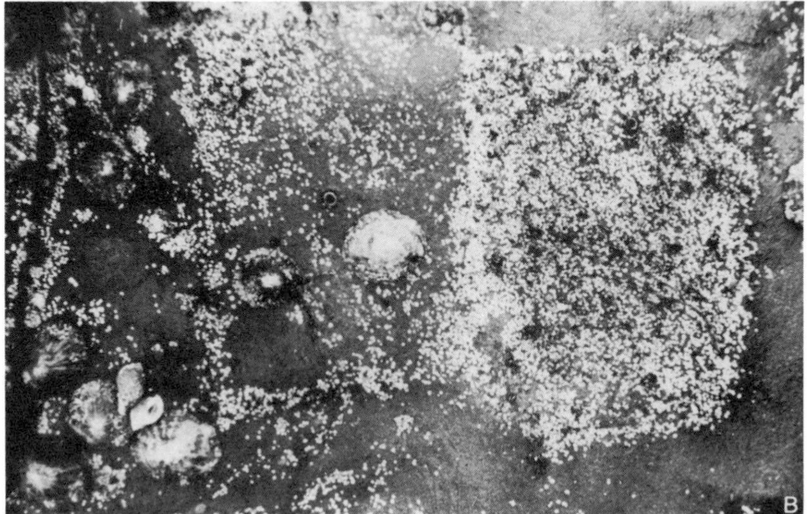

Fig. 6. The effect on barnacle settlement of excluding large grazing limpets (*Patella vulgata*) at Millport, Scotland. (A) Shows the stainless-steel mesh cages in place. (B) Shows the same rock with the cages removed. On the left, two limpets which were inside the cage for 10 weeks during the settlement season greatly reduced the density of barnacles. On the right, with no limpets, the density of barnacles was much greater. Predatory snails (*Thais lapillus*) were excluded from both cages (Connell 1961a).

vironment and not due to inherent differences in the two populations of fish.

3. *Ecological Communities*

Most of the field experiments dealing with whole communities have been concerned with the development of mixed-species assemblages by colonization of cleared substrates. One of the earliest of these was done by Wilson (1925), who observed the rate of colonization of algae on concrete blocks. He suspended the blocks at different depths and scraped some clean at intervals, while leaving others undisturbed. Thus he could separate the effect of both seasonal variations in settlement and of depth, in the development of an algal community. Hewatt (1935) removed mussel beds from intertidal surfaces and observed the recolonization; he did no further experimental manipulation. Other such experiments are described in the treatise on marine fouling (Woods Hole Oceanographic Inst., 1952).

Few experiments have been done with organisms inhabiting soft substrates. Reynaud-Debyser (1958) followed the recolonization of depopulated beach sand. Boaden (1962) did the same, but as described earlier, first sorted the sand into 6 classes based upon particle size. It is interesting that with a restricted range of grain sizes, fewer individuals and fewer species colonized the tubes. This seems to indicate that reduced complexity leads to reduced diversity. However, this is confounded by the smaller number of individuals, which also leads to fewer species.

Many studies have been made of "community metabolism," but evidently none have included experimental manipulation in the field. Studies similar to that of Hargrave (1970) in fresh water could profitably be done in the sea. He showed that experimental variations in the population density of amphipods had a great effect on the metabolism of microorganisms in the bottom sediments.

The study of species diversity in marine communities has been mainly confined to description and speculation. I have performed one field experiment on a coral reef, where the diversity increases from the inner reef flat toward the outer edge. I transplanted colonies of about 20 species from the outer region of high diversity to the inner flat. As a control for the disturbance caused by transplanting, I split each colony in half and carried one half of each back to the original area near the outer edge of the reef. The survival over 2 years was almost as good in the region of low diversity as in the source area of high diversity. Thus the low diversity on the inner reef flat does not seem to be due to higher mortality of adult colonies in periods of normal weather. Several other possible mechanisms could account for this pattern of diversity.

IV. Discussion

It seems clear from this survey that field experiments have proved to be very useful in marine ecology. They have shown that the physical environment directly determines the distribution of marine organisms mainly when it reaches extreme values. For example, species can tolerate relatively great increases in the temperature of seawater (Naylor, 1965). When extremes are reached, e.g., very low salinities, great wave action, long exposure to air, etc., marine organisms are killed. But in less extreme circumstances, they can tolerate variations of their physical environment, and under such conditions biological interactions become critical.

Although it has long been recognized that such factors as population density, interspecific competition, grazing, etc., must be important in the ecology of marine organisms, it is difficult to make a convincing case using laboratory results to calculate the presumed effect in the field. But the field experiments described above have clearly demonstrated that the distribution and abundance of some marine organisms are directly determined by biological interactions.

It should also be clear from the survey in this chapter that there is much room for improvement in field experimentation. Rather than criticize particular instances, I will outline my views on the proper methods, from which it will be clear that most studies so far have been deficient in one or more respects.

The chief value of field experiments is that the results can be more directly applied to natural ecosystems than can those from laboratory experiments. This is because the interactions with other organisms, and the natural variations in the abiotic environment, are included in the experiment. Therefore, the best field experiments are those done close to the natural community and with the least disturbance to it.

For example, removing a species by hand should cause less disturbance than excluding it with fences or cages. Likewise, to measure the rate of settlement of larvae from the plankton onto a rocky substrate it would be preferable to scrape a patch of rock bare, or to place a panel on the rock, surrounded by the existing organisms, rather than to hang it from a raft nearby. The reason is that in natural situations, the existing members of the community have a great effect on the settlement of larvae [as indicated by Goodbody's (1961) work], so that they must be included. Some of the field experiments described were either too isolated from the natural communities or may have involved too much disturbance.

Another important requirement is that control populations must be set up at the same time and closely adjacent to the experimental ones, so that the environmental conditions vary in the same way in both sets. In many cases it is more difficult to set up adequate controls than to apply the experimental treatment. This is especially true when some sort of artificial apparatus is being used. For example, cages may cast shade, slow down water movements, restrict the movement of mobile animals inside, provide surfaces for the growth of algae, etc. As controls I have usually used roofs extending out from the cage (Fig. 5); these modify the physical environment to about the same degree as cages. However, it is difficult always to be sure that the controls are fully adequate, and it is better to try to establish several sorts of controls. One of the best examples of a controlled field experiment is Boaden's (1962) on the interstitial fauna of a sandy beach. A general rule is that controls should be established in which the experimental apparatus is duplicated in every respect except for the factor under study.

Because environmental conditions vary in space as well as in time, replicate plots of both experimental and control treatments need to be established. To allow a proper statistical analysis of variance, the plots should ideally be randomly arranged, within the habitat being studied. Few if any of the field experiments done so far have been fully satisfactory in these respects. The reasons are, of course, that if adequate controls are included, there are not enough resources of material, labor, and time to set up enough replicates. Fortunately, the differences between the experimental treatments and the controls are often so much greater than the differences between replicates, that extensive replication has not been necessary. However, it is always advisable to have some replication; many of the studies described had none, but one hopes that future ones will improve on this low standard.

Clearly the experimental areas must be accessible, and the population density great enough so that a reasonable number of individuals can be included in each replicate. These requirements usually restrict the use of field experiments to nearshore shallow areas and to species which live on or near the bottom. Most field experiments have so far been done in the intertidal zone or on reefs in shallow water. With present techniques, such experiments cannot be done with the plankton and deeper water organisms. Biological oceanographers should view this as a challenge to their ingenuity.

In preparing this chapter I made a rapid survey of the contents, since 1960, of 11 journals concerned with marine biological subjects. They are probably fairly representative of the recent work on marine ecology published in Europe and the United States. While I undoubtedly missed some relevant papers, it was clear that few marine biologists have used field experiments to test their hypotheses. This seems to be due to the impression that experimentation is an activity which can only be carried out in the laboratory under controlled conditions. What is not generally recognized is that if proper attention is paid to establishing control areas, and if sufficient replication is done, that valid "controlled experiments" can be conducted under natural conditions.

Since laboratory experiments have been so successful in solving problems at the lower levels of organisation (molecular, cellular, organismal), there is evidently the feeling that they will be equally successful at the population level. I hope that the examples in this chapter will illustrate the differences between these levels and show that new techniques must be applied to studying populations and communities in natural conditions. If questions are asked at the individual or lower levels, e.g., "to what stimulus does an individual larva respond in selecting its substrate?", then laboratory experiments are completely appropriate. It is only when the question concerns the behavior of the population in nature, e.g., "What determines the distribution of species A on the beaches?", that laboratory experiments are less useful than field experiments.

Laboratory experiments probably also appeal because they can be done in a tidy fashion, in well-controlled conditions, often using sophisticated and expensive apparatus which is undoubtedly more impressive than the sealing-wax and string methods of many field experiments.

The question remains for marine ecology, are field experiments always superior to laboratory experiments? My answer would be yes, if it is possible to do field experiments which satisfy the requirements outlined above. On the other hand, if adequate controls cannot be established, or if they would create great disturbance, or if adequate replication is impossible, then field experiments cannot be done and laboratory experiments are the next best thing. For these reasons I have often used laboratory experiments, as have most ecologists. But if the conditions are such that satisfactory field experiments can be done, they are to be preferred to laboratory experiments.

References

Barnes, H., Crisp, D. J., and Powell, H. T. (1951). Observations on the orientation of some species of barnacles. *J. Anim. Ecol.* **20**, 227–241.

Barnes, H. (1967). Ecology and experimental biology. *Helgoland, Wiss. Meeresunters.* **15**, 6–26.

Behrens, S. (1971). The distribution and abundance of the intertidal prosobranch *Littorina scutulata* (Gould 1899) and *L. sitkana* (Philippi 1845). M.Sc. thesis, Univ. of British Columbia.

Berry, A. J. (1961). Some factors affecting the distribution of *Littorina saxatilis* (Olivi). *J. Anim. Ecol.* **30**, 27–45.

Blake, N. J., and Jeffries, H. P. (1971). The structure of an experimental infaunal community. *J. Exp. Mar. Biol. Ecol.* **6**, 1–14.

Boaden, P. J. S. (1962). Colonization of graded sand by an interstitial fauna. *Cah. Biol. Mar.* **3**, 245–248.

Carlisle, J. G., Turner, C. H., and Ebert, E. E. (1964). Artificial habitat in the marine environment. *Calif. Dept. Fish Game Fish Bull.* **124**, 1–93.

Castenholz, R. W. (1961). The effect of grazing on marine littoral diatom populations. *Ecology* **42**, 783–794.

Castenholz, R. W. (1963). An experimental study of the vertical distribution of littoral marine diatoms. *Limnol. Oceanogr.* **8**, 450–462.

Clarke, T. A. (1970). Territorial behavior and population dynamics of a pomacentrid fish, the Garibaldi, *Hypsypops rubicunda*. *Ecol. Mon.* **40**, 189–212.

Connell, J. H. (1961a). Effects of competition, predation by *Thais lapillus*, and other factors on natural populations of the barnacle *Balanus balanoides*. *Ecol. Mon.* **31**, 61–104.

Connell, J. H. (1961b). The influence of interspecific competition and other factors on the distribution of the barnacle *Chthamalus stellatus*. *Ecology* **42**, 710–723.

Connell, J. H. (1963). Territorial behavior and dispersion in some marine invertebrates. *Res. Pop. Ecol.* **5**, 87–101.

Connell, J. H. (1970). A predator-prey system in the marine intertidal region. I. *Balanus glandula* and several predatory species of *Thais*. *Ecol. Mon.* **40**, 49–78.

Crisp, D. J. (1964). Racial differences between North American and European forms of *Balanus balanoides*. *J. Mar. Biol. Ass. U.K.* **44**, 33–45.

Crisp, D. J. (1965). Surface chemistry, a factor in the settlement of marine invertebrate larvae. *Proc. 5th Mar. Biol., Symp. Bot. Gothoburgensia*, **III**, 51–65.

Crisp, D. J. (1968). Differences between North American and European populations of *Balanus balanoides* revealed by transplantation. *J. Fish. Res. Bd. Can.* **25**, 2633–2641.

Crisp, D. J., and Barnes, H. (1954). The orientation and distribution of barnacles at settlement with particular reference to surface contour. *J. Anim. Ecol.* **23**, 142–162.

Crisp, D. J., and Patel, B. (1961). The interaction between breeding and growth rate in the barnacle *Elminius modestus* Darwin. *Limnol. Oceanogr.* **6**, 105–115.

Curl, H., and McLeod, G. C. (1961). The physiological ecology of a marine diatom, *Skeletonema costatum* (Grev.) Cleveland. *J. Mar. Res.* **19**, 70–88.

Dayton, P. K. (1970). Competition, predation, and community structure: the allocation and subsubsequent utilization of space in a rocky intertidal community. Ph. D. thesis, Dept. of Zool., Univ. of Washington, Seattle.

Dayton, P. K. (1971). Competition, disturbance, and community organization: The provision and subsequent utilization of space in a rocky intertidal community. *Ecol. Monogr.* **41**, 351–389.

Ebert, T. A. (1968). Growth rates of the sea urchin *Strongylocentrotus purpuratus* related to food availability and spine abrasion. *Ecology* **49**, 1075–1091.

Edwards, R. R. C., Steele, J. H., and Trevallion, A. (1970). The ecology of O-group plaice and common dabs in Loch Ewe. III. Prey-predator experiments with plaice. *J. Exp. Mar. Biol. Ecol.* **4**, 156–173.

Fager, E. W. (1971). Pattern in the development of a marine community. *Limnol. Oceangr.* **16**, 241–253.

Foster, B. A. (1971). On the determinants of the upper limit of intertidal distribution of barnacles (Crustacea: Cirripedia) *J. Anim. Ecol.* **40**, 33–48.

Frank, P. W. (1965). The biodemography of an intertidal snail population. *Ecology* **46**, 831–844.

Friedrich, H. (1961). Physiological significance of light in marine ecosystems. *In* "Oceanography" (M. Sears, ed.), Publ. 67, pp. 257–270. Amer. Assoc. Advance Sci., Washington, D.C.

Gaarder, T., and Gran, H. H. (1927). Production of plankton in the Oslo Fjord. *Rapp. Proc. Verb. Cons. Perm. Int. Explor. Mer.* **42**, 1–48.

Gauld, D. T. (1950). A fish cultivation experiment in an arm of a sea loch. III. The plankton of Kyle Scotnish. *Proc. Roy. Soc. Edinburgh* **64** (B), 36–64.

Gersbacher, W. M., and Dennison, M. (1930). Experiments with animals in tide pools. *Publs. Puget. Sound Mar. Biol. Stn.* **7**, 209–215.

Ghelardi, R. (1960). Structure and dynamics of the animal community found in *Macrocystis pyrifera* holdfasts. Ph. D. thesis, Univ. of California, San Diego.

Ghelardi, R. (1971). "Species" structure of the animal community that lives in *Macrocystis pyrifera* holdfasts. *In* "The Biology of Giant Kelp Beds (Macrocystis) in California" (ed., W. J. North). *Beihefte zur Nova Hedwigia*, Heft **32**, 381–420.

Glynn, P. W. (1968). Mass mortalities of echinoids and other reef flat organisms coincident with midday, low water exposure in Puerto Rico. *Mar. Biol.* **1**, 226–243.

Goodbody, I. (1961). Inhibition of the development of a marine sessile community. *Nature (London)* **190**, 282–283.

Gross, F., Marshall, S. M., Orr, A. P., and Raymont, J. E. G. (1947). An experiment in marine fish cultivation I-V. *Proc. Roy. Soc. Edinburgh* **63** (B), 1–95.

Gross, F., Nutman, S. R., Gauld, D. T., and Raymont, J. E. G. (1950). A fish cultivation experiment in an arm of a sea-loch. I-V. *Proc. Roy. Soc. Edingurgh* **64** (B), 1–135.

Hall, D. J., Cooper, W. E., and Werner, E. E. (1970). An experimental approach to the production dynamics and structure of freshwater animal communities. *Limnol. Oceanogr.* **15**, 839–928.

Harger, J. R. E. (1968). The role of behavioral traits in influencing the distribution of two species of sea mussel, *Mytilus edulis* and *Mytilus californianus*. *Veliger* **11**, 45–49.

Harger, J. R. E. (1970a). The effect of wave impact on some aspects of the biology of sea mussels. *Veliger* **12**, 401–414.

Harger, J. R. E. (1970b). The effect of species composition on the survival of mixed populations of the sea mussels *Mytilus californianus* and *Mytilus edulis*. *Veliger* **13**, 147–152.

Harger, J. R. E. (1971). Variation and relative "niche" size in the sea mussel *M. edulis* in association with *M. californianus*. *Veliger* **14**, 275–282.

Hargrave, B. T. (1970). The effect of a deposit-feeding amphipod on the metabolism of benthic microflora. *Limnol. Oceanogr.* **15**, 21–30.

Hatton, H. (1938). Essais de bionomie explicative sur quelques especes intercotidales d'algues et d'animaux. *Ann. Inst. Oceanogr. Monaco* **17**, 241–348.

Haven, S. B. (1973). Competition for food between the intertidal gastropods *Acmaea scabra* and *Acmaea digitalis*. *Ecology* **54**, 143–151.

Hewatt, W. G. (1935). Ecological succession in the *Mytilus californianus* habitat as observed in Monterey Bay, California. *Ecology* **16**, 244–251.

Holme, N. A. (1950). Population-dispersion in *Tellina tenuis* Da Costa. *J. Mar. Biol. Ass. U.K.* **29**, 267–280.

Isaacs, J. D. (1969). The nature of oceanic life. *Sci. Amer.* **221**, 146–162.

Jones, N. S. (1948). Observations and experiments on the biology of *Patella vulgata* at Port St. Mary, Isle of Man. *Proc. Liverpool Biol. Soc.* **56**, 60–77.

Jones, N. S., and Kain, J. M. (1967). Subtidal algal colonization following the removal of *Echinus*. *Helgol. Wiss. Meeresunt.* **15**, 460–466.

Jones, W. E., and Demetropoulos, A. (1968). Exposure to wave action: measurements of an important ecological parameter on rocky shores of Anglesey. *J. Exp. Mar. Biol. Ecol.* **2**, 46–63.

Kensler, C. B., and Crisp, D. J. (1965). The colonization of artificial crevices by marine invertebrates. *J. Anim. Ecol.* **34**, 507–516.

Kinne, O. (1963). The effects of temperature and salinity on marine and brackish-water animals. I. Temperature. *Oceanogr. Mar. Biol. Ann. Rev.* **1**, 301–340.

Kinne, O. (1964). The effects of temperature and salinity on marine and brackish-water animals. II. Salinity and temperature-salinity relations. *Oceanogr. Mar. Biol. Ann. Rev.* **2**, 281–339.

Kitching, J. A., and Ebling, F. J. (1961). The ecology of Lough Ine. XI. The control of algae by *Paracentrotus lividus* (Echinoidea). *J. Anim. Ecol.* **30**, 373–383.
Kitching, J. A., and Ebling, F. J. (1967). Ecological studies at Lough Ine. *Advan. Ecol. Res.* **4**, 197–291.
Kitching, J. A., Muntz, L., and Ebling, F. J. (1966). The ecology of Lough Ine. XV. The ecological significance of shell and body forms in *Nucella*. *J. Anim. Ecol.* **35**, 113–126.
Klugh, A. B., and Martin, J. R., (1927). The growth-rate of certain marine algae in relation to depth of submergence. *Ecology* **8**, 221–231.
Knight-Jones, E. W. (1953). Laboratory experiments on gregariousness during setting in *Balanus balanoides* and other barnacles. *J. Exp. Biol.* **30**, 584–598.
Knight-Jones, E. W., and Stevenson, J. P. (1950). Gregariousness during settlement in the barnacle *Elminius modestus* Darwin, *J. Mar. Biol. Ass. U.K.* **29**, 281–297.
Landenberger, D. E. (1967). Studies on predation and predatory behavior in Pacific starfish (*Pisaster*). Ph. D. dissertation, Univ. of California, Santa Barbara.
Landenberger, D. E. (1968). Studies on selective feeding in the Pacific starfish (*Pisaster*) in southern California. *Ecology* **49**, 1062–1075.
Lasker, R. (1970). Utilization of zooplankton energy by a Pacific sardine population in the California current. *In* "Marine Food Chains" (J. H. Steele, ed.), University of California Press, Berkeley, pp. 265–284.
Lewis, J. R. (1964). "The Ecology of Rocky Shores." English Univ. Press, London.
Luckens, P. A. (1970). Breeding, settlement and survival of barnacles at artificially modified shore levels at Leigh, New Zealand. *N.Z. J. Mar. Freshwater Res.* **4**, 497–514.
Marshall, S. M., and Orr, A. P. (1928). The photosynthesis of diatom cultures in the sea. *J. Mar. Biol. Ass. U.K.* **15**, 321–360.
Mayer, A. G. (1918). Ecology of the Murray Island coral reef. *Pap. Dept. Mar. Biol. Carnegie Inst. Wash.* **9**, 1–48.
McAllister, C. D., Parsons, T. R., Stephens, K., and Strickland, J. D. H. (1961). Measurements of primary production in coastal sea water using a large-volume plastic sphere. *Limnol. Oceanogr.* **6**, 237–258.
McIntyre, A. D., Munro, A. L. S., and Steele, J. H. (1970). Energy flow in a sand ecosystem. *In* "Marine Food Chains" (J. H. Steele, ed.), University of California Press, Berkeley, pp. 19–31.
Menge, B. A. (1972). Competition for food between two intertidal starfish species and its effect on body size and feeding. *Ecology* **53**, 635–644.
Mullin, M. M. (1969). Production of zooplankton in the ocean: the present status and problems. *Oceanogr. Mar. Biol. Ann. Rev.* **7**, 293–314.
Murdoch, W. W. (1969). Switching in general predators: experiments on predator specificity and stability of prey populations. *Ecol. Monogr.* **39**, 335–354.
Naylor, E. (1965). Effects of heated effluents upon marine and estuarine organisms. *Advan. Mar. Biol.* **3**, 63–103.
Neushul, M. and Powell, J. H. (1964). An apparatus for experimental cultivation of marine algae. *Ecology* **45**, 893–894.
Newell, R. C. (1970). "The Biology of Intertidal Animals." Logos Press, London.
Paine, R. T. (1966). Food web complexity and species diversity. *Amer. Natur.* **100**, 65–75.
Paine, R. T. (1971). A short-term experimental investigation of resource partitioning in a New Zealand rocky intertidal habitat. *Ecology* **52**, 1096–1106.
Paine, R. T., and Vadas, R. L. (1969). The effects of grazing by sea urchins, *Strongylocentrotus* spp. on benthic algal populations. *Limnol. Oceanogr.* **14**, 710–719.
Petipa, T. S. (1966). On the energy balance of *Calanus helgolandicus* in the Black Sea. *In* "Physiology of Marine Animals," pp. 60–81. Oceanogr. Comm. Sci. Publ. Ho., Moscow. (English transl.).

Pomerat, C. M., and Reiner, E. R. (1942). The influence of surface angle and of light on the attachment of barnacles and other sedentary organisms. *Biol. Bull.* **82**, 14–25.

Pomerat, C. M., and Weiss, C. M. (1946). The influences of texture and composition of surface on the attachment of sedentary marine organisms. *Biol. Bull.* **91**, 57–65.

Randall, J. E. (1961). Overgrazing of algae by herbivorous marine fishes. *Ecology* **42**, 812.

Randall, J. E. (1965). Grazing effect on sea grasses by herbivorous reef fishes in the West Indies. *Ecology*. **46**, 255–260.

Raymont, J. E. G. (1966). The production of marine plankton. *Advan. Ecol. Res.* **3**, 117–205.

Reynaud-Debyser, J. (1958). Contribution a L'etude de la faune interstitielle du Bassin d'Arcachon. *Proc. Int. Congr. Zool., 15th* 323–326.

Rodella, T. D. (1971). An ecological study of tide pool fish parasites from two areas near Santa Barbara, California. Ph.D. thesis, Univ. of California, Santa Barbara.

Round, F. E., Sloane, J. F., Ebling, F. J., and Kitching, J. A., (1961). The ecology of Lough Ine. X. The hydroid *Sertularia operculata* (L.) and its associated flora and fauna: effects of transference to sheltered water. *J. Ecol.* **49**, 617–629.

Sandison, E. E., and Hill, M. B. (1966). The distributions of *Balanus pallidus stutsburi* Darwin, *Gryphaea gasar* ((Adanson) Dautzenberg), *Merceriella enigmatica* Fauvel and *Hydroides uncinata* (Philippi) in relation to salinity in Lagos Harbour and adjacent creeks. *J. Anim. Ecol.* **35**, 235–250.

Scanland, T. B. (1971). Effects of predation on epifaunal assemblages in a submarine canyon. Ph.D. thesis, Univ. of California, San Diego.

Seed, R. (1968). Factors influencing shell shape in the mussel *Mytilus edulis*. *J. Mar. Biol. Ass. U.K.* **48**, 561–584.

Smith, F. G. W. (1946). Effect of water currents upon the attachment and growth of barnacles. *Biol. Bull.* **90**, 51–70.

Southward, A. J. (1964). Limpet grazing and the control of vegetation on rocky shores. *In* "Grazing in terrestrial and marine environments (ed., D. J. Crisp), Blackwell Sci. Publ. Ltd., Oxford, pp. 265–273.

Staiger, H. (1957). Genetical and morphological variation in *Purpura lapillus* with respect to local and regional differentiation of population groups. *Annee Biol.* **61**, 251–258.

Stephens, J. S., Johnson, R. K., Key, G. S., and McCosker, J. E. (1970). The comparative ecology of three sympatric species of California blennies of the genus *Hypsoblennius* Gill (Teleostomi, Blenniidae). *Ecol. Mon.* **40**, 213–233.

Stephenson, W., and Searles, R. (1960). Experimental studies on the ecology of intertidal environments at Heron Island. I. Exclusion of fish from beach rock. *Aust. J. Mar. freshwater Res.* **11**, 241–267.

Stimson, J. S. (1968). The population ecology of the limpets *Lottia gigantea* (Gray) and several species of *Acmaea* (Eschsholtz) coexisting on an intertidal shore. Ph.D. thesis, Univ. California, Santa Barbara.

Stimson, J. S. (1970). Territorial behavior in the owl limpet, *Lottia gigantea*. *Ecology* **51**, 113–118.

Trevallion, A., and Ansell, A. D. (1967). Preliminary experiments in enriched sea water. *J. Exp. Mar. Biol. Ecol.* **1**, 257–270.

Trevallion, A., Edwards, R. R. C., and Steele, J. H. (1970). Dynamics of a benthic bivalve. *In* "Marine Food Chains" (J. H. Steele, ed.), University of California Press, Berkeley, pp. 285–295.

Wilson, O. T. (1925). Some experimental observations of marine algal successions. *Ecology* **6**, 303–311.

Woods Hole Oceanographic Inst. (1952). "Marine Fouling and its Prevention." U.S. Naval Inst., Annapolis, Maryland.

Chapter 3

Behavior: *In Situ* Approach to Marine Behavioral Research

William F. Herrnkind

I. Introduction

Our contemporary view of animal behavior includes the concept that behavior, as a dynamic process through which animals interact with the environment, is adaptive (Hinde, 1970; Eibl-Eibesfeldt, 1970). Implicit in this statement, then, is the value of understanding through repeated observation and experimentation, the behavior of animals in their natural environment. This knowledge is derivable only inferentially from studies conducted in the laboratory or in some simulated condition. It is most appropriately and validly determined from direct study *in situ*. Field research has contributed enormously to the understanding of behavior in terrestial–aerial biotopes but a perusal of the literature shows a relative lack of such information for marine (and aquatic) animals.

Behavioral research in the marine environment is limited in large part by the difficulties imposed by the medium on the scientist. First, visualization, the primary method of witnessing and in turn recording and representing behavior, is limited or distorted by the air–water interface, water clarity, and depth. Second, even mildly turbulent sea conditions preclude the presence of the observer. Third, under the most optimal conditions, the sea imposes severe depth and duration limitations which are further magnified by cold water, strong currents, wave surge and other common physical perturbations. These limits are in some part imposed both on the scientist-in-the-sea and his technological sensory extensions.

Past underwater research was also hampered by the lack of a discrete methodology and the lack of appropriately trained scientists (Starck, 1968). For example, the experimentalist avoided field research because he was ingrained with the misconception that valid scientific data requires absolute controls on all but one or a few experimental variables. The validity of this is dependent on the question being asked and, of course, applies to both marine and nonmarine research. Until recently, few marine behaviorists conducted *in situ* research and the research approaches and programs for training new scientists have reflected this.

Within the past two decades *in situ* studies have increasingly contributed to knowledge of marine animal behavior. For example, the late Conrad Limbaugh using diving techniques described and documented the diversity, behavioral mechanisms, and functional significance of cleaning symbionts on tropical and subtropical reefs (Limbaugh, 1961; Limbaugh *et al.*, 1961). Interestingly, Limbaugh was expanding upon observations first noted by William Beebe (1928), a scientist prominent for emphasizing the *in situ* approach to obtain data on marine organisms. A wealth of unique and significant observations on reef fishes and invertebrates were made by Lindberg (1955), Randall (1958), Clark (1961), Schroeder and Starck (1964), Hobson (1965), and others using diving techniques. Underwater observations contributed significantly to Breder's (1959) thorough analysis of fish schooling also during this same period. The importance of information obtained by these and other workers clearly demonstrated the applicability and desirability of *in situ* research.

The impetus from the above works together with the availability of SCUBA (self-contained underwater breathing apparatus) produced an increasing interest in underwater studies as evidenced by the number of studies by diving scientists which began to appear during the 1960's (Feddern, 1965; Starck and Davis, 1966; Randall, 1964; Fager *et al.*, 1966). *In situ* SCUBA studies have also recently become important for sampling and long term observations on populations of economically important species. The National Marine Fisheries Service, NMFS, (formerly Bureau of Commercial

Fisheries) has used SCUBA techniques for such purposes even in Alaska and the northeast Atlantic coast for some time now (Dr. R. A. Cooper, Personal Communication). Recent contributions to the fisheries ecology of palinurid lobsters in South Africa stemmed directly from diving studies which avoid many of the biases and weaknesses of purely trap based programs (Heydorn, 1969).

In addition to SCUBA gear, basically unchanged over the past 20 years, the development of new undersea technology has extended the capacity for gathering and recording data. The use of mixed gas rebreathers, submersibles and habitats extend the depth and time at depth for the scientist-in-the-sea (Miller *et al.*, 1971). Human sensing capabilities are expanded by the use of underwater television (UTV) (Myrberg *et al.*, 1969), acoustic monitoring systems (Steinberg *et al.*, 1965; Cummings *et al.*, 1964) and miniaturized ultrasonic tags (Hasler, 1966; Johnson, 1971).

Because the application of new underwater techniques is accelerating along with the interest in animal behavior in general, it seems appropriate to examine the rationale underlying research on the behavior of marine animals and to examine various *in situ* techniques.

Why study marine animal behavior? Most of the present research in ethology centers on the study of freshwater and terrestrial vertebrates. For example, the fishes which are the most thoroughly studied are generally freshwater species easily held in aquaria (e.g., cichlids, characins, cyprinodonts). However, marine vertebrates and invertebrates are receiving increasing attention as indicated by the recent literature (Reese, 1964; Eibl-Eibesfeldt, 1970; Winn and Olla, 1972). This trend will surely continue as marine species become more available to observation.

Marine animals are behaviorally interesting because of their phylogenetic and biological diversity and because they present a variety of behavioral adaptations to conditions found in no other environment. The marine realm is phyletically the richest and certain animal groups are represented only in that environment. For example, all ctenophores, cephalopods, echinoderms, most crustaceans, cnidarians and fishes (particularly elasmobranchs) are marine. Continued study of these groups is necessary for a thorough understanding of the relationship of behavior to the different types of nervous organization they represent. The intensive analysis of behavior and neurosensory processes in octopods by Young, Wells and others is one of the few significant contributions in this area (Wells, 1966). (However, it should be added that the ethology of octopods has not yet been adequately studied.) Even the basic knowledge of orientational responses mediating locomotory movements, feeding, reproductive aggregation, and substrate selection are known for only a relatively small number of invertebrates. This type of information is required to support such attempts as that of Jander (1965) to comprehend the ontogeny and evolution of arthropod orientation mechanisms. In general, further research on diverse marine invertebrates seems necessary to understanding the phylogenetic evolution of animal behavior and neurobiological mechanisms.

A wealth of behavioral problems are presented by the inhabitants of uniquely marine habitats including the coral reef, the deep sea, and intertidal zone. The study of behavioral ecology of the tropical coral reef is especially formidable since it is one of the most diverse and complex of all animal communities. Behavioral phenomena associated with animal symbioses and vertical migration are also primarily subject to study in the marine environment. In addition to behaviors peculiar to marine animals, there exist many activities and mechanisms parallel in function to those represented in the terrestrial—aerial environment. Thus there are marine examples of periodic migration, homing, mimicry, territoriality, agonistic and sexual communication, parental care, dominance hierarchies, etc., many of which need comparative examination with terrestrial—aquatic examples (Hazlett, 1972). The operation of behavioral mechanisms found in both marine and terrestrial animals is of particular interest to determine the equivalent functional components and those modifications necessary for operation in one medium or the other, e.g., acoustic echolocation and communication (See Tavolga 1964a, 1967; Adler, 1971), sun-compass orientation (Winn *et al.*, 1964; Hasler, 1966), pheromone mediated behavior (Ryan, 1966; Atema and Engstrom, 1971).

Why study behavior *in situ*? The rationale for *in situ* research on marine animals was eloquently presented by Starck (1968) who points out three major advantages of direct observation. First, aquaria are often inadequate representations of the marine environment. Conditions including hydrodynamics, lighting, *Lebensraum*, depth, and complexity of intra- and interspecific interaction cannot always be reasonably duplicated.

Second, the natural environment often presents more valid conditions for behavioral experimentation. That is, instead of artificially eliminating possibly important stimuli, the experimentalist allows the natural conditions to act as the control and the effect of some experimental change is noted. Studies of acoustic attraction on free ranging sharks (Nelson and Gruber, 1963; Myrberg *et al.*, 1969; Banner, 1972) and fishes (Richard, 1968; Fish, 1972; and Myrberg, 1972) successfully demonstrate this approach. In these studies, the distances involved and the alterations of sound caused by depth and the other limitations in artificial enclosures preclude laboratory simulation. Moreover, one is not left with the question, "... but does it work that way in nature?"

Another valuable approach is to observe and record behavior while monitoring certain environmental factors which can later be correlated with the behavioral events. This approach was taken by Stevenson (1972) in establishing the controlling influence of such factors as light and current on the activity and feeding of pomacentrid fishes. Studies of orientational mechanisms guiding long distance movements in spiny lobsters have a similar methodology and are discussed later in this chapter (Herrnkind and McLean, 1971).

Starck's third point is the *a priori* discovery of new phenomena; we simply have not yet begun to investigate the wealth of new problems available to the long term, on-site observer. Furthermore, the restrictive nature of aquaria may prevent the occurrence of certain behaviors or produce behavioral artifacts. For example, the early studies on soniferous fishes included prodding a captive fish until it made (or failed to make) a sound. This is analagous to beating a bird and recording the squawk. The true character of a complex communication mechanism is not readily apparent in this artificial situation. Presently, much of marine acoustic communication research is conducted *in situ* to remedy this shortcoming (Cummings *et al.*, 1964; Myrberg, 1972; Winn, 1972).

The discovery of certain animal symbiotic interactions (Limbaugh, 1961) and behaviorally mediated hermaphroditic reproduction in the serranid fish, *Serranus subligarius* (Clark, 1961), was initially established through field study. These phenomena were subsequently examined under laboratory conditions dictated by knowledge derived *in situ* (Clark, 1965; Johnson, 1969). Mass migrations by the spiny lobster, *Panulirus argus* involving thousands of individuals in head-to-tail queues was witnessed only fortuitously in the field by scientists using SCUBA and would be an unlikely discovery to take place in a seawater tank (Herrnkind and Cummings, 1964; Herrnkind 1969, 1970). Thus discovery, experimentation, and corroboration of events, as well as development of valid laboratory models are tasks appropriate to the *in situ* approach.

II. Methods

Methods applied to undersea behavioral research fall under the general headings of (1) direct observation by the scientist, (2) indirect observation by remote sensing, or (3) an integration of the two. The following discussion describes various available techniques, examines their utility in light of research results derived from them, and points out the inherent drawbacks in each.

A. DIRECT OBSERVATION

1. *Introduction*

Direct underwater observation by the scientist confers several unique advantages unattainable in unmanned systems (Starck, 1968). No machinery is capable of the rapid integration, selectivity, and adjustment to circumstances inherent in a human observer. The direct contact with aspects of the environment outside a researcher's specific academic discipline force upon him a natural interdisciplinary approach providing an enriched perspective. The scientist-in-the-sea also possesses open input channels to stimuli and peripheral events beyond the range of specialized remote sensors, thus increasing his opportunity for new discoveries. The scientist, with time, accumulates information which may yield insights unavailable to the lab-based worker. Additionally, in reasonable seriousness, a human is often more reliable and less subject to failure than gadgetry. Finally, the only effective means usually available to most investigators is to enter the sea and study the organisms directly.

Upon entering the medium, the relatively unrestricted diver can maneuver easily and adjust his position relative to the subjects under observation. The entire visual field from nearly any angle is available and very slow movements, helpful when approaching animals, are facilitated by the three-dimensional freedom available to the neutrally buoyant observer. (To appreciate this, imagine slowly descending upon a bird nest at an oblique vertical angle!) Given reasonable water conditions the buoyant snorkeler or the appropriately weighted diver has both hands free to record notes or photograph. A properly insulated observer in warmer waters (20°–25°C) can operate for several hours at depths from the surface to 10 m or so (depending on air supply, exertion, etc.). Chances are good that during this time the underwater observer will be less fatigued by maintaining body position and posture than a terrestial observer subject to the full measure of gravity (typically brought to bear on a sensitive area resting on a protrusion).

Close-up observation is to some extent facilitated in the sea as in the terrestial–aerial environment by habituation on the part of the animals observed. Initially, upon a diver's approach, fishes, crustaceans, and cephalopods frequently seek cover or cease activities but resume their behavior after several minutes if the diver remains still. After a relatively short time, it is often possible to move about more vigorously without apparent disruptive effect.

2. *The Scientist-in-the-Sea*

a. *Observations Using Mask and Snorkel.* Studies in shallow, clear waters where excursions below the surface are needed only occasionally can often be performed with mask and snorkel (flippers optional depending on depth, distance from shore, etc.). Any good swimmer in reasonable health can usually master the breathing tube and short surface dives to 2 m depth. Below this depth, the diver must have the ability to "clear" his ears, (i.e., equalize pressure between the oral cavity and inner ear across the eustachian tubes). Individuals vary in this capacity according to anatomy, sinus condition, condition of the eardrum, and practice, and it is dangerous to overextend oneself. The capable snorkeler has several advantages over the SCUBA diver. Being unencumbered by air tanks, the snorkeler has more freedom of movement, can remain stationary in a vertical plane (as long as the surface is calm), and can communicate by voice to companions.

An example of research done while snorkelling is the field experimentation by Snyder and Snyder (1970) on alarm responses of the long-spined urchin, *Diadema antillarum* to intra- and interspecific body fluids. The Snyders conducted their studies in a shallow (1–2 m) lagoon where the urchins tended to clump by day. The protocol was to pulverize a *Diadema*, other test organism or substrate (control) upcurrent from a clump of urchins, then record their behavior. Since *D. antillarum* is unresponsive to a diver (so long as he does not cast a shadow over them or cause excessive turbulence), the experimenters moved about with considerable freedom. The experimental results showed the *D. antillarum* respond to the body fluid of conspecifics (from up to 8 m away) by "racing" away on their ventral spines. They were much less responsive to juices of other local echinoderms although they showed an alarm response to contact from the helmet conch, *Cassis tuberosa*, a predator of *Diadema*. The investigators suggested that the alarm response is an adaptation to reduce predation.

This work further illustrates the desirability of a field approach for experimental work because the results directly apply to the operation of the behavioral mechanism under natural conditions. The information obtained is also useful for designing laboratory studies requiring more stringent controls. Other examples of research conducted in some part by free-diving are found in Nelson *et al.* (1969), Fishelson (1971), Clarke (1970), and Reed (1971).

The usefulness of free-diving from the surface is sharply reduced as the need for sustained periods below the surface increases. It is inefficient and somewhat ineffective even at shallow depths for detailed observation since the diver can usually remain submerged only for periods of less than one minute before needing to resurface for air. In addition, this intermittent commotion often disturbs the animals under observation. The simplest solution is the adoption of SCUBA, permitting longer periods at depth with less exertion.

b. *Observations Using SCUBA.* A basic SCUBA outfit consists of single or twin 38, 58, or 72 ft^3 tanks holding air at 1800–2500 psi and a two-stage, single or double-hose regulator with mouthpiece. Mask, flippers, and safety vest complete the basic outfit. The degree of attendant paraphernalia, such as snorkel, wet suit, weightbelt, knife, compass, depth gauge, watch, etc., depends on the water conditions and task at hand. A wet suit for thermal insulation is strongly recommended when the observer is inactive for long periods. Even in water of 25°C, the temperature differential between body and medium will result in chills after several hours exposure.

In addition to items strictly for diving, an investigator usually carries some device for recording notes. A plastic slate with attached lead or wax pencil is the simplest method although underwater tape recorders using microphones mounted against the throat (Losey, 1971), in the mouthpiece (Fager *et al.*, 1966), or in a full-face mask (Hollien and Rothman, 1971) are available. Such devices as portable TV cameras, manual event recorders, still and cine-cameras, directional hydrophones, and flood and strobe lights in pressureproof or pressurizable, leakproof cases are also marketed or can be adapted by the researcher for his specific needs (Losey, 1971; Fager *et al.*, 1966; Starck and Davis, 1966).

The most important remaining addition for the undersea scientist is at least one competent diving "buddy." Behavioral work by its nature requires a high degree of concentration by the investigator who is often unaware of potentially troublesome situations. Classic is the experience of a colleague, engrossed in observing an animal close-up, who was approached to within a few meters by a 4 m long shark. When asked by his buddy why he didn't respond to the enormous predator he replied, "What shark?" I find it valuable to assign the task of observation to one partner while the other assumes responsibility of watching the observer and monitoring the situation. It hardly needs elaboration, but it is essential that a scientist receive adequate training and develop his skills appropriate to the demands of his diving tasks. A poorly prepared diver is at best inefficient scientifically, and at worst, dangerous to himself and his partner.

The recent literature includes numerous studies using SCUBA, so my examples are chosen mainly to demonstrate the range of applications. Present knowledge of the behavioral ecology of coral reef animals, especially fishes, has been significantly enriched by the studies of Randall, Starck, Hobson, and their respective associates (Randall, 1958, 1967; Schroeder and Starck, 1964; Starck and Davis, 1966; Hobson, 1965). These investigators performed their observations using standard SCUBA, underwater photography, and underwater lights. The SCUBA enabled them to work hundreds of hours at depths down to 40 m (although mostly at shallower depths). The photography permited successful documentation of important behavioral events, and the portable lights allowed night studies. This latter aspect is particularly significant since these workers discovered the reef ecosystem operates on a spectacular diurnal–nocturnal shift in activity levels and behavior of the inhabitants.

Some major points presented by the collective studies are briefly stated below. A crepuscular transition occurs in the presence and activities of most species of fishes and invertebrates (Collette and Talbot, 1971) reflecting the diversity of niches and behavioral activities. After dark, dominant diurnal reef fishes including pomadasyids (grunts), lutjanids (snappers), labrids (wrasses), pomacentrids (damsel fishes), scarids (parrot fishes), and acanthurids (surgeon fishes) either seek cover in the reef interstices (e.g., the latter four groups), or move off the reef to feed (e.g., the grunts and snappers) (Starck and Davis, 1966). Fishes which either school over the reef by day or occupy crevices, emerge at dusk to feed and engage in other activities. This is also true of such invertebrates as the coral polyps, spiny lobsters (Cooper and Herrnkind, 1971), octopods (Hochberg and Couch, 1971), arrow crabs (Barr, 1971), and *Diadema antillarum* (Birkeland and Gregory, 1971). Associated with the day–night changes in behavior, many fishes undergo dramatic color changes (Schroeder and Starck, 1964; Starck and Davis, 1966). Obviously, this range of information on such a diversity of species and involving so many behavioral activities is available only through *in situ* observation, in this case facilitated by SCUBA. Other examples of observational studies employing standard SCUBA are found in Clarke (1970), Myrberg *et al.* (1967), Randall (1964, 1967), Heydorn (1969); Feddern (1965), Larcombe and Russell (1971), Russell (1971); Chapman and Rice (1971), Cobb (1971).

Experimental studies are also facilitated by SCUBA (e.g., Hartline *et al.*, 1971; Herrnkind and McLean, 1971; Waterman and Forward, 1972). Our research group initially examined the orientational behavior of spiny lobsters, *Panulirus argus*, through underwater field experiments conducted by divers (Herrnkind, 1970; Herrnkind and McLean, 1971; Cooper and Herrnkind, 1971). Experiments under simulated conditions were precluded by the difficulty of reproducing the spatial requirements, hydrodynamic surge and current forces, and substrate conditions present during directed locomotory movements witnessed in the field (Creaser and Travis, 1950; Herrnkind and Cummings, 1964; Clifton *et al.*, 1970). Thus we experimentally examined the directional orientation of spiny lobsters in the field while monitoring possible guidance cues.

Briefly, a SCUBA diver released a lobster with its longitudinal body axis oriented along a given compass angle (Fig. 1). Its path and bearing at the vanishing point (limit of underwater visibility, usually 10–30 m) was recorded. The lobster was then retrieved and released at a 45° angle clockwise or counter-clockwise from the original release angle. The procedure was repeated a total of eight times (0°, 45°, 90° . . ., through 360°) to control for

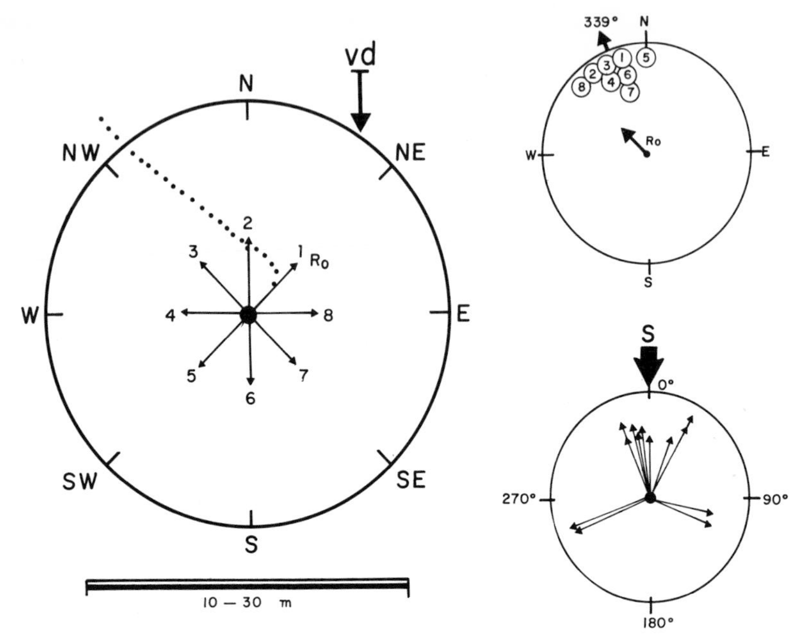

Fig. 1. (left) Field technique for testing orientation of spiny lobsters (*Panulirus argus*) on open, level sand substrate at depths of 5–10 m. A lobster is first released (1, Ro) heading in some arbitrary direction (e.g., NE) and observed to the vanishing distance (vd) or some specified point. The lobster is then retrieved and released at successive angles of 45° either clockwise or counterclockwise, a total of eight times. A typical pathway is indicated by the dotted line. (Right) Upper scatter diagram shows typical results of eight runs by one lobster. The lower diagram shows the performance of 13 lobsters by mean vectors calculated from the r_i values. In this case, most lobsters moved in the direction from which waves approached and passed over the test site. The resultant surge (S) force (a back and forth water displacement) is suggested as the orienting cue. (Figure modified from Herrnkind and McLean, 1971.)

potential bias of body axis orientation at each release. Numerous individuals were tested with and without sensory restrictions.

Results were analyzed by comparing the orientation of each lobster with the directional components of potential guidance cues measured at the time of testing; e.g., the maximum bottom downslope and the approach direction of current and wave surge. Well oriented lobsters often travelled in the direction from which waves approached, independent of slope angle or current direction. We hypothesized that some aspect of wave surge provides a directional guidepost to spiny lobsters under certain conditions. On the strength of our observations we are designing experiments for use in a wave test-tank to determine the response of lobsters to wave forces while controlling for other factors.

A major drawback of standard SCUBA, frequently mentioned (if not always in print), is the optical and acoustical artifact of the bubble stream. While animals appear in many cases to ignore or habituate to the diver, it is sometimes obvious that this is not the case. In some situations, it is not possible to determine if the organism's behavior is somehow being modified (High and Beardsley, 1971).

A second limitation is the reduced time at depth as depth increases due to the increasing volume of air inspired and then lost upon expiration. Thus a diver's expiration volume of 1 liter of air at the surface becomes an expiration of approximately 2 liters when it reaches the surface from 10 m (approximately 2 × atmospheric pressure) and 8 liters from 60 m (approximately 8 × atmospheric pressure). Thus, a volume of air sufficient for 1 hour at the surface lasts only $\frac{1}{2}$ hour at 10 m.

c. Observations Using Rebreathers. Nearly complete avoidance of the bubble problem and a partial solution to the time-at-depth limitation is provided by mixed gas rebreathers developed within the past decade (not to be confused with pure oxygen rebreathers using by the armed services for many years, which are limited to use in depths of less than 10 m).

The physical principles of rebreather operation are simple (Schmidt, 1971). Instead of exhausting the exhalant, it is passed through a carbon dioxide absorbent and electronically injected with sufficient additional pure O_2 to bring the oxygen partial pressure back to some preprogrammed level as monitored by redundant oxygen sensors. The unused oxygen in each exhalation is then "rebreathed." Depending on the make and model, currently marketed rebreathers (e.g., General Electric and Bio Marine) provide sufficient air or mixed gas for four or more hours at virtually any depth. This latter capability is provided by the fact little gas escapes and human metabolic requirements do not greatly increase as a result of hyperbaric conditions imposed in moderate depths. The rebreathers are generally packaged in a removable housing and are worn much like a standard SCUBA. The weight of these devices on land varies with make and model; e.g., the GE Mark-X Model 3 weighs approximately 65 lb, or some 15 lb less than twin 72 ft^3 air tanks.

Only a few behavioral studies facilitated by rebreathers have appeared thus far. Collette and Talbot (1971) examined the crepuscular changeover of reef fishes at 15 m depth off St. John, U.S. Virgin Islands. During approximately 100 diving hours, the investigators identified 107 species in 36 families and established times of activity change and other information on 35 species. This work nicely complements the day–night comparison studies by Hobson (1965) and Starck and Davis (1966).

Smith and Tyler (1971) conducted studies on space resource sharing in a coral reef fish community in the same general area. They confirmed the general hypothesis that the fishes characteristic of small patch reefs truly interact to establish a community rather than being merely random assemblages. More specifically, they determined some of the niche relationships of certain important species as related to feeding areas, prey, nest sites, territoriality, social patterns, and home range. The authors of both studies pointed out the value of long term (up to 4 hr) observation periods and the absence of the effect of bubbles on the organisms while using rebreathers.

Other behavioral research using rebreathers is reported by Earle (1971) and Cooper and Herrnkind (1971).

Working at depths exceeding 25 m is considerably more difficult than in shallower conditions. The first problem, mentioned earlier, is maximum time at depth. Closed circuit rebreather systems or increased capacity open-circuit rigs (by using double or triple tanks) can provide increased time at depth. The latter are considerably more cumbersome and less efficient, but also less expensive, than the former. However, once the diver has exceeded certain critical periods under compression, nitrogen in the system begins to saturate the blood and tissues requiring decompression before returning to the surface. Failure to decompress results in nitrogen bubble formation and in the blood stream and may result in the "bends."

The rate of saturation increases with depth in such a way that a diver has 50 min time at 20 m, 10 min at 40 m and only 5 min at 60 m before decompression is necessary. At depths below 40 m, the small amount of on-site time obviates many studies. Should the time limit at depth be exceeded, the decompression time increases markedly in comparison to the bottom time. Decompression can take place *in situ* at progressively shallower depths or by transferring the diver to a pressurized chamber aboard ship or ashore. Neither situation is satisfactory in that both require valuable research time. In addition, the saturated diver is in great potential danger until decompression is completed.

The second problem is the narcotic effect of nitrogen under pressure. Nitrogen narcosis may affect different individuals at different depths but becomes noticeable to most divers by 40 m. This narcosis problem is solved by using a helium–oxygen mixture for work of long duration at deep depths. However, "heliox" is more expensive and cumbersome to handle than compressed air.

Yet another problem in deep diving for extended periods is oxygen poisoning, since 20% O_2 (a partial pressure of 0.2 atm) in the breathing mixture causes severe physiological distress under high pressures. Hence, the level of oxygen in the gas cylinder or the programmed setting in a rebreather must be properly adjusted depending on depth. Transcending all of this, is the increased risk of injury or death in event of accidents at extreme depths, including human error as well as equipment failure. The formidable physiological limitations, aptitude, and experience restrictions, training, expense, and logistics are barriers hurdled by a relatively few diving scientists at present. The depths below 100 m are presently more amenable to *in situ* study by means of remote sensing. This will be discussed in a later section.

d. *Use of Undersea Habitats.* Within the past decade the concept of undersea habitats which divers may freely enter and leave has been realized. The habitat occupants live in a gas atmosphere under the ambient hydrostatic pressure for that depth and can leave the habitat through a hatch directly into the sea. The limits here are vertical since the individuals are totally saturated with the inert gas (nitrogen or helium) at the habitat's depth-determined pressure and are subject to the bends if they approach too near the surface.

The primary advantages of such habitats are the enormous increases in the ratio of useful work time to decompression time. Once totally saturated, the inhabitants can remain at depth indefinitely without increasing the necessary decompression time. For example, the decompression time for aquanauts submerged for 14–20 days during the Tektite-2 program was approximately 22 hr, the same time required after being down only a few days at that depth (approximately 12 m). In theory, the concept applies to any depth, since once 100% saturation occurs, the same decompression time will be required, regardless of the length of time at that depth.

The use of saturation mode habitats began in the United States with the Navy's SEALAB program. This was initiated during the late 1950's under the guidance of Capt. George Bond (MD). The first program, SEALAB-1, took place off Bermuda in 1964 at the depth of 60 m and established the feasibility of such operations. This was followed by a more ambitious program in 1966, SEALAB-2, in which teams of aquanauts lived and performed research tasks at a depth of 65 m off San Diego, California. Also about this time, Jacques Cousteau successfully operated the CONSHELF habitat at 100 m in the Mediterranean. The 1968 SEALAB-3 operation designed to take place at over 180 m depth was aborted upon the accidental death of one of the aquanauts. Since that time, the United States Navy has temporarily, at least, given up the idea of manned undersea habitats and the most ambitious, deep habitat research program in the world has been halted.

Most of the other recent habitat programs, some developed by private industry, are aimed at shallower depths where nitrogen atmospheres are practical and the problems associated with great depth can be avoided.

Among these operations are the EDELHAB (FLARE Program 1971–1972), Perry HYDROLAB, European HELGOLAND experiment, the ongoing Texas Tektite and PRINUL (Puerto Rico International Undersea Laboratory) programs. To date, the most successful from the standpoint of marine science, particularly behavioral research, were the Tektite-1 (1969) and Tektite-2 (1970) Programs (Fig. 2 and 3; Clifton *et al.*, 1970; Miller *et al.*, 1971).

The general plan of Tektite-2 involved 2 or 3 week habitations by two teams of two scientists each (usually each team had a different study objective). In addition, each mission included an engineer who was responsible for the technical operation of the habitat. Each individual was given several

Fig. 2. (top) Artists cutaway depiction of the Tektite habitat (from Clifton *et al.*, 1970).

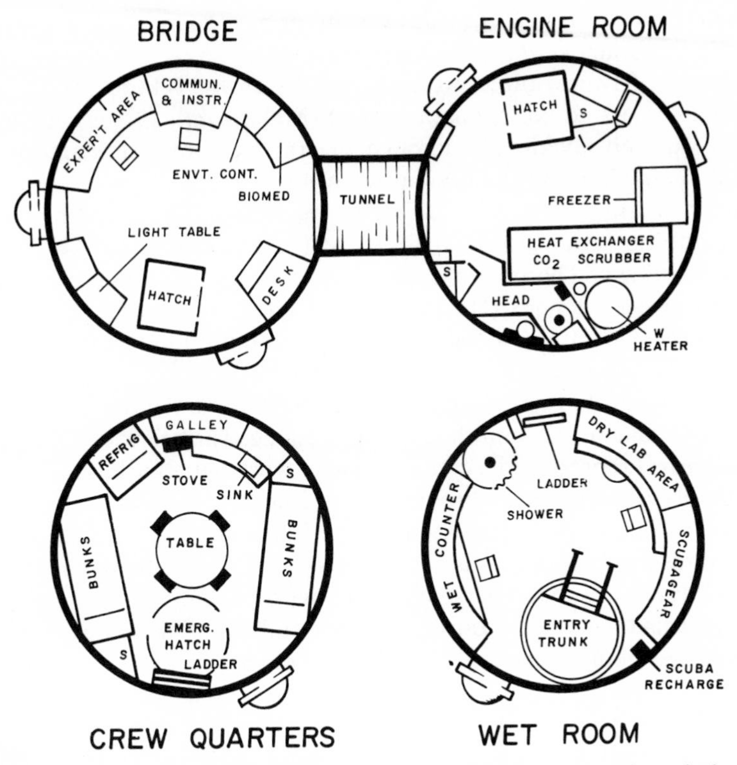

Fig. 3. Layout of the Tektite habitat interior. The aquanauts enter through the trunk in the wet room. Immediately overhead is the engine room housing the atmospheric control system. Crossing through the tunnel leads to the bridge with monitoring equipment and communications. Immediately below are the crew quarters with sleeping and food preparation areas. (Figure modified from Miller *et al.*, 1971, p. II–3).

weeks of training and familiarization with the undersea on-site topography prior to each mission. Rather than being a strictly experimental program, one of the primary objectives of Tektite was to increase the research capability of the participating scientists.

The program took place at Great Lameshur Bay on the south shore of St. John, U.S. Virgin Islands (for details see Clifton *et al.*, 1970; and Miller *et al.*, 1971). The mild weather, protected bay, and clear, warm water (20–30 m visibility, 80°F temperature) were all well suited for undersea work. The habitat was situated at a depth of approximately 15 m on a sand substrate adjacent to a live coral reef. The area available to the aquanauts varied in depth from the 8 m upper excursion limit to a depth of approximately 23 m about $\frac{1}{4}$ mile away on the sand–algae plain seaward of the reef. The biological character of this tropical reef area and biota are described by Kumpf and

Randall (1961), Clifton *et al.* (1970), and in various papers from the Tektite-2 final report (Miller *et al.*, 1971).

Thirteen of twenty-three projects undertaken by Tektite-2 researchers involved some study of animal behavior (Miller *et al.*, 1971). Most of the studies were observational, although several incorporated *in situ* experimental manipulations. The descriptive work included observations on the behavior of fishes relative to different trap designs (High and Beardsley, 1971), feeding strategies of certain reef invertebrates (Birkeland and Gregory, 1971) and herbivorous fishes (Earle, 1971), the ranging and activity patterns of cephalopods (Hochberg and Couch, 1971), fishes (Smith and Tyler, 1971; Collette and Talbot, 1971), and spiny lobsters (Cooper and Herrnkind, 1971; Herrnkind *et al.*, in press).

By design, most projects were limited to the duration of the mission, plus whatever could be accomplished during training. The behavioral project on spiny lobsters represented a continuation of work initiated during Tektite-1 and involved a year-long program including three, 3-week missions on Tektite-2 (Cooper and Herrnkind, 1971). Both this long term project and some of the single mission studies involved coordination between the aquanauts and surface support personnel, although other projects were conducted entirely from the habitat. Coordination and collaboration between teams on a given mission frequently occurred even though each had separate overall objectives. The variety of projects, techniques, and approaches allow a preliminary appraisal of the applicability of undersea habitats to behavioral research (as well as to other fields).

The advantages most often mentioned in the researcher's reports were the long length of time available at depth and the ease with which investigators operated from the habitat as compared to working from the surface. For example, cumulative in-water time of 50–86 hr per individual for the 2-week missions are reported (Collette and Talbot, 1971; Earle, 1971). My partner and I accumulated a total of 162 hr during approximately 20 available working days; an average of 4 hr per day each. On certain days, as many as 8 diving hours were logged by individual aquanauts at depths to 20 m. Perhaps even more significant was the daily regularity of dives even during turbulent weather that restricted surface diving operations. Brief storms did not usually limit diving from the habitat and one team even took the opportunity to examine the effect of storm-caused turbidity on fish activity patterns (Collette and Talbot, 1971).

Besides providing long periods *in situ* under a wide range of conditions, operation from the habitat was more convenient and probably safer than surface operations. For instance, surface problems such as locating an underwater site from a boat, weighing anchor, loading, unloading and preparing equipment, donning diving gear on a tossing deck, long trips enroute, engine breakdown, long periods at decompression stops, etc., were eliminated or reduced. Night work was especially convenient since preparations took place on a stable, well-lighted platform and water entry was a simple process. The capacity for sustained underwater work without undue fatigue, a major cause of diving mishaps, was a major contribution of this type of habitat operation that cannot be equalled by most other diving techniques. In addition, the time saved in avoiding logistic and decompression problems was available for scientific endeavor. The advantages cited above will become even more important in future habitat operations situated at long distances from shore and/or at greater depths.

Any appraisal based on the preliminary reports thus far published may be somewhat preliminary, although several scientifically significant points are already apparent. First, considering the brevity of the missions, the work reported is amazingly thorough, due largely to the extensive dive time. The studies by Collette and Talbot (1971); Smith and Tyler (1971), and Earle (1971) are particularly notable in this regard. Collette and Talbot examined the crepuscular change-over in reef fish activity during a 2-week mission. They consider their results comparable to portions of surface based studies of much longer duration by Hobson (1965) and Starck and Davis (1966). Earle's behavioral observations on herbivorous fishes, while not quantitatively equivalent to similar studies by Randall (1967), are at least qualitatively comparable especially when considering the shorter length of time involved. Barr's (1971) descriptive study of the arrow crab, *Stenorhynchus seticornis*, provides another interesting example since he accomplished the study largely on spare air-time available after completing his primary research work associated with the lobster project. This could be relatively easily done simply by studying the arrow crabs living with 10 m of the habitat entrance.

The Tektite studies of homing behavior and habitation patterns of spiny lobsters, *Panulirus argus*, provided much new information about a species on which millions of research dollars have been spent and hundreds of papers have been written (Cooper and Herrnkind, 1971; Herrnkind and McLean, 1971; Herrnkind *et al.*, in press). The undersea habitat approach greatly facilitated the continual daily and semidaily surveys of the reef to monitor the location and activity patterns of individually identifiable lobsters.

For example, we found that reef lobsters typically reside by day in specific dens from which they emerge at night to feed and to which they return by the following dawn. The den sites most frequently occupied often contained more than one individual. Certain lobsters maintained residency in one or a few dens for weeks at a time while others, after leaving the area for periods of up to several months, eventually returned to the same dens.

Displacement experiments, discussed in a later section, showed that most lobsters homed to the vicinity of the den from which they were captured (Herrnkind and McLean, 1971). Other data suggest that lobsters subjected to disturbance within the den during the day leave at night and take up residence in other dens. This information dispels the commonly held notion that spiny lobsters fortuitously choose dens and adds the knowledge that older, mature lobsters are selective in their habitation rather than being loosely migratory, as are the younger individuals (Dawson and Idyll, 1951). These results emphasize that the dynamics of a lobster population may be more clearly understood on a basis of daily observations of identified individuals rather than through inference from intermittent trap collections.

Although there are many advantages to undersea habitats, they are not without their own problems, including such things as personnel and equipment logistics, technical difficulties, and cost effectiveness. The habitat-based aquanaut, being fully saturated, can not surface and is in constant danger of this eventuality should an accident occur. While the surface-based diver can usually decompress (short term) anywhere, before returning to the surface, the aquanaut must return to the habitat (thus, the benefit of the mixed gas rebreathers for habitat operations).

The capacity to home to the habitat efficiently, day or night, requires familiarization with surrounding topography, and competency in compass navigation, acoustic or sonar tracking devices, or a combination of these (Hollien and Rothman, 1971). A reasonably serious problem occasionally occurs when an aquanaut working in areas shallower than the habitat depth cannot equalize (clear) the inner ear pressure in order to descend back to the entryway. In addition, excursions from the habitat are limited by air supply, fatigue, and water conditions. Aquanauts are also denied access to shallow water areas which may be important to their study.

The problem of personnel and equipment transport can be partly alleviated at the added expense of a surface based support team ready to complement the capacities of the saturated workers. Such cooperation requires effective communication, both from surface-to-diver and between divers. Efficient interdiver communication systems are presently being perfected and their absence during previous habitat operations is regrettable.

Undersea transport can be facilitated by diver delivery vehicles such as the Rebikov Pegasus, Perry Reef Hunter, and one-man scooters commercially marketed for sport divers. Another approach to facilitate logistics is the development of movable habitats, perhaps selfpowered and somewhat independent of surface support. The EDELHAB habitat used in conjunction with a surface support ship for power, gas supply, and transportation and the newly designed habitat for PRINUL incorporate some of these approaches. Future designs will likely approach more optimal capabilities as the interest in habitats grows.

Regardless of design and utility, habitats and their operation are expensive in comparison to surface-based diver programs. The argument becomes, "Is it worth it?" In my opinion, the work accomplished during Tektite-2 could have been accomplished from the surface at less cost although it would have taken more scientific personnel working a much longer time. This is true only because the working depths were generally less than 25 m and the location was only a few hundred meters from the shore facilities. However, at depths greater than 50 m, or further offshore, the work would have been virtually impossible except by use of a habitat or lock-out type submersibles. It is under these conditions then that the habitat concept will achieve its optimal value in undersea research.

The Tektite program successfully demonstrated the advantages and adaptability of this approach for scientific research and provided a wealth of new information for future applications. The SEALAB programs suggest that the human and technical problems of saturation at deeper depths can be overcome when we consider the results to be worth the price. The space program is a case in point.

3. *The Scientist-Outside-the-Sea*

a. *Observation through the Air–Sea Interface.* Surprising contributions to our knowledge of the behavior of marine organisms have come from observations through the water surface from above. The conditions under which this technique is useful are dictated by water clarity, depth, and degree of surface ripple. Under any condition, the ability to see is improved when glare is reduced using Polaroid glasses and locating the best vantage point to avoid reflection. Larger areas and gross details at greater depths can be observed by raising the viewpoint as high as practical from a platform or tower. The main advantages of such aerial observation are convenience, safety, and extended time.

The technique is especially valuable because it maximizes the observable area in shallow murky water. The study of acoustic attraction of young lemon sharks (*Negaprion brevirostrus*) to sounds of struggling fish by Banner (1972) illustrates this nicely. The work was conducted in shallow, (less than 1 m depth) sand–grass flats where the sharks spend much of their juvenile period. Depending on wind and tidal factors, visibility seldom exceeded a few meters and was often less than 1 m, making underwater observation over a wide area impossible. To record the effect of sound signals on the orientation of a distant shark, Banner observed from a boat anchored an appropriate distance from a transducer. He found that young sharks detected and oriented to struggling fish (or acoustic simulations) from distances of over 10 m, thus far exceeding the underwater visibility.

The value of high vantage points is shown in the work of Yarnall (1969) in establishing the ranging and activity patterns of *Octopus cyanea*. The diurnally active octopods, confined within a large seminatural pool, were observed from a tower mounted high enough to avoid most sun reflection and observe influence in addition to providing a panoramic view of the pool. A similar approach was taken by Hodgson and Mathewson (1971) in an experimental study of the orientation of sharks to current-borne chemicals. The movements of the untethered experimental animals were recorded visually and cinematographically from a platform 6 m above a 12 m × 24 m enclosure approximately 3 m deep. Other similar examples are found in Aronson (1971), and Arnold (1965).

Nelson and Gruber (1963) used an airborne observer to record the acoustic orientation of sharks to an underwater sound source on a coral reef from several hundred meters away. Observations from aircraft are also useful in tracking migrations, locating fish schools, and in studying the behavior of marine mammals.

The drawbacks of observation through the air–water interface relate in part to the limiting factors such as depth, turbidity, and the water surface conditions. Thus Hodgson and Mathewson were able to monitor their sharks at 3 m depth through the clear waters at Bimini, but they would be unable to do so at that depth in the murky waters of Biscayne Bay (Miami) where Banner worked. Further restrictions of this technique relate to the size of organisms and degree of detail required. The higher the vantage point, the more difficult it is to see small animals or details of posture, appendage movements, rapid body movements, etc. Also, being above the animal prevents visualizing changes in color patterns or other actions most readily seen from a lateral view. Movements from above may induce fright responses or otherwise interrupt the behavior of many marine species. The obvious usefulness of this technique is in tide pools, shallow turbid waters, and in reasonably clear waters where entry may be impractical or dangerous.

b. *Observation through a Viewport*. Another obvious way to maintain the air-bound observer in his natural habitat while providing a direct subsurface view is the use of a glass-bottom bucket or a viewport in a ship, submersible, or raft. The inherent values of this approach are convenience in note taking, photography and other manipulations, as well as relative safety and long observation time. The main advantages over the previous method are the absence of distortion and the availability, in many cases, of a lateral view of animals.

This method was used with excellent results by researchers associated with the National Marine Fisheries Service (then Bureau of Commercial Fisheries) Biological Laboratory in Hawaii during the 1960's. Their general objective was the accumulation of knowledge on the behavior of pelagic fishes, particularly the tunas (Nakamura, 1972; Gooding and Magnuson, 1967). The equipment and techniques, described by Nakamura (1972), included a retractable caisson lowered from a ship, subsurface viewports in research ships, a 360° view-cylinder suspended below a free-drifting raft, and both wet and dry submersibles.

An example of the value of such studies is the new knowledge obtained on the ecological significance of drifting objects to pelagic fishes (Gooding and Magnuson, 1967). A free-drifting raft with subsurface viewing cylinder was employed for this work; the raft serving as the drift object around which the pelagic animals naturally accumulate and interact. The support ship with its attendant chemical, acoustic, and hydrodynamic artifacts remained in the distance. The investigators concluded that floating objects attracted pelagic fishes, and served both to concentrate the food supply of a number of species and, at the same time, acted as protective cover for the prey. In addition, the previously unreported interspecific cleaning behavior of pelagic fishes was recorded for the first time. It seems unlikely that such information could have been obtained except through an *in situ* approach.

A major drawback of the viewport approach is the passive role of the observer since one must wait for action to take place in the vicinity. A problem specific to most small floating vehicles is the motion induced by wave action which can disrupt visualization directly by gyrating the visual field or indirectly by causing motion sickness in the observer. Such problems were not insignificant in the studies mentioned above.

B. INDIRECT OBSERVATION: REMOTE SENSING

1. *Introduction*

The use of remote sensors to obtain behavioral data, or concurrent information about the physical environment, can sometimes replace the scientist-in-the-sea or provide complementary information. A remote sensor may extend one of our senses, such as the acoustic or optic, into otherwise inaccessible areas, or may amplify energy levels which are below the human threshold (as with photo-multiplying devices), or it may convert some form of information-containing energy into a useful or recognizable form (as in thermal recording or acoustical holography). Long duration recording devices, and those placed at deep depths or in other situations which preclude a human presence, are of primary significance for undersea behavioral research. As stated earlier, depths below a few hundred meters or regions of excessive turbidity and water motion are accessible only by remote means.

In other situations, the nature of the information desired may not require an on-site observer. For example, a simple photocell device can adequately monitor the number of times an animal leaves or enters a den while direct human observation would be a waste of effort.

Finally, breakdowns, while perhaps more frequent in remote sensors than in man, are only mentally, rather than physically, painful and such equipment can usually be repaired or replaced.

2. *Underwater Photography*

Photography is invaluable in studying animal behavior in any environmental setting (Eibl–Eibesfeldt, 1970; Winn and Olla, 1972).

Among the important applications of photography are: (1) the permanent recording for purposes of documentation and description of events, (2) the recording of actions for later detailed analysis after enlargement, color filtering, contrast modification, stereoscopy, etc., (3) recording of photic events outside the human perceptive range by magnification, special films (e.g., infra red, ultraviolet sensitive), slowing or speeding of cine film to analyze temporal events beyond human perception, and (4) the capability of recording in darkness and at deep depths using high speed strobes. Modern camera gear (including 35 mm and larger formats, cine 8 mm, super-8 mm, 16 mm and larger) is adapted for underwater use by encasing the camera in separate pressurizable or pressure and leak-proof housings, although some equipment is designed for direct immersion. This applies also to accessories such as strobes (including slaves), flash bulbs, and light meters. Films and filters in ranges appropriate to the light frequency modification by water are available.

The techniques needed for good underwater photography are probably no more complicated than many applications in air but referral to any number of basic books on the subject saves effort for the novice (e.g., Starck and Brundza, 1966). Photographic and skin-diving (as well as scientific) journals are useful in keeping pace with new advances (e.g., *Photographic Applications in Science, Technology and Medicine*). With the advent of small portable cameras, such as the Nikonos 35 mm, as well as numerous housings for a variety of small still and cine cameras, combined with the new high speed films, the diving researcher is relatively unencumbered by the equipment, but is capable of professionally recording significant behavioral events.

Scanning the literature reveals many interesting photo-applications so the examples given here are only suggestive of the range. The analysis of the spatial–temporal character of fish schools requires (for larger species, at least) *in situ* photo recording (McLean and Herrnkind, 1971). This is apparent in the review by Breder (1959) describing features of schools in relation to behavioral stress and changes in physical conditions. The careful photographic analysis of interfish distance enabled Shaw (1969) to develop formulae predicting such spatial relationships. The use of suspended particles in association with photo techniques to record water displacement and turbulence has further extended this approach.

In many cases, photographic documentation is essential in the recording and documentation of unique undersea behaviors. For example, without photographic documentation, it would have been difficult to record the mass single-file migrations of the spiny lobster, *Panulirus argus*, and impossible to convince skeptics that such a phenomenon could occur. (Fig. 4; Herrnkind and Cummings, 1964; Herrnkind, 1969).

Field photo documentation of fish color changes (Starck and Davis, 1966; Nakamura, 1972), cleaning behavior (Limbaugh, 1961), thoracic collapse in deep diving porpoises (Ridgway *et al.*, 1969), new shell acquisition by hermit crabs at gastropod predation sites (McLean, in prep.), and the escape responses of sea urchins (Snyder and Snyder, 1970) all constitute examples of behavioral events first recorded photographically *in situ*. Waterman and Forward (1972) used underwater photography to obtain the first clear-cut evidence of fish orientation to polarized light. In some cases, extensive photo analyses are better performed under lab conditions after the initial field evidence is secured.

Some immediate drawbacks of photography are the relatively short duration for continuous recording, as compared to direct observation, and the lack of rapid feedback in analyzing the data so obtained. Otherwise, the

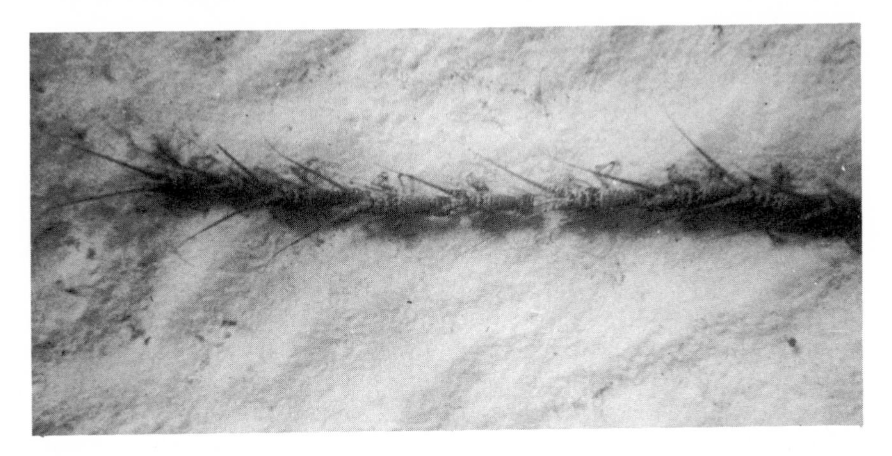

Fig. 4. A queue, or single-file line, of spiny lobsters, *Panulirus argus*, during a mass migration off Bimini, Bahamas in 1969. Each trailing lobster maintains contact with the lobster ahead using its antennules and anterior pereiopods. Hundreds of such queues ranging from 2 to 50 lobsters move in the same general heading over periods of up to several days, usually following autumn storms. (Photo by W. Herrnkind)

limitations of photography, in the marine environment, as one would expect, follow those of human vision. Cloudy water restricts distance and occludes detail as well as reducing available light. In some situations detail may be resolved with Polaroid filters (Lythgoe and Hemmings, 1967). Similarly, as depth increases, light intensity decreases making necessary the use of artificial light sources to locate and record subjects. Such illumination, however, may modify behavior and care must be exercised in analyzing such records. Lens and film choice may cause problems since in most cameras they cannot be changed underwater. Focusing by standard view finders is made difficult by interference from the divers face plate although the new optical viewfinders are a significant improvement.

Another problem with increasing depth is the change in the wave length character of ambient light. Red is quickly filtered out even in clear oceanic water and the turbidity of inshore waters may cause a high absorbence of red or other wave lengths which might cause distortion or undesirable background color (Luria and Kinney, 1970). Filters (e.g., to reduce the amount of blueness) or artificial light are needed to achieve good color results even near the surface in clear sunlit water.

At extreme depths and other situations demanding remote controls, the photographer has to rely on fate to capture the desired event. In addition, the chance of equipment failure is greatly increased.

Finally, underwater camera equipment is considerably more expensive than that used on land. For example, the underwater housing for the popular Bolex 16 mm cine camera costs more than the camera itself. Regardless of limitations, photography is still one of the most useful ways of recording behavioral events *in situ.*

3. *Underwater Television (UTV)*

Television, either portable or permanently mounted, has been successfully used for a variety of underwater studies over the past 20 years (Barnes, 1963; Kumpf, 1964; Cummings *et al.*, 1966; Myrberg *et al.* 1969). Reduction in size and cost of cameras, recorders, and power supplies make television suitable for a variety of applications to undersea work. UTV has the great advantage over photography of providing an instantaneous and continuous picture plus a videotape replay capability. Also a remotely focused stationary camera is less likely than a human observer to disturb subjects under study.

Along with the capacity for continuous observation, UTV makes possible long term, detailed study of behavior at depths not optimally available to SCUBA divers. This approach was successfully used for the last decade at the video–acoustic installation in Bimini, Bahamas. The system at various stages of refinement is described by Kronengold *et al.* (1964), Stevenson (1967), and Myrberg (1972). The most successful model (Fig. 5)

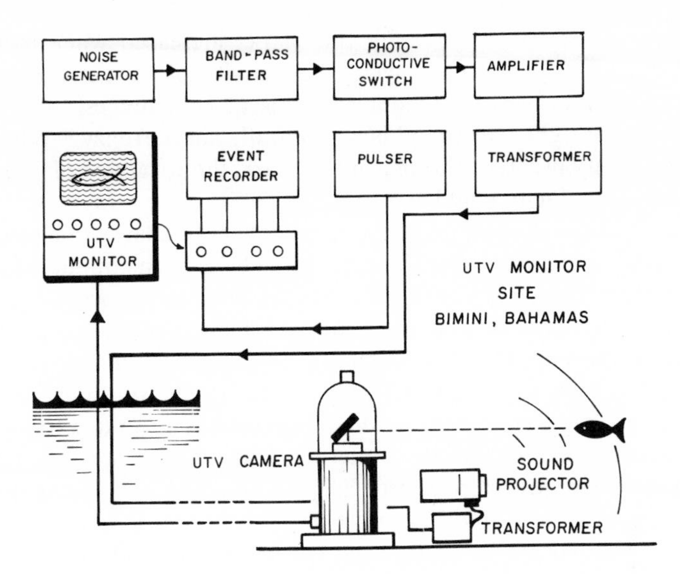

Fig. 5. An underwater television (UTV) installation 20 m deep off Bimini, Bahamas. Panning and tilting are accomplished through a movable prism system. In addition the camera has a zoom lens. The UTV was used in conjunction with an acoustic monitoring and playback system allowing remote experimental manipulation and recording. (Modified from Richard, 1968.)

was placed at approximately 20 m depth near a reef outcrop off the west coast of Bimini where visibility ordinarily ranged from 20 to 30 m. The camera was enclosed in an opaque housing and was remotely focused, zoomed, panned, and tilted via a small movable prism (mounted above the camera lens) in a transparent hemispherical housing. This eliminated much of the movement that might cause startle responses of nearby animals. The monitoring station included a video tape with acoustical channels from hydrophones near the camera and from voice microphones operated by an observer viewing the monitor.

Myrberg *et al* (1969) used this system to study the attraction of free ranging sharpnose sharks (*Rhizoprionodon* sp.) to sounds of various frequency and pulse rate. By television scanning the underwater horizon during alternating periods of acoustic signals (experimental) and silence (control), the researchers were able to demonstrate that low frequency sounds, projected as an irregular pulsed signal, attracted sharks from beyond visual range within minutes to the area of the speaker. They further found that frequencies below 1000 Hz, especially in the 20–100 Hz range, were attractive to this species. Frequent playback of the signals in the absence of reinforcement eventually resulted in apparent habituation by the sharks in the area.

By testing under various visual conditions, they found that tighter circling and criss-crossing of the area by the sharks increased when light level decreased. Additionally, the swimming activity of attracted sharks increased as the number of individuals increased. They suggest that this may provide some insight into the phenomenon of "shark frenzy." While most observed activity involved the sharpnose shark, the attraction of three other species was also documented.

Although useful studies on acoustic responses of sharks (Nelson, 1967; Banner, 1967) and bony fishes (Iversen, 1967) have been conducted in large tanks or pools, the results suffered from the presence of acoustic artifacts or spatial restrictions. Other applications of UTV in relation to experimental acoustical studies are found in Kumpf (1964), Steinberg *et al.* (1965), Cummings *et al.* (1966), Richard (1968), Stevenson (1972), Myrberg (1972).

A fixed system, such as used in the Bimini studies, is limited to studies of those organisms attracted to the camera or in the immediate area. Portable systems have been developed, especially as an adjunct to diving and submersible operations. A good example of this is the work on the behavioral ecology of the Norway lobster, *Nephrops norvegicus* (Rice and Chapman, 1971; Chapman and Rice, 1971). The researchers also used SCUBA and submersibles to obtain *in situ* information on distribution, habitat characteristics and population dynamics. A portable UTV camera was used to observe the characteristic patterns of emergence from dens and other aspects of den-related and social behavior thereby reducing the possible influence of a diver and permitting long term viewing under conditions restrictive to a human observer (e.g., darkness, depth–air supply, cold temperature). This study did not require a permanent camera site. In addition, the long distance to any shore facility made this impractical. Other studies using portable UTV systems are discussed in Barnes (1963), Sand (1957), Steinberg and Koczy (1964) and Nakamura (1972).

UTV is costly for some applications where photography might suffice. Permanent installations such as the Bimini system are especially costly in requiring high quality, durable housings, cable, etc. which must function during months of submersion without deteriorating. Even then, anchor snaggings, fouling, electronic or mechanical failures, and adjustments may require all too frequent attention from divers and technicians. Naturally, the equipment must be surfaced for repair, often under less than ideal sea conditions.

4. *Passive Acoustic Monitors*

Mechanisms of acoustic intraspecific communication (Winn, 1964; Frings, 1964; Fish, 1972), orientation (Banner, 1972), and echolocation (Kellogg, 1961) have evolved in numerous marine animals. Monitoring discriminable signals such as hydrodynamic sounds indicating the presence of an animal or acoustic communications associated with some behavioral act, provide a useful technique for "observing" marine animals under conditions precluding vision. Furthermore, depending on background noise levels, acoustic events can be constantly monitored somewhat independently of water or weather conditions.

Another major advantage is the capability of detecting significant information and signals which are not based on visual systems (Winn, 1964). For example, Cummings *et al.* (1966) found that escape responses by wrasses occurred in the absence of any detectable visual stimuli but were most probably caused by hydrodynamic sounds of accelerating carangids (jacks) and other fish predators.

Acoustic equipment consists basically of a hydrophone, amplifier(s), and speaker. Short distances between hydrophone and speaker may require little amplification, filtering, and receiver elaboration. Since each system is usually tailored to the particular conditions and problem, investigators generally assemble the components as necessary. Studies on echoranging signals of cetaceans, for example, may require a sensitivity range of frequencies from 10 to 150,000 Hz (Kellogg, 1961), whereas examination of communication in many fish species involves only the lower 10% of this range (Tavolga, 1964b). At the speaker end of the system the requirements also vary; some researchers require only earphones or a loudspeaker, others need multichannel recorders, sound level meters, calibration systems, etc. (see discussion by Tyrrell, 1964). Sonograms, where required, necessitate sonographic analysis on fairly expensive and sophisticated equipment (Winn, 1972). Also it is desirable for bio-acousticians to maintain a reference library of identified recordings to familiarize new observers or to compare with possible variants or new sounds.

The development of acoustic and related equipment for defense related purposes, popular applications (e.g., musical recording), telecommunications, and biological communications research have provided a variety of items usefully adopted to marine bio-acoustical purposes. In recent years, the number of submersible items available commercially has also increased. Thus, acoustical monitoring systems useful for a variety of applications are available over a wide price range to those investigators with the technical expertise to design, build, and operate them.

The field of marine bio-acoustics grew rapidly after development of adequate underwater detecting and recording apparatus in the 1940's (Tavolga, 1964b). Reports on recent research and extensive bibliographies are found in Tavolga (1964a, 1967), Cahn (1967), Winn and Olla (1972), and Adler (1971).

Among the more ambitious bio-acoustic systems was the permanent installation operated off the coast of Bimini by Steinberg, Cummings, Myrberg, and their associates during the period from 1960 to 1971 in conjunction with the video system described previously (Fig. 5) (Myrberg, 1972). The primary capability of the installation was the sustained recording from an array of inshore hydrophones located at 15–20 m depth and from another hydrophone at approximately 400 m depth in the Florida Straits (Kronengold *et al.*, 1964; Steinberg *et al.*, 1965). By recording for 24 hr of each 72 hr period year round, the daily, monthly, and seasonal patterns of oceanic sound production were characterized. Sonic activity was recorded on magnetic tape for aural identification of specific sounds and graphically to determine the overall ambient acoustic levels.

Perhaps the most striking observation of this program was the long-term rhythmic nature of sound production by marine animals. This included sounds both produced by some biological mechanism and those resulting from the incidental physical factors associated with moving or feeding (such as scraping on the substrate) (Cummings *et al.*, 1964, 1966). Certain species exhibited strong crepuscular peaks of sound production which varied in time of onset or peak level according to sunrise and/or sunset at different times of year. Some sounds occurred either diurnally or nocturnally reflecting the crepuscular change in active species and behavior. Lunar periodicity of movement in the area of the hydrophones apparently occurred in the queen conch (*Strombus gigas*), while seasonal trends appeared in the frequency of occurrence of certain other sounds (Cummings *et al.*, 1964).

In addition to documenting the rhythmic nature of soniferous activity, movements of sound producers were determined by triangulation from an equilateral array of three hydrophones approximately 50 m apart. The sound produced by a moving fish was recorded on tape and printed out on a fast moving chart strip and the time of arrival difference to the three hydrophones was used to estimate the position of the animal. Both this technique and the recording of specific sound categories can be analyzed by computer directly from the hydrophones although such a system was not used in the studies reported here.

A particularly fruitful approach to experimental analysis of sonic communication is the use of sound playback. The necessary precursor to such studies is a thorough knowledge of the behavioral acts and act sequences associated with the acoustic repertoire of a species. Such information is typically obtained from temporal correlation (e.g., by sequential contiguity analysis) between any class of individual actions associated with a sound. Once the appropriate baseline is achieved, the predictability of behavior modification through sound playback can be used to test hypothetical models and the resultant data compared to acoustic communication in other species. This approach was applied to evaluations of the intraspecific communication systems of the toadfish, *Opsanus tau* (Winn, 1967; Fish, 1972);

bicolor damsel fish, *Eupomacentrus partitus* (Myrberg, 1972); goby, *Bathygobius soporator* (Tavolga, 1958); fiddler crabs, *Uca* spp. (Salmon and Horch, 1972). Interspecific responses to acoustic signals and artifacts were examined in cetaceans and in various fish species as reviewed by Myrberg (1972).

Many acoustic phenomena consequential to invertebrates involve reception of low frequency water particle displacements or substrate vibrations effective only over very short distances (Banner, 1967; Salmon and Horch, 1972). Because of spatial restrictions, only a few investigators have examined the acoustic behavior of marine invertebrates *in situ* (e.g., Salmon and his co-workers studying terrestrially active ocypodid crabs; Salmon and Horch, 1972).

5. *Underwater Telemetry*

The problem of locating and tracking the movements of specific organisms by visual means is especially difficult in the marine environment. Even when the observer can maintain visual contact with a subject, his presence may modify its behavior. The development over the past 15 years of miniaturized high frequency (30–150 KHz) sonic transmitters constitutes a major advance in field studies of migration, orientation, and behavioral ecology.

Most sonic transmitter "tags" are used as simple acoustic beacons and project a pulse of high frequency sound at periodic intervals. The ultrasonic frequencies, although attennuated more rapidly than low frequencies, carry a considerable distance in the sea (100 m–5 km). Theoretically, the lower the frequency, the greater the range when the appropriate power is available. They are apparently inaudible to most fishes and invertebrates. By intermittently pulsing the sound, battery life is extended and a distinct signature can be given each tag by varying the pulse rate and/or frequency.

Each tracking situation dictates the specific characteristic of the tags needed and a wide range of frequencies, signal strength, and duration can be achieved. For example, we used tags with a 200 m range and 4–5 day life for short-term orientation studies of spiny lobsters, *Panulirus argus*, in situations where several tagged individuals were released in the same area (Herrnkind and McLean, 1971). In this way, we avoided overlapping of signals that might occur with long-lasting, more powerful tags. For longer studies on migratory movements we used larger tags with a range of 1 km and an expected life of 30–60 days. This facilitated relocating the individuals several weeks later after they had dispersed and perhaps moved several kilometers away.

In addition to sending a fixed signal, the pulse rate of the tag can be controlled through electronic linkage to measure pressure, temperature, light or other physical parameters, as well, providing physiological information (MacKay, 1968). Examples of the type of information transmitted are external temperature, ambient light level, pressure (depth), compass bearing, swimming speed (relative to water), EKG, and internal temperature (Standora *et al.*, 1972; Johnson, 1971). The capabilities of transmitters are continually increasing as the state of the art advances.

Past ultrasonic tagging studies largely involved fishes and, to a lesser extent, other marine vertebrates and crustaceans (Stasko, 1971). The range of applications and results from fish migration research are reported by Stasko along with a useful comparative coverage of other approaches to tracking fish movements. A useful reference list of ultrasonic telemetry studies on aquatic/marine organisms, including fishes, is published in the various issues of *Underwater Telemetry Newsletter* (Stasko and Johnson, eds.).

The tagging process in fish may require forcing the tag into the stomach, inserting it into the body cavity through a closable incision (Hasler, 1966) or attaching it externally by inserting an anchor under the skin (Standora *et al.*, 1972). Crustacea are tagged externally with the transmitter either wired (Lund and Lockwood, 1970) or glued to the carapace (Herrnkind and McLean, 1971). Tagging internally seems desirable in long term studies to reduce drag, injury, or chance of snagging. Externally anchored transmitters allow attachment to free swimming subjects, such as sharks, and eventual retrieval by a diver or through a timed release mechanism. In addition, telemetry of light intensity, water temperature, and swimming speed (relative to the water) requires external sensors.

Upon release, the animal is followed by a boat or diver using a directional receiver, sometimes in conjunction with an omnidirectional receiver. The experienced tracker can usually estimate his approximate distance from the tag and thereby maintains sufficient distance to avoid affecting the speed or course of the animal. The position of the animal is periodically determined from compass references, visible landmarks, radar, LORAN, RDF, bottom contour, and occasionally, "dead reckoning." Data on currents, weather conditions, visibility, etc., are taken depending on the need to correlate such conditions with the pathway and orientation of the subjects.

The technique of continuous tracking while recording consequential physical information has provided new insights into behavior associated with homing and migration as well as suggesting possible orientational guideposts in some long-studied species (Stasko, 1971). The research on salmon (*Oncorhynchus* spp.) migration by Hasler, McCleave, Stasko, and their colleagues presents an example of the value of this approach. Among the major goals of the salmon study was the characterization of the migratory pathway for both the open sea phase and home river movements. Extensive coordinated tracking studies were conducted using numerous boats and shore monitoring stations (Stasko, 1971).

The researchers discovered considerable individual variability in pathways since some salmon moved from the open sea toward the home river mouth with the tidal currents and others moved against it. Some individuals oriented in accord with water currents while others moved in an oriented fashion only in the presence of the sun. These results suggest that multiple or redundant sources of directional information are available to orienting salmon (and perhaps other fish) during migration and argue against a single guidepost mechanism such as sun compass or rheotaxis. Other fairly extensive tracking studies are reported for sturgeon (*Acipenser*; Poddubny, 1969), white bass (*Roccus*; Hasler *et al.*, 1969), tuna (*Katsuwonus*; Yuen, 1970).

Ultrasonic telemetry is particularly well suited to diver (as well as surface) tracking of benthic crustaceans and gastropods which range over a relatively small area and do not move at a rapid pace (Clifton *et al.*, 1970). Saddle-shaped transmitters (70 KHz, 200 m range) fitted to the dorsal cephalothoracic carapace of spiny lobsters, *Panulirus argus*, by high tensile-strength glue (e.g., Eastman 910 or Permabond) permitted divers to locate individuals during nocturnal forays and to establish the temporal and spatial character of movements between dens and feeding areas (Fig. 6). The sonic tags also

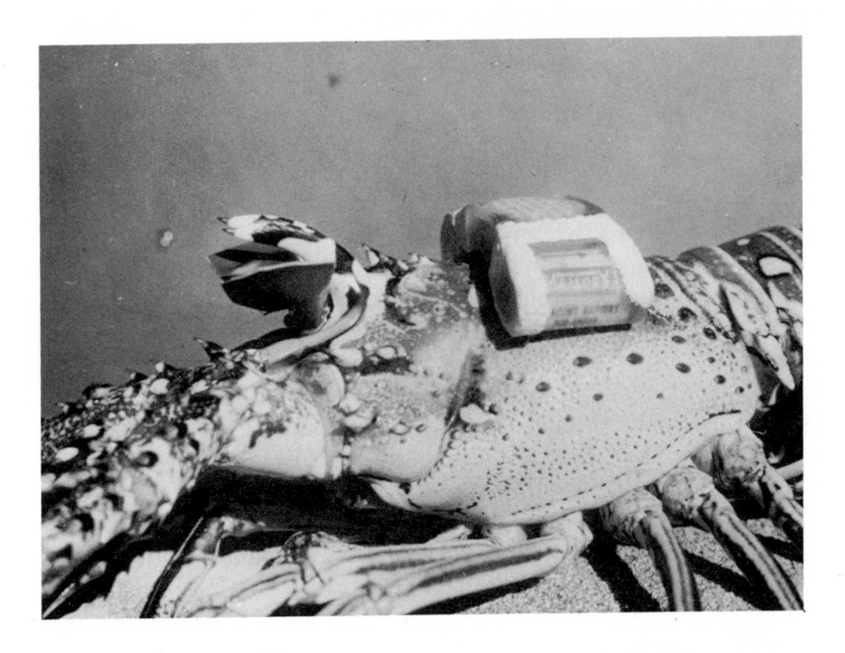

Fig. 6. (Top) Saddle-shaped sonic tag mounted on the dorsal cephalothorax of a spiny lobster *Panulirus argus*. The acoustic signal (70 KHz) and physical characteristics of the tag have no apparent effect on the behavior of the lobster. (Photo by J. G. Halusky.)

facilitated relocating individual lobsters during the day when they were in dens, although the signal range was markedly reduced except when in line with the den aperture.

The tracking studies established that lobsters moved onto the reef edge or onto adjacent sand–algal plains as far as 300 m to forage and returned before dawn to the reef, and often to the original den. This homing ability was further examined in displacement experiments by hand-carrying tagged lobsters to an arbitrary release location approximately 200 m from their den (Herrnkind and McLean, 1971; Herrnkind *et al.*, in press). All individual lobsters tested oriented in the appropriate bearings and most returned within 30 m of the capture site before the following morning, some to the original den. In all daylight tests the lobsters remained quiescent shortly after release, or stopped upon reaching the reef edge, then moved the remaining distance by night. Individuals demonstrated different pathways, some moving in a bee-line across the reef, others circumnavigating the reef by moving along the edge and through surge channels (Fig. 7).

This detail of documentation was permitted only by the capabilities afforded by integrating telemetry with concurrent diving observation and would not be possible from the surface or by any of the other techniques described. Other sonic tagging studies of invertebrates are reported by Clifton *et al.* (1970; *Strombus gigas*, the queen conch) and Lund *et al.* (1971; *Homarus americanus*, the American lobster). The movements of many other large crustaceans and gastropods are readily subject to examination by this approach.

More sophisticated telemetry systems have been successfully used or are being tested at present. The use of sonic transmitters with signal modulation by internal and ambient temperature enabled Carey *et al.* (1971) to document the thermal regulatory ability of tunas. Standora *et al.* (1972) reported success with a multiplexing circuit in an ultrasonic tag (on angel sharks) with sensors providing data on light level, velocity relative to the water, compass bearing, depth (pressure), and temperature.

The development of shore-based, buoyed and shipboard monitoring stations, particularly those capable of triangulating distance and bearing is also proceeding (Stasko, 1971), although this aspect lags behind transmitter technology. In addition to ultrasonic techniques; radio telemetry techniques are available with a much longer range. These are restricted, however, in marine application to surfacing animals such as cetaceans (Evans, 1971), pinnipeds, and turtles (Carr, 1967).

A major drawback of ultrasonic telemetry lies in the nature of the information transmitted; the signal indicates only position vs time or some physical or physiological condition from which behavior must be inferred. Thus, one does not know whether the sudden change in course of an animal

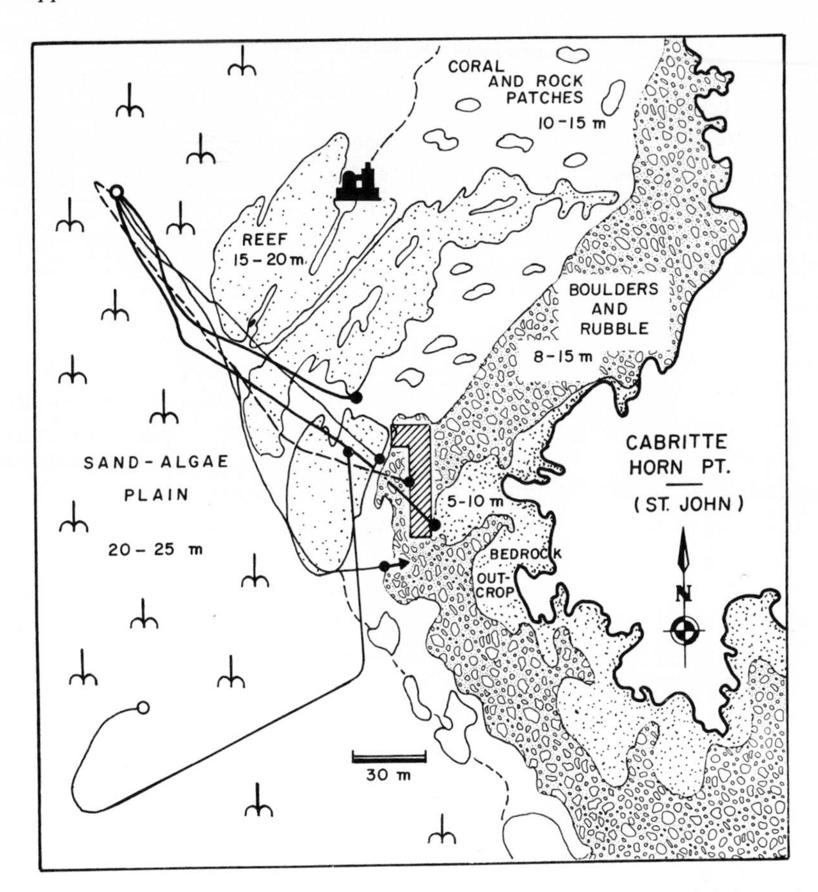

Fig. 7. Pathways of 6 spiny lobsters (*Panulirus argus*) displaced from their home dens (hatched area), brought to the Tektite habitat (silhouette) for attachment of sonic tags, and released at points indicated by open circles. Lobsters were tracked by divers using directional receivers. Lobsters typically homed to the den area during the night following release. The closed circles indicate their location at dawn after release the previous afternoon. (Modified from Herrnkind and McLean, 1971).

represents feeding, attack by a predator, disorientation, or social interaction. Even when the main goal is merely tracking, only a few individuals can be tracked simultaneously, especially if they diverge, move at different rates or if the signals overlap. Also, natural sounds, such as water turbulence and snapping shrimp, and artifacts, such as propeller and motor noise, frequently mask the transmitted signal. Accumulating sufficient data on which to generalize takes much time and effort. Boat and diver operations for long term tracking require considerable manpower and are subject to hardships imposed by the weather and sea conditions. Most of the past and ongoing open sea tracking studies require heavy funding as in the case of the salmon and cetacean studies.

Despite the range of transmitter characteristics available, adequate battery life and signal strength requires tags too large for many smaller species. Additionally, only a few companies market ultrasonic gear for marine use. Finally, the lack of pinpoint determination of position reduces the accuracy of tracking data.

6. *Sonar*

Generally, sonar devices operate by sending ultrasonic signals and receiving echoes reflected from objects of variable acoustic impedance. Depth recorders and so-called fish finders can indicate the presence, size, or number of free swimming animals under some conditions. Such information is important in studies of horizontal and vertical movements of aggregates, or schools, and large individual animals. With the appropriate equipment, Doppler shifts caused by the body flexures and motion of an animal can be detected and discriminated (Cummings, 1964; Hester, 1967). The return signals are monitored either by their aural characteristics or by a visual display indicating the frequency shifts (Lenarz and Green, 1971). Systems are presently under study which translate the acoustic echoes into actual visual images (acoustical holography) therefore optimizing the analysis of information by a human observer. One has only to think of the remarkable range, acuity and discriminating capabilities of porpoise "sonar" (Kellogg, 1961) to appreciate the potential of such a technological reality for human undersea observation.

Sonar systems have the basic advantages of other acoustic techniques over vision as regards useful range, and share the advantage of sonic telemetry over passive monitoring in providing a periodic signal. In addition, continuously transmitting systems, as opposed to pulsed sonar, provide a high rate of received information permitting sustained contact with fast moving targets (Hester, 1967). This latter system is capable of distinguishing not only size and distances of targets but can be used to identify certain fish species by the character of their body undulations and tail beat frequencies.

Most past biological applications of sonar involved monitoring the size and position of fairly large aggregates such as fish schools or planktonic scattering layers. This information is particularly useful to studies on horizontal migrations of large free-swimming animals, such as cetaceans, as well as to characterizing vertical migration. Some recent research on salmonids serves to illustrate the type of useful information obtained by sonar techniques. The daily pattern of horizontal and vertical movements and associated feeding activities of sockeye salmon *Onchorhynchus nerka*, were determined by Narver (1970) who located suspected aggregations by sonar and then sampled by trawl to assure the species composition. The sockeyes during summer tended to cluster by day at depths of 20–35 m and begin ascending at about 1.5 hr before sunset to 3 m below the surface. The descent occurred about 1.5 hr after sunrise. Heaviest feeding took place during the period of ascent and descent as determined from stomach content analysis. Sonar was particularly valuable to establishing the relative proportion of the population involved in the migration over the course of the day. No other technique could readily provide this information.

Broader studies were undertaken in 1964 (IIOE program—Bradbury *et al.*, 1970) and in 1965 (the SOND program—Currie *et al.*, 1969) to study the vertical distribution of organisms in the upper 1000 m of a selected oceanic area. The three major concurrent thrusts involved biological sampling, hydrography, and acoustic characterization. The latter task used sonar to characterize planktonic communities by their acoustic structure. The presence and movements of planktonic organisms were determined from their sonar signature and correlated with the prevailing physical conditions such as ambient light and temperature.

Cummings (1964) demonstrated the utility of Doppler shift detection for determining both the character and the degree of movement in a study on damsel fish. He recorded visual descriptions of certain actions, such as feeding, simultaneously while recording acoustically with a fish-finder. In this way, he was later able to aurally monitor 24-hr tape recordings and establish the behaviors associated with daily activity rhythms.

The drawbacks of present sonar devices center around the problem of determining the identity and activity of the signal producer. Preliminary capture, visual observation, or photography (Cushing, 1968) is generally necessary but often difficult in open sea situations. Complications also arise from excessive or variable movements of the transducer as may occur in a violently rolling vessel (Hester, 1967). Echo strength varies with the acoustic impedance of the target so that fishes with swim bladders provide a strong signal while animals without bones, scales, or gas bladders give a much weaker signal. High frequency sounds from cetaceans may jam or mislead the sonar operator since the frequencies overlap. Finally, the more sophisticated transducing and analysis systems are complex and expensive thus restricting their general availability.

III. Discussion

The past two decades have seen a sharp rise in the level of knowledge of marine animal behavior. The primary information has been derived by the *in situ* application of many of the techniques discussed above. Among the areas in which our knowledge has advanced tremendously in recent years are marine bio-acoustics and animal communication, the study of pelagic organisms, biorhythmic functions, inter-, and intraspecific associations within the coral reef community and life history characterizations of economically important marine animals.

With the important characterizations of behavior partially accomplished for some species, study of mechanisms, causation, and development is now beginning. Even in these latter areas, where the experimental methods may primarily require a laboratory approach, experimentation and the testing of hypotheses may still often be best accomplished in the field.

The integration of many of the above techniques has been and will continue to be an important approach. For example, the scope of knowledge from bio–acoustic communication studies involved observational and experimental studies in both field and laboratory using remote sensing as well as scientist-in-the-sea applications (Tavolga, 1964a, 1967). The behavioral ecology of the Norwegian lobster, *Nephrops norvegious*, was elucidated under difficult *in situ* conditions through the concerted effort of diving scientists, submersible operation and UTV (Chapman and Rice, 1971). The behavior of pelagic fishes was first studied by submersible, drift raft, and other viewport techniques in the open ocean and through further observation and experimentation under controlled, simulated field conditions (Nakamura, 1972.) The continued integration of methods will likely characterize much of the future research in this field.

The technological need for future *in situ* marine behavioral research falls into three main categories: (1) increased application of available techniques, (2) refinement of certain techniques and equipment in order to improve performance and to increase availability and reliability, and (3) development of new techniques, equipment, and applications for research in presently restrictive conditions. In the same manner that increased research activity by scientists using face mask and standard SCUBA has resulted in much new information since 1950, it is expected that the increased recruitment of scientists-in-the-sea and use of readily available adjuncts such as photography, cinematography, acoustic monitors, portable UTV, and ultrasonic telemetry will result in much new and important knowledge about the activities of marine organisms. The fulfillment of this need depends primarily on expansion in training of diving scientists, an objective further discussed below.

Improving the operation and applicability of technology applies to such diving hardware as mixed gas rebreathers, underwater communication systems, lock-out submersibles, and habitats. Effective aids to communication, navigation, diver transport, and thermal insulation are particularly necessary to habitat programs as the operating depth is increased, or as more difficult environments are examined. Virtually all remote sensing equipment needs improvement in performance and reduction in cost. Ultrasonic telemetry, especially, shows great promise should it achieve further transmitter miniaturization, as well as increased range and life, a greater array of sensors, and precise position estimation. Also needed are sonar systems with increased acuity and discrimination capabilities.

Illuminating the behavioral biology of deep sea animals *in situ* will likely occur only with development of improved sonar systems and submersibles. Perhaps the creation of noninjurious capture techniques for deep sea animals and large high pressure chambers to bring the deep sea environment into the laboratory will represent a necessary concomittant approach. Past advances in such technology benefited from the input by the fisheries, defense and space program as well as from the interdisciplinary interests of oceanographers. Hopefully, this will continue and expand in the future through such agencies as the Sea Grant and Manned Undersea Science and Technology (MUST) programs of the National Oceanic and Atmospheric Administration (NOAA), as well as NMFS, the Office of Naval Research (ONR) and the National Science Foundation (NSF).

The need to train scientists in a technical and physically rigorous endeavor, i.e., diving, in order for them to ply their profession, has analogies with the training required for the space program. Sport-type training is not necessarily adequate, although many scientific institutions frequently leave the diving education of scientists to YMCA, NAUI (National Association of Underwater Instructors), and other similar courses aimed at the nonscientist. Even in institutions using diving in research and where excellent organization and regulation of diving is provided, training in advanced technology and scientifically applicable techniques is seldom given. The National Marine Fisheries Service solves this need, in part, through its own diving programs and the obvious encouragement of its personnel to take part in outside programs.

A particularly appropriate program is a 12-week multidisciplinary graduate level course called Scientist-In-The-Sea given by the State University System of Florida and the Naval Coastal Systems Laboratory at Panama City, Florida. The course provides advanced diving technology and theory as well as relevant scientific principles and methods from the marine sciences. Graduate students are selected from various disciplines on a basis of academic performance and scientific potential, as well as being required to meet medical and diving qualifications. The faculty includes physicians specializing in diving medicine, underwater communications researchers, psychologists, oceanographers, ocean engineers, diving technologists, marine biologists, geologists, habitat aquanauts, and numerous visiting lecturers, expert in relevant fields. Virtually all the faculty are themselves experienced diving scientists or diving technicians. The curriculum includes classroom and field application of techniques in navigation, search, communication, underwater photography, advanced open and closed circuit SCUBA, advanced umbilical techniques, priciples of saturation diving, submersibles, habitats, safety, planning, and the implementation of diving programs for scientific purposes. Principles and research methods from physical, chemical, and geological oceanography, marine ecology and behavior, engineering science, diving medicine, and psychology make up a major portion of the program. The future will likely see the expansion of this type of course with the ultimate objective of providing a cadre of scientists skilled in the application of *in situ* methods to research in their professional disciplines.

Acknowledgments

First I must acknowledge the influence by William C. Cummings, William P. Davis, Walter A. Starck II, and John G. VanDerwalker who, each in his way, helped me to see beneath the "surface." I am philosophically in their debt. I thank my graduate research assistants James Farr, Joseph Halusky, Paul Kanciruk, and Greg Bill for critical reading of portions of the manuscript.

Aspects of research in which I had a direct role, and portions of which I have discussed in the paper, were supported as follows: Bimini video–acoustic program, University of Miami, 1961–1964,—Office of Naval Research and Bureau of Ships. Spiny lobster research, 1969,—Florida State University and Office of Naval Research through the American Museum of Natural History; Tektite-2, 1970—Department of Interior (and numerous agencies listed in Miller *et al.*, 1971); 1971–Sea Grant (NOAA); 1971–1972–National Science Foundation; 1972, Scientist-In-The-Sea Program—Sea Grant (NOAA). Facilities provided by the Psychobiology Research Center at Florida State University are gratefully acknowledged.

References

Adler, H., ed. (1971). Orientation: Sensory Basis. *Ann. N.Y. Acad. Sci. Vol.* **188**.

Arnold, J. (1965). Observations on the mating behavior of the squid *Sepioteuthis sepioidea*. *Bull. Mar. Sci.* **15**, 216–222.

Aronson, L. (1971). Further studies on orientation and jumping behavior in the gobiid fish, *Bathygobius soporator*. *Ann. N.Y. Acad. Sci.* **188**, 378–392.

Atema, J., and Engstrom, D. (1971). Sex pheromone in the lobster. *Nature* (*London*) **232**, 261–263.

Banner, A. (1967). Evidence of sensitivity to acoustic displacements in the lemon shark, *Negaprion brevirostris* (Poey). *In* "Lateral Line Detectors" (P. Cahn, ed.), pp. 265–273. Indiana Univ. Press, Bloomington, Indiana.

Banner, A. (1972). Use of sound in predation by young lemon sharks, *Negaprion brevirostris* (Poey). *Bull. Mar. Sci.* **22**, 251–283.

Barnes, H. (1963). Underwater television. *Oceanogr. Mar. Biol. Ann. Rev.* **1**, 115–128.

Barr, L. (1971). Observations on the biology and behavior of the arrow crab, *Stenorhynchus seticornis* (Herbst), in Lameshur Bay, St. John, Virgin Islands. *In* "Scientists-In-The-Sea" (J. Miller, J. Van Derwalker, and R. Waller, eds.), pp. VI 213–220. Dept. of Interior, Washington, D. C.

Beebe, W. (1928). "Beneath Tropic Seas." Putnam, New York.

Birkeland, C., and Gregory, B. (1971). Feeding behavior of a tropical predator *Cyphoma gibbosum* Linnaeus. *In* "Scientists-In-The-Sea" (J. Miller, J. Van Derwalker, and R. Waller, eds.), pp. VI 58–69. Dept. of Interior, Washington, D.C.

Bradbury, M., Abbott, D., Bovbjerg, R., Mariscal, R., Fielding, W., Barber, R., Pearse, V., Proctor, S., Ogden, J., Wourms, J., Taylor, L. Jr., Christofferson, J., Christofferson, J., P., McPhearson, R., Wynne, M., and Stromberg, P. Jr. *et al.* (1970). Studies of the fuana associated with the deep scattering layers in the equatorial Indian Ocean, conducted on R/V Te Vega during October and November 1964. *Proc. Int. Symp. Biolog. Sound Scattering in the Ocean* pp. 409–452. U.S. Naval Oceanographic Office.

Breder, C. (1959). Studies on social groupings in fishes. *Bull. Amer. Mus. Natur. Hist.* **117**, 397–482.

Cahn, P., Ed. (1967). "Lateral Line Detectors." Indiana Univ. Press, Bloomington, Indiana.

Carey, F., Teal, J., Kanwisher, J., Lawson, K., and Beckett, J. (1971). Warm bodied fish. *Amer. Zool.* **11**, 135–144.

Carr, A. (1967). Adaptive aspects of the scheduled travel of *Chelonia*. *In* "Animal Orientation and Navigation" (R. Storm, ed.), pp. 35–56. Oregon State Univ. Press, Corvallis, Oregon.

Chapman, C., and Rice, A. (1971). Some direct observations on the ecology and behavior of the Norway lobster *Nephrops norvegicus*. *Mar. Biol.* **10**, 321–329.

Clarke, E. (1961). Functional hermaphroditism and self-fertilization in serranid fish. *Science* **129**, 215–216.

Clark, E. (1965). Mating of groupers. *Natur. Hist.* **74**, 22–25.

Clarke, T. (1970). Territorial behavior and population dynamics of a pomacentrid fish, the garibaldi, *Hypsypops rubricunda*. *Ecol. Monogr.* **40**, 189–212.

Clifton, H., Mahnken, C., Van Derwalker, J., and Waller, R. (1970). Tekite 1, man-in-the-sea project: Marine science program. *Science* **168**, 659–663.

Cobb, J. (1971). The shelter related behavior of the lobster, *Homarus americanus*. *Ecology* **52**, 108–115.

Collette, B., and Talbot, F. (1971). Activity, patterns of coral reef fishes with emphasis on nocturnal-diurnal changeover. *In* "Scientists-In-The-Sea" (J. Miller, J. Van Derwalker, and R. Waller, eds.), pp. VI 257–260. Dept. of Interior, Washington, D.C.

Cooper, R., and Herrnkind, W. (1971). Ecology and population dynamics of the spiny lobster, *Panulirus argus*, of St. John Island, U.S. Virgin Islands. In "Scientists-In-The-Sea" (J. Miller, J. Van Derwalker, and R. Waller, eds.). pp. VI-24–57. U.S. Dept. of Interior, Washington, D.C.

Creaser, E., and Travis, D. (1950). Evidence of a homing instinct in the Bermuda spiny lobster. *Science* **112**, 169–170.

Cummings, W. (1964). Using the Doppler effect to detect movements of captive fish in behavior studies. *Trans. Amer. Fish Soc.* **92**, 178–180.

Cummings, W., Brahy, B., and Herrnkind, W. (1964). The occurrence of underwater sounds of biological origin off the west coast of Bimini, Bahamas. *In* "Marine Bio-acoustics" (W. Tavolga, ed.), pp. 27–43. Pergamon, Oxford.

Cummings, W. C., Brahy, B. D., and Spires, J. (1966). Sound production, schooling, and feeding habits of the Margate, *Haemulon album* Cuvier, off North Bimini, Bahamas. *Bull. Mar. Sci.* **16**, 626–640.

Currie, R., Boden, B., and Kampa, E. (1969). An investigation on sonic-scattering layers: the

R.R.S. 'Discovery' SOND cruise, 1965. *J. Mar. Biol. Ass. U.K.* **49**, 489–514.

Cushing, D. (1968). Direct estimation of a fish population acoustically. *J. Fish Res. Bd. Can.* **25**, 2359–2364.

Dawson, C., and Idyll, C. (1951). Investigations on the Florida spiny lobster, *Panulirus argus* (Latreille). *Mar. Lab. Miami Tech. Ser. No. 2* 1–39.

Earle, S. (1971). The influence of herbivores on the marine plants of Great Lameshur Bay. *In* "Scientists-In-The-Sea" (J. Miller, J. Van Derwalker, and R. Waller, eds.), pp. VI 132–200. Dept. of Interior, Washington, D.C.

Eibl-Eibesfeldt, I. (1970). "Ethology." Holt, New York.

Evans, W. (1971). Orientation behavior of delphinids: radio telemetric studies. *Ann. N.Y. Acad. Sci.* **188**, 142–160.

Fager, E., Flechsig, A., Ford, R., Clutter, R., and Ghelardi, R. (1966). Equipment for use in ecological studies using SCUBA. *Limnol Oceanogr.* **11**, 503–509.

Feddern, H. (1965). The spawning, growth and general behavior of the bluehead wrasse, *Thalassoma bifasciatum* (Pisces: Labridae). *Bull. Mar. Sci.* **15**, 896–941.

Fish, J. (1972). The effect of sound playback on the toadfish. *In* "Behavior of Marine Animals" (H. Winn and B. Olla, eds.), Vol. 2, pp. 386–434.

Fishelson, L. (1971). Ecology and distribution of the benthic fauna in the shallow waters of the Red Sea. *Mar. Biol.* **10**, 113–133.

Frings, H. (1964). Problems and prospects in research on marine invertebrate sound production and reception. *In* "Marine Bio-Acoustics" (W. Tavolga, ed.), pp. 155–174. Pergamon, Oxford.

Gooding, R., and Magnuson, J. (1967). Ecological significance of a drifting object to pelagic fishes. *Pacific Sci.* **21**, 486–497.

Hartline, P., Hartline, A., Szmant, A., and Flechsig, A. (1971). Escape response in a pomacentrid reef fish. *In* "Scientists-In-The-Sea" (J. Miller, J. Van Derwalker, and R. Waller, eds.), pp. VI-201–208. Dept. of Interior, Washington, D.C;

Hasler, A. (1966). "Underwater Guideposts." Univ. of Wisconsin Press, Madison, Wisconsin.

Hasler, A., Gardella, E., Horrall, R., and Henderson, H. (1969). Openwater orientation of white bass, *Roccus chrysops*, as determined by ultrasonic tracking methods. *J. Fish. Bd. Can.* **26**, 2173–2192.

Hazlett, B. (1972). Ritualization in marine Crustacea. *In* "Behavior of Marine Animals" (H. Winn and B. Olla, eds.), Vol. 1, pp. 97–125. Plenum Press, New York.

Herrnkind, W. (1969). Queuing behavior of spiny lobsters. *Science* **164**, 1425–1427.

Herrnkind, W. (1970). Migration of the spiny lobster. *Natur. Hist.* **79**, 36–43.

Herrnkind, W., and Cummings, W. (1964). Single file migrations of the spiny lobster, *Panulirus argus* (Latreille). *Bull. Mar. Sci.* **14**, 123–125.

Herrnkind, W., and McLean, R. (1971). Field studies of homing, mass emigration, and orientation in the spiny lobster, *Panulirus argus*. In "Orientation: Sensory Basis" (H. Adler, ed.) *Ann. N.Y. Acad. Sci.* **188**, 359–377.

Herrnkind, W., Van Derwalker, J., and Barr, L. (in press). Population ecology and dynamics of spiny lobsters, *Panulirus argus*, of St. John, U.S.V.I.: (4) habitation, patterns of movement and general behavior.

Hester, F. (1967). Identification of biological sonar targets from body-motion Doppler shifts. *In* "Marine Bioacoustics" (W. Tavolga, ed.), Vol. 2. pp. 59–73. Pergamon, Oxford.

Heydorn, A. (1969). The rock lobster of the South African west coast *Jasus lalandii* (H. Milne-Edwards), 2. Population studies, behavior, reproduction, moulting, growth and migration. *Div. Sea Fisheries Investigational Rep.*, No. 71, 1–52.

High, W., and Beardsley, A. (1971). Observations of fish behavior in relation to fish pots. *In*

"Scientists-In-The-Sea" (J. Miller, J. Van. Derwalker and R. Waller, eds.), pp. VI-4–14. Dept. of Interior, Washington, D.C.

Hinde, R. (1970). "Animal Behavior." McGraw-Hill, New York.

Hobson, E. (1965). Diurnal-nocturnal activity of some inshore fishes in the Gulf of California. *Copeia* 291–302.

Hochberg, R. and Couch, J. (1971). Biology of cephalopods. *In* "Scientists-In-The-Sea" (J. Miller, J. Van Derwalker, and R. Waller, eds.), pp. VI-221–228. Dept. of Interior, Washington, D.C.

Hodgson, E. and Mathewson, R. (1971). Chemosensory orientation in sharks. *Ann. N.Y. Acad. Sci.* **188**, 175–182.

Hollien, H., and Rothman, H. (1971). Studies of diver communication and retrieval. *In* "Scientists-In-The-Sea" (J. Miller, J. Van. Derwalker, and R. Waller, eds.), pp. X-1–9. Dept. of Interior, Washington, D.C.

Iversen, R. (1967). Response of yellow fin tuna (*Thunnus albacares*) to underwater sound. *In* "Marine Bio-acoustics" (W. Tavolga, ed.), Vol. 2, pp. 105–119. Pergamon, Oxford.

Jander, R. (1965). Die Phylogenie von Orientierungsmechanismen der Arthropoden. *Verh. Deutsch. Zool. Ges. Jena* 266–306.

Johnson, J. (1971). A brief history of ultrasonic tracking. *Underwater Telemetry Newsletter* **1**, 2–4.

Johnson, V., Jr. (1969). Behavior associated with pair formation in the banded shrimp *Stenopus hispidus* (Olivier). *Pac. Sci.* **23**, 40–50.

Kellogg, W. (1961). "Porpoises and Sonar." Univ. of Chicago Press, Chicago, Illinois.

Kelly, M., and Conrod, A. (1969). Aerial photographic studies of shallow water benthic ecology. *In* "Remote Sensing in Ecology" (P. Johnson, ed.), pp. 173–184. Univ. of Georgia Press, Athens, Georgia.

Kronengold, M., Dann, R., Green, W., and Lowestein, J. (1964). Description of the system. *In* "Marine Bio-acoustics" (W. Tavolga, ed.), pp. 11–26. Pergamon, Oxford.

Kumpf, H. (1964). Use of underwater television in bio-acoustic research. In "Marine Bio-acoustics" (W. Tavolga, ed.), pp. 45–57. Pergamon, Oxford.

Kumpf, H., and Randall, H. (1961). Charting the marine environments of St. John, U.S. Virgin Islands. *Bull. Mar. Sci.* **11**, 543–551.

Larcombe, M., and Russell, B. (1971). Egg-laying behavior of the broad squid, *Sepioteuthis bilineata*. *N.Z. J. Mar. Fw. Res.* **5**, 3–11.

Lenarz, W., and Green, J. (1971). Electronic processing of acoustical data for fishery research. *J. Fish. Res. Bd. Can.* **28**, 446–447.

Limbaugh, C. (1961). Cleaning symbiosis. *Sci. Amer.* **205**, 42–49.

Limbaugh, C., Pederson, H., and Chase, F., Jr. (1961). Shrimps that clean fishes. *Bull. Mar. Sci.* **11**, 231–257.

Lindberg, R. (1955). Growth, population dynamics and field behavior in the spiny lobster, *Panulirus interruptus* (Randall). *Univ. Calif. Publ. Zool.* **59**, 157–248.

Losey, G. Jr. (1971). Communication between fishes in cleaning symbiosis. *In* "Aspects of the Biology of Symbiosis" (T. Cheng, ed.), pp. 45–76. Univ. Park Press, Baltimore, Maryland.

Lund, W., and Lockwood, R. (1970). Sonic tag for decapod crustaceans. *J. Fish. Res. Bd. Can.* **27**, 1147–1151.

Lund, W., Stewart, L., and Weiss, H. (1971). Investigation on the lobster. Final Rep. Commercial Fish. Res. Develop. Act No. 3, 1–105.

Luria, S. and Kinney, J. (1970). Underwater vision. *Science* **167**, 1454–1461.

Lythgoe, J. and Hemmings, C. (1967). Polarized light and underwater vision. *Nature* (*London*) **213**, 893–894.

MacKay, R. (1968). Aquatic animals. *In* "Bio-Medical Telemetry," pp. 296–330. Wiley, New York.

McLean, R., and Herrnkind, W. (1971). Compact schooling during a mass movement by grunts. *Copeia* 328–330.

Magnuson, J., and Prescott, J. (1966). Courtship, locomotion, feeding and miscellaneous behavior of Pacific bonito (*Sarde chiliensis*). *Anim. Behav.* **14**, 54–62.

Mariscal, R. (1972). Behavior of symbiotic fishes and sea anemones. *In* "Behavior of Marine Animals" (H. Winn and B. Olla, eds.), Vol. 2, pp. 327–360. Plenum Press, New York.

Miller, J., Van Derwalker, J. and Waller, R. (eds.) (1971). "Scientists-In-The-Sea." U.S. Dept. of Interior, Washington, D.C.

Myrberg, A., Jr. (1972). Using sound to influence the behavior of free-ranging marine animals. *In* "Behavior of Marine Animals" (H. Winn and B. Olla, eds.), Vol. 2, pp. 435–468. Plenum Press, New York.

Myrberg, A., Jr., Brahy, B., and Emery, A. (1967). Field observations on reproduction of the damselfish, *Chromis multilineata* (Pomacentridae), with additional notes on general behavior. *Copeia* 819–827.

Myrberg, A., Jr., Richard, J., and Banner, A. (1969). Shark attraction using a video-acoustic system. *Mar. Biol.* **2**, 264–276.

Nakamura, E. (1972). Development and uses of facilities for studying tuna behavior. *In* Behavior of Marine Animals" (H. Winn and B. Olla, eds.), Vol. 2, pp. 245–277. Plenum Press, New York.

Narver, D. (1970). Diel vertical movements and feeding of underyearling sockeye salmon and the limnetic zooplankton in Babine Lake, British Columbia. *J. Fish. Res. Bd. Can.* **27**, 281–316.

Nelson, D. (1967). Hearing thresholds, frequency discrimination, and acoustic orientation in the lemon shark, *Negaprion brevirostris* (Poey). *Bull. Mar. Sci.* **17**, 741–768.

Nelson, D., and Gruber, S. (1963). Sharks: attraction by low-frequency sound. *Science* **142**, 975–977.

Nelson, D., Johnson, R., and Waldrop, L. (1969). Responses in Bahamian sharks and groupers to low-frequency, pulsed sound. *Bull. S. Calif. Acad. Sci.* **68**, 131–137.

Norris, K. (1964). Some problems of echolocation in cetaceans. *In* "Marine Bio-acoustics" (W. Tavolga, ed.), pp. 317–336. Pergamon Press, New York.

Poddubny, A. (1969). Sonic tags and floats as a means of studying fish response to natural environmental changes and to fishing gears. *FAO Fish. Rep.* **62**, 793–801.

Randall, J. (1958). A review of the labrid fish genus *Labroides*, with descriptions of two new species and notes on ecology. *Pac. Sci.* **12**, 327–347.

Randall, J. (1964). Contributions to the biology of the queen conch, *Strombus gigas*. *Bull. Mar. Sci.* **14**, 246–295.

Randall, J. (1967). Food habits of reef fishes of the West Indies. *In* "Studies in Tropical Oceanography," No. 5, pp. 665–847. Univ. of Miami, Inst. of Mar. Sci., Miami.

Reed, R. (1971). Underwater observations of the population, density and behavior of pumpkinseed, *Lepomis gibbosus* (Linnaeus) in Cranberry Pond, Massachusetts. *Trans. Amer. Fish. Soc.* **100**, 350–353.

Reese, E. (1964). Ethology and marine zoology. *Oceanogr. Mar. Biol. Ann. Rev.* 455–488.

Rice, A. and Chapman, C. (1971). Observations on the burrows and burrowing behavior of two mud-dwelling decapod crustaceans, *Nephrops norvegicus* and *Goneplax rhomboides*. *Mar. Biol.* **10**, 330–342.

Richard, J. (1968). Fish attraction with pulsed low-frequency sound. *J. Fish. Res. Bd. Can.* **25**, 1441–1452.

Ridgway, S., Scronce, B., and Kanwisher, J. (1969). Respiration and deep diving in the bottlenose porpoise. *Science* **166**, 1651–1654.

Russell, B. (1971). Underwater observations on the reproductive activity of the demoiselle *Chromis dispilus* (Pisces: Pomacentridae). *Mar. Biol.* **10**, 22–29.

Ryan, E. (1966). Pheromone: evidence in a decapod crustacean. *Science* **151**, 340–341.

Salmon, M. and Horch, K. (1972). Acoustic signalling and detection by semiterrestrial crabs of the family Ocypodidae. *In* "Behavior of Marine Animals" (H. Winn and B. Olla, eds.), Vol. 1, pp. 60–95. Plenum Press, New York.

Sand, R. (1957). Application of underwater television to fishing gear research. *Trans. Amer. Fish. Soc.* **86**, 158–160.

Schmidt, R. Jr., (1971). Mark X closed circuit rebreather training course. *In* "Scientists-In-The-Sea" (J. Miller, J. Van Derwalker and R. Waller, eds.), pp. IV-7–14. Dept. of Interior, Washington, D.C.

Schroeder, R., and Starck, W. II. (1964). Photographing the night creatures of Alligator Reef. National Geographic Magazine, Vol. 125, pp. 128–154.

Shaw, E. (1969). Some new thoughts on the schooling of fishes. *FAO Fish. Rep.* **62**, 217–231.

Smith, C., and Tyler, J. (1971). Space resource sharing in a coral reef community. *In* "Scientists-In-The-Sea" (J. Miller, J. Van Derwalker, and R. Waller, eds.), pp. VI 261–269. Dept. of Interior, Washington, D.C.

Snyder, N., and Snyder, H. (1970). Alarm response of *Diadema antillarum*. *Science* **168**, 276–278.

Standora, E., Jr., Sciarrotta, T., Ferrel, D., Carter, H., and Nelson, D. (1972). Development of a multichannel, ultrasonic telemetry system for the study of shark behavior at sea. California State Univ., Long Beach Foundation, *Tech. Rep. No. 5*.

Starck, W., II (1968). Editorial. *Undersea Biol.* **1**, 1–3.

Starck, W., II, and Brundza, P. (1966). "The Art of Underwater Photography." Amer. Photogr. Book Publ., New York.

Starck, W., II, and Davis, W. (1966). Night habits of fishes of Alligator Reef, Florida. *Ichthyologia* **38**, 313–356.

Stasko, A. (1971). Review of field studies on fish orientation. *In* "Orientation: Sensory Basis" (H. Adler, ed.). pp. 12–29. *Ann. N.Y. Acad. Sci.* **188**.

Steinberg, J., Cummings, W., Brahy, B., and MacBain (Spires), J. (1965). Further bio-acoustic studies off the west coast of North Bimini, Bahamas. *Bull. Mar. Sci.* **15**, 942–963.

Steinberg, J. and Koczy, F. (1964). Objectives and requirements. *In* "Marine Bio-acoustics" (W. Tavolga, ed.), pp. 1–9. Pergamon, Oxford.

Stevenson, R. Jr. (1967). Underwater television. *Oceanol. Int.* **2**, 30–35.

Stevenson, R. Jr., (1972). Regulation of feeding behavior of the bicolor damselfish (*Eupomacentrus partitus* Poey) by environmental factors. *In* "Behavior of Marine Animals" (H. Winn and B. Olla, eds.), Vol. 2, pp. 278–302. Plenum Press, New York.

Tavolga, W. (1958). The significance of underwater sounds produced by the males of the gobiid fish, *Bathygobius soporator*. *Physiol. Zool.* **31**, 259–271.

Tavolga, W. (ed.) (1964a). "Marine Bio-acoustics." Pergamon, Oxford.

Tavolga, W. (1964b). Sonic characteristics and mechanisms in marine fishes. *In* "Marine Bio-acoustics" (W. Tavolga, ed.), pp. 195–211. Pergamon, Oxford.

Tavolga, W. (ed.) (1967). "Marine Bio-acoustics," Vol. II. Pergamon, Oxford.

Tyrrell, W. (1964). Design of acoustic systems of obtaining bio-acoustic data. *In* "Marine Bio-acoustics" (W. Tavolga, ed.), pp. 65–86. Pergamon, Oxford.

Waterman, T., and Forward, R., Jr. (1972). Field demonstration of polarotaxis in the fish *Zenarchopterus*. *J. Exp. Zool.* **180**, 33–54.

Wells, M. (1966). The brain and behavior of cephalopods, *In* "Physiology of Mollusca" (K. Wilbur and C. Yonge, eds.), pp. 547–589. Academic Press, New York.

Winn, H. (1964). The biological significance of fish sounds. *In* "Marine Bio-acoustics" (W. Tavolga, ed.), pp. 213–232. Pergamon, Oxford.

Winn, H. (1967). Vocal facilitation and the biological significance of toadfish sounds. *In* "Marine Bio-acoustics" (W. Tavolga ed.), Vol. 2, pp. 283–304. Pergamon, Oxford.

Winn, H. (1972). Acoustic discrimination by the toadfish with comments on signal systems. *In* "Behavior of Marine Animals" (H. Winn and B. Olla, eds.), pp. 361–385. Plenum Press, New York.

Winn, H., and Olla, B. (eds.) (1972). "Behavior of Marine Animals," Vols. 1 and 2. Plenum Press, New York.

Winn, H., Salmon, M. and Roberts, N. (1964). Sun-compass orientation by parrot fishes. *Z. Tierpsychol.* **21**, 798–812.

Yarnall, J. (1969). Aspects of the behavior of *Octopus cyanea* Gray. *Anim. Behav.* **17**, 747–754.

Yuen, W. (1970). Behavior of skipjack tuna, *Katsuwonus pelamis*, as determined by tracking with ultrasonic devices. *J. Fish. Res. Bd. Can.* **27**, 2071–2079.

Chapter 4

Comparative Physiology: Neurophysiology of Marine Invertebrates

M. S. Laverack

I. Introduction

The classic approach to a problem in natural history is to observe and describe. You may look at where an animal lives and describe its habitat, the

*This review was completed in 1971 and already advances in intracellular dye injection, whole animal preparations, and telemetry have overtaken the tentative and conservative predictions made here.

trees it lives in, the temperature of the water, or what it feeds on, or some other of the multitudinous aspects of its life. By studying these various habits, one gradually accumulates information about the natural world and the parts played in it by the various inhabitants.

If in the light of previous experience, either published or unpublished, one particular example takes your fancy, however, you concentrate on that. Or if one system of organs is the interesting feature, you concentrate upon that. In this type of investigation as many examples as possible of the species, or of the families within an order, are examined. Gradually there appears an indication of the range, evolutionary patterns, and variability that exist within the population. Such is the study of morphology in which the basic principles that govern the design of animals are elucidated. The shape of the whole or its component parts is the basic fabric of zoology and in the late nineteenth and early twentieth centuries it was *à la mode*. The more detailed the description the better and the more valuable. Taxonomy as a subsidiary science based upon morphologic characteristics became fashionable and in the present day the use of the scanning electron microscope has revealed even greater structural details than the most sophisticated light microscopy. Thus we can now base our taxonomy upon setal arrangements previously unsuspected because they were too small for resolution by the light microscope. In the same way, biochemistry and the various analytical methods can now reveal differences of blood or tissue chemistry that are valuable and informative but require expensive equipment.

Nonetheless the classic approach to a problem is via the habitat and then the morphology of the individual. Teleologically speaking the morphology, as demonstrated in the living animal, reveals its ancestry and also the adaptability of the species in changing conditions. We may then begin to look for the functional value of the organs that are exhibited.

Study of function begins with the necessity to understand the mechanism underlying the usual range of activities exhibited by the animal in question. To understand the function of a steam engine you need to know what it does; in the case of a radio, the valves and the resistors mean little until you have heard the music emanating from the loudspeaker.

Physiology as the symptom of function is too extensive a topic to cover in its entirety. In the case of animal studies the behavior of the animal is all important and behavior per se is dealt with by Herrnkind in this volume. My own purpose will be to indicate the lines that studies have taken on the underlying physiological mechanisms within the animal body that are apparent to the observer as some facet of behavior. Marine examples are of great value in this study as the comparative approach enables conclusions to be reached as to fundamental mechanisms, evolutionary development, variation, convergence, and adaptability of organisms. The variety of marine animals, encompassing nearly every major taxonomic group, makes available as wide a spectrum of possible physiologic types as can be envisaged. Very few have yet received adequate attention.

II. The Past

A. ANATOMY

The history of physiology depends to a considerable extent upon the broad and substantial basis of an understanding of morphology. The study of the nervous system is no exception to this rule. The structure of the nervous system must be understood in order to explain physiological characteristics, although it is unwise perhaps to suggest that all physiological characteristics will be apparent by reference solely to the morphology of the system. Enough is now known about the variability of electrical responses of nervous tissue and of the various interconnexions between nervous elements to be aware that structural differences and connexions, if any, may be extremely subtle. Nonetheless the initial approach to the nervous system has always been a morphological one.

The anatomist is content perhaps with a dissection that lays open the animal and makes clear the cartography of the organ system. In the case of the nervous system of, say, an annelid the dissection is a simple one in which the cerebral ganglia and the various ganglia of the ventral chain are revealed by straightforward means. The various segmental nerves that originate from the ganglia may also be seen—although if their trajectory soon takes them into the musculature of the body walls the precise pattern of innervation may be obscured by other tissues. Dissection therefore enables the gross outline to be obtained with relatively little effort. Further detail demands a more precise examination, and in this case the examination takes the form of tracing tracts of nerves or perhaps individual nerves by the use of various histological procedures. Sectioning material from the nervous system can be carried out by the use of paraffin wax sections and by the staining of such sections with general and specific stains. General mapping stains such as Mallory's or Heidenhain, Azan, and various other polychromatic stains yield considerable information regarding the passage of nerves and, to a certain extent, the fibers within those nerves to their various destinations. It was, however, discovered early by morphologists that more information could be obtained by the use of special techniques.

Anatomical and morphological investigations then became dominated by the valuable and still more useful methods of specific staining for nerves, although specificity is often only a relative term. The use of silver by Cajal and by Golgi revolutionized the knowledge of anatomy and the individual geography of nerve cells through the early part of this century. Reduced silver enabled the individual nerve cell to be identified and characterized and its dendritic and axonic branches typified in some detail. These methods still have considerable use as seen in the work of Strausfeld (1970) and Strausfeld and Blest (1970). Thick sections can be dealt with in this manner in order to obtain considerable detail about the nerve cell. Silver staining by Holmes' method or the more refined Rowell method applied to sectioned material of known nervous nature gives greater information on the fiber arrangement and their disposition within ganglia and in peripheral nerves (see Heimer, 1971). Nonetheless they are dependent upon the vagaries of pathways of the growing nerve fibers. Sections cut in one plane do not necessarily give information regarding the disposition of fibers in other planes, and so serial reconstruction of a three-dimensional nature must be carried out using a number of sections. This method can be refined even further by the use of light microscopy using thin araldite sections cut at approximately 1 μm and stained in toluidine blue; but the method is very laborious and time-consuming. In the past, however, these are the sorts of methods that have given immense detail and still will reveal items of interest in new or re-studied systems.

Another method much favored by anatomists, both past and present, has been the use of methylene blue. Methylene blue staining is by no means an easy or easily repeatable method, and many repetitions of the same investigation are usually advisable in order to be sure that the details obtained are correct and complete. Methylene blue stain is taken up selectively under certain conditions by neurones and their processes and can give considerable detail of the anatomy of individual cells. In the hands of such masters as Retzius and Alexandrowicz a great deal of information has become available for certain receptor systems and of pathways in the central nervous system and of the innervation of muscles. It probably could not be claimed that every detail is given by this method but it is an extremely valuable tool in the neuroanatomist's locker, and will remain so for a long time in the future. It is, however, surprising that it is not equally useful for study of all animal phyla. It is particularly straightforward and useful for the study of anthropods, especially decapod Crustacea, but is less useful and simple to use on mollusks.

For a mine of information on such subjects and an indication of the range and variety of the many animals involved, the reader is referred to Bullock and Horridge (1965).

B. PHYSIOLOGY

Physiology is not a new science. Studies have been made for hundreds of years, especially, of course, in respect to humans and mammals since the work serves the field of medicine. Many of the methods and techniques used in the study of physiology were devised for use on mammals and needed large often clumsy apparatus that could be manipulated only with some difficulty. Also they were quite inappropriate to other groups of animals. Respiration studied by the use of a Douglas bag could hardly be modified for use in investigations on fish.

Nonetheless the kymograph became a standard tool since it was reliable in use and variable over a great range in terms of speed so that information regarding events taking either several hours or a few milliseconds were equally amenable to mechanical analysis. The contractions of skeletal muscles in isolation, the activity rhythms of whole animals, and the beating of hearts were all phenomena approachable by such means. They still are.

The days of string and sealing wax physiology when simple instruments and the ingenuity of the experimenter were all that was available have now largely passed, but one must pay tribute to what was accomplished. In the field of marine physiology the modification of mammalian-inspired apparatus was brought about due to an upsurge of interest in marine animals at about the turn of the century.

The systematic background had been laid, and at least for Europe and much of America, most animals had been identified, classified, labeled, catalogued, and descriptions published. This meant that with a solid background of morphological work to stand on, the characteristics and individual peculiarities of marine forms could now be studied. The aquatic habitat presents a set of problems with which man is unfamiliar by virtue of his evolution, and which therefore makes certain "normal" viewpoints untenable. Swimming below the surface with an aqualung is not quite the same exercise as being a lobster with gills or a sea urchin with none.

In the same way it is perhaps difficult to maintain perspective about sensory attributes in a watery medium, about osmoregulation when the question is one of desiccation by extraction of water against a concentration gradient, about excretion of ammonia and trimethylamine oxide, and about other aspects of life with which man is not familiar.

Thus in the early stages, the design of experiments necessitated rethinking the equipment needed for the job. The uptake of oxygen from solution became a chemical problem solved by the Winkler titration; the excretion of ammonia into water was estimated by various chemical and biochemical techniques. Microscopy became more and more refined as the question became more precise, and again involvement of new chemical methods became apparent.

On the physical side, the development of more and better recording techniques that introduced electrical methods have the attributes of speed, precision, and cleanness of working to facilitate experiments.

III. The Present

A. STRUCTURE

The advances in microscopy that have revolutionized the approach to neurophysiology and an understanding of the anatomy of the nervous system, leading eventually, one hopes, to a knowledge of the cellular level of behavior, interaction, and integration, may be classified under two headings. One relates to the tools and the second to the techniques.

1. *Tools of the trade*

a. *Transmission Electron Microscopy.* There can be no doubt that the classic morphologists felt the need for a more detailed look at the material they worked with (the light microscope having certain theoretical and predictable limitations beyond which it cannot perform). The invention of the transmission electron microscope therefore opened new doors and its use has led to a flood of papers as the "New Anatomy" acquired adherents. For students of the nervous system it has given enormous detail where little was available previously.

To give one example, annelids of the three major classes (Oligochaeta, Polychaeta, and Hirudinea) all possess central nervous system arrangements of ganglia with a double set of connectives that link each ganglion to its neighbor. From these ganglia, there arise peripheral nerves that extend to the body wall and the musculature. Within these nerve trunks are found nerve fibers. The question arises as to how many nerve fibers and where do they run?

Earlier anatomists counted assiduously the number of fibers and argued over whether or not there were sufficient to account for the multiplicity of end organs and muscles that required innervation. Langdon (1895, 1900) has described large numbers of sense organs in earthworms and *Nereis* that apparently were not represented singly in the peripheral nerves. This led to discussion of peripheral fusion or synaptic interaction as a means of explaining the apparent decrease in numbers of fibers in the segmental nerves (Smallwood, 1926; Ogawa, 1934). Large numbers of sensory neurones have been confirmed in *Nereis* by Dorsett (1964).

Electron micrography (Horridge, 1963) revealed many axons in polychaete peripheral nerves, including very fine ones that were unresolvable by optical methods. Since that time, further investigations have shown that innumerable fine fibers exist where previously a general unstructured background was reported. This does not mean that all the problems are solved and Lawry (1970) has described synaptic structures at the periphery in *Aphrodite*, thus suggesting that integration may indeed take place at the periphery. Horridge (1963) reported efferent sensory impulses in annelid nerve fibers (*Harmothoë*) that may have been the result of curious synaptic interaction within the CNS. Mill and Knapp (1967) have since reported similar events in earthworms, but Lawry is dissatisfied with the explanation. Another possible interpretation was suggested by the work of Nicholls and Purves (1970) on leeches. In these animals certain sensory cells with somata located in the central ganglia have bifurcate axons running to the skin and body wall. Impulses can travel in both branches presumably and may allow transmission to move from one to the other. Nicholls and his colleagues do not believe this happens, but Günther (1970) working on earthworms shows that efferent sensory impulses of touch receptors can result from the bifurcate nature of similar nervous elements.

Thus the anatomical basis of the knowledge of nerve systems may solve problems of numbers, but cannot demonstrate the complexity of the arrangements or connexions of neurons. The neuropil of a ganglion in which nervous elements meet and synapse is complex, but nonetheless organized in particular ways. Synaptic junctions are not totally random but while the precise position at which interaction occurs may be variable, the precision with which one unit contacts another cannot be doubted.

Axon fusion may not occur at all in peripheral sensory nerves. Steinbrecht (1969), using insects, demonstrates enough primary sensory axons to account for all known receptors on the insect (*Bombyx* and *Rhodnius*) antenna; a site where axon fusion was long suspected.

The anatomical complexity of cellular interaction that can be described is further complicated by the manner in which nerve cells react to stimulation, respond to long term changes in the environment (internal and external), and affect one another by direct and indirect means. Ideas that neurons function solely on the all or none principle are now overthrown in the light of recent work. Some of the phenomena observed may be explicable in terms of structures (especially membrane structure), but we cannot go further at present (see Section III, B, 8).

b. *Scanning Electron Microscopy.* The second major anatomical advance of recent times has been made possible by the advent of the scanning electron microscope. This machine enables surfaces to be observed in fine detail and in several different planes at the same sitting. It reveals the exact disposition of component parts without dissection on a virtually intact and unmolested portion of an animal.

For sensory physiologists, this must be regarded as a boon of the highest order. The surface of animals, be they soft or hard bodied, is a highly sensitive region served by numerous sense organs of differing types. In the case of arthropods, especially decapod Crustacea, the sense organs are essentially cuticularized caps to projections of the hypodermis toward the exterior. In the case of aquatic Crustacea, it seems very unlikely that any part of the dermal layers makes actual contact with the external environment, with the possible exception of some chemoreceptors. The information obtained from the environment therefore is garnered by sense organs that incorporate cuticular projections of some form. The form may be critical to their function and hence an intimate knowledge is required.

Mechanoreceptors are aligned at the surface as hairs that are innervated (presumably in an analagous fashion to insects in which the nervous elements insert at the base of the hair). These hairs are of several sorts, being elongate and stout, or feathery, or short and broad (see Debasieux, 1949). Hairs can be arranged in preferred directions on the cuticle and this may confer properties of directionality in the responses of the sensory nerves (Mellon, 1963, Laverack, 1963). Physiological investigations alone do not reveal reasons for directionality, but anatomy can give good clues.

Chemoreceptors are also arranged on the surface of animals. In crabs and lobsters there may or may not be a pore at the end of the hair through which external fluids may make contact with the sensory elements within (Ghiradella *et al.*, 1968a, b; Laverack and Ardill, 1965). At present, in marine decapods there is some confusion over the existence of a pore, or it may be that there are genuine differences and whilst some have a pore others do not (see also Thomas, 1970, on crayfish). The scanning EM is the sort of instrument that can clear up points of this kind.

2. *Methods and Techniques*

Recent advances in staining techniques have concentrated on specificity and exactitude rather than on general overall staining. Precision in methods such as those achieved by histochemistry are valuable and important in localizing substances within certain cell systems, or within certain cells. The Falck-Hillarp fluorescence method for the location of catecholamines and indolamines has in the last few years become a standard tool (Falck, 1962; Falck and Owman, 1965; Björklund *et al.*, 1972). The localization of 5-HT (5-hydroxytryptamine) and catecholamines (adrenaline, noradrenaline, and dopamine) has been thoroughly examined in a number of phyla, including examples from marine fauna (coelenterates, arthropods, mollusks) (see Cottrell and Laverack, 1968). Taken together with EM studies, pharmacological experiments, autoradiography, and chromatography, much information is obtained at the single cell level (e.g., Rude *et al.*, 1969; Osborne and Dando, 1970).

Refinement and sophistication of existing techniques takes us further. Chromatography is now possible using individual cells isolated from ganglia. It has been successfully carried out on large neurons from *Helix pomatia*, and the giant cells and axons of mollusks such as nudibranchs (*Tritonia*, *Aplysia*) could be readily subjected to this kind of analysis (Osborne *et al*., 1971a, b).

Histochemical techniques can probably be devised for many substances, provided the biochemistry of the product is known. The value of the fluorescence method lies in its applicability to one group of substances that are suspected transmitters in many nervous systems. Similar methods for acetylcholine, γ-aminobutyric acid, glutamic acid, and other possible transmitters would obviously facilitate studies on precise localization of their occurrence. Otsuka *et al*. (1967) demonstrated GABA in sundry central neurons of lobsters by microdissection methods. Answers might now be more easily obtained. Such is the pace of advance in methodology that what is thought of as a major step this year is already outmoded within a year or two.

Staining of a different kind is that provided by the use of fluorescent dyes. The best known of these is procion yellow. Some years ago potassium ferricyanide injected into nerve cells enabled single cell bodies to be identified and located within ganglia (Kerkut and Walker, 1962). Reaction to prussian blue within the cell and subsequent clearing and microscopical examination showed the experimental cell clear of all others. The method only proved valuable for the cell body region, however, and gave no information regarding dendrites and axon pathways. The value of procion yellow and its chemical relatives is that it diffuses more readily in the cytoplasm of the cell and consequently reaches into fine cell processes (Stretton and Kravitz, 1968). By injection with electrophoretic or pressure methods (see Davis and friends, 1970), procion yellow is ejected from microelectrodes into single cells. It then may be left to diffuse for some hours and later examined either in sectioned material or in whole cleared ganglia. The anatomy of the cell is then revealed in surprising detail, although one must remain dubious about the resolution of the finest dendritic terminals. These are often so small as to be unresolvable in the light microscope anyway, so care must be taken in interpretation of the neuropil, for example. The drawing of the course of individual axons now begins to take on great precision as the result of this method and maps are now available of several sets of neurons and their endings as in lobster abdominal ganglia (Selverston and Kennedy, 1969; Davis, 1970), leeches (Nicholls and Purves, 1970), earthworms (Mulloney, 1970), mollusks and others. It is obvious that the future will see many more maps of this kind for animals of different phyla, marine forms among them, although possibly echinoderms may be difficult because of the generally small size of the axons. The only class in that phylum that seems even remotely possible to approach may be the ophiuroids where there are reported to be giant axons (Pentreath and Cottrell, 1971). Procion yellow dyeing is not of course histochemistry, but only a means of demonstrating the anatomy of a nerve cell. When specific stains are developed for all the transmitters it will be of great value, as also will specific stains for active peptides and polypeptides such as are found in certain neurosecretory systems.

The general Gomori and paraldehyde fuchsin stains utilized by endocrinologists are only indicators of secretory activity without any specificity as to carrier or active principle. A stain that could distinguish active from inactive compounds would facilitate investigations, though other cellular contents might also be stainable. The immunofluorescence techniques in use in studies on sympathetic neurones (Livett *et al.*, 1971a) could well be adapted to other systems, as in the study of endocrines in mammals (Livett, *et al.*, 1971b).

3. *Electrophysiology*

The major revolution in physiological analysis of the nervous system came about with the introduction of the oscilloscope. Prior to that time, although it was realized that nerve activity was essentially an electrical phenomenon, the response characteristics of galvanometers were too slow to allow detailed accounts to be prepared. The oscilloscope changed this because of its ease of use and the clarity of the record obtained. In the hands of Adrian and his contemporaries it soon became "The" tool for nerve system work. The development of high gain amplifiers, stimulators, and pen recorders was of less significance, although equally important.

The oscilloscope has now become more refined and by changes in ancillary equipment more adaptable, but basically remains the premier tool in this field of research. The use of wire electrodes, platinum, silver, tungsten, stainless steel, alloy, and so on did, and still does, provide a method of obtaining small signals from the axon for display, but does not usually allow for individual identification of nerve elements except by signal size. Isolation has to be accomplished by either prior dissection, by teasing nerve tracts into single fiber preparations, by selection of tracts in which only fiber remains active, or by more sophisticated means.

The second major step in electrophysiology was undoubtedly the final development by Graham and Gerard (1946), and Ling and Gerard (1949) of the intracellular electrode. First used for the study of the transmembrane potentials of frog muscles the use of fine drawn glass saline-filled electrodes has since been expanded into a variety of different approaches to the nervous system. Associated with this has been the development of the unity gain pre-amplifier as an associated piece of equipment.

Marine animals provide many suitable preparations for analysis by the most refined techniques presently in use.

The modern era in electrophysiology of marine animals may be said to start with the realization by Young (1939) that the giant axon of squids was indeed an axon and not a skeletal structure. This system of axons is located in the mantle of squids and the individual fibers may be of immense size (up to 2 mm in diameter for large squids) and hence the manipulation of the nerve becomes relatively easier. The insertion of a microelectrode into an axon is straightforward, though in the early developmental stages it no doubt seemed extremely difficult. The size of the axon also allows for the extrusion of the axoplasm by physical means so that the internal composition of the cell may be changed to one of known values. The exchange for known salines enables better examination of the effects of changing ionic concentrations on the activity of the membrane. Work on this preparation won the Nobel Prize for Hodgkin (1963). Squids still provide material for many workers.

With the greater and greater refinement of the intracellular electrode more and more cells have become prey to the marauding electrophysiologist. With neuroanatomists describing more and more cells of medium to large size, for example in *Aplysia*, nudibranch mollusks, lobsters, and crabs, more and more studies are being made by intracellular means.

B. PHYSIOLOGY

1. *Cell to Cell Conduction without Benefit of Nerves*

A number of instances have been reported in the last few years in which cells interact electrically directly with the next cell and without the interpolation of nerves. The phenomenon occurs widely in epidermal layers and is not restricted to invertebrates vs vertebrates, nor to terrestrial vs aquatic animals.

Some particularly interesting examples exist among marine animals that indicate the possible importance of such interactions in areas where nervous integration might otherwise be suspected. The simplest case, and certainly that which occurs earliest in the life of the animal, was reported by Tupper *et al.* (1970) studying the development of the starfish *Asterias forbesi*. In the 2, 4, 8, and 16 cell stages no electrical coupling between cells is noted, but at 32 cells (fifth cleavage) injected current passes between the cells. That is to say a low resistance pathway is established by this stage and this is concurrent with the formation of desmosomes at this time (Wolpert and Mercer, 1963). This does not indicate that information flow occurs between cells, nor that the function of the nervous system is usurped in such cases, but the very fact that electrical changes are transferred so readily (though with attenuation) indicates that subtle changes of milieu may occur rapidly and continuously. There is no definitive nervous system of course at these early stages. Strathmann (1971) has recently shown ciliary patterned beating on the surface of later echinoderm larvae. He postulates that such activity may be coordinated by a general field effect from cell to cell. Moores (1970, unpublished) in observations in the author's laboratory has shown bundles of presumptive small diameter axons in certain areas of the pluteus of *Echinocardium cordatum* which may also be concerned with coordination of the larva.

Coordination between cells without the intervention of nerves demonstrated by correlated changes in potential in contiguous cells has been termed neuroid conduction. Some information content distribution is necessary for this definition to be acceptable, since simple coupling is not sufficient. Neuroid conduction implies that some transmission of meaningful information is accomplished.

Neuroid conduction has been shown in sundry coelenterates, both medusae and polyp in form (see for example, McFarlane, 1969). Mackie (1970) gives a good account of the range of occurrence and types. Perhaps the most interesting example is in ctenophores (Horridge, 1965a,b), where a double control system apparently exists for the modification of comb plate beat. The compound cilia of the combs are synchronized to beat together and in a distinct phase relation to each plate. Nervous transmission takes place between the elements of the nerve net that run over the surface of the animal and under the ciliated cells. This system seems inhibitory in function. Waves of ciliary activity that act in locomotion on the other hand are associated with depolarizing waves in the underlying tissue, and transmission is by neuroid or cell to cell conduction.

Similar phenomena of cell to cell conduction are shown in the hearts of tunicates. This is not neuroid, but myoid (or muscle cell to muscle cell) conduction. Integration of muscle contraction occurs directly from cell to cell. Nervous coordination of the muscles of the heart is not in evidence here. The pacemaker activities of one end of the heart are dominant for a while, and then dominance reverts to the opposite end. Conduction of the contractile wave along the surface of the heart is via the muscle cells (Kriebel, 1968a,b; Anderson, 1968). This type of coordination ensures a smooth contraction over the heart without interruption.

Some discontinuity and interruption of current flow is demonstrated in crustacean heart. In *Homarus americanus* the muscle fibers are of limited length, ending at a disc which abuts the contiguous fiber. It is not known if these are rectifying in nature. Contraction of the muscle fibers is graded and activated by local excitatory potentials (ejp's). Propagation of ejp's across a disc is probably not physiologically significant (Anderson and Smith, 1971).

These questions of cell to cell conduction are possibly important in consideration of the evolution of the nervous system from its earliest beginnings (Horridge, 1968; Mackie, 1970) and further work is required to elucidate the basic features of these unusual examples.

2. *Motor Neurons*

The function of motor neurons is to convey information from the CNS to the muscles. A great deal is known in some marine animals about the patterns of motor innervation, and about the patterns of impulse firing, and types of musculature affected. Fast and slow muscles, inhibition, and neuromuscular junctions are well described (see Kennedy *et al.*, 1966).

The use of dye injection and sophisticated careful analysis of electrical responses now enables us to make a generalization about motor neurons that was not possible a few years ago. Sandeman (1969) showed for the crab (*Carcinus maenas*) that the motor neuron involved in reflex eye withdrawal is not a simple unit acting in a simple way, but a rather complicated morphological and physiological entity. He showed that the proximal region of the axon, closest to the cell soma, was of large diameter while the conducting portion is of smaller diameter. The "fatter" portion is concerned with the integration of signals received from external sources (sensory input) (Fig. 1).

Integration is a function of the input because those inputs ending closest to the spiking portion of the axon will be more effective in modifying propagated activity than those further away. Sandeman suggests that this may be an answer to the question of differing effects of differing inputs. Information derived from mechanoreceptors around the eye is more likely to affect the motor neuron than similar information from the carapace; he hypo-

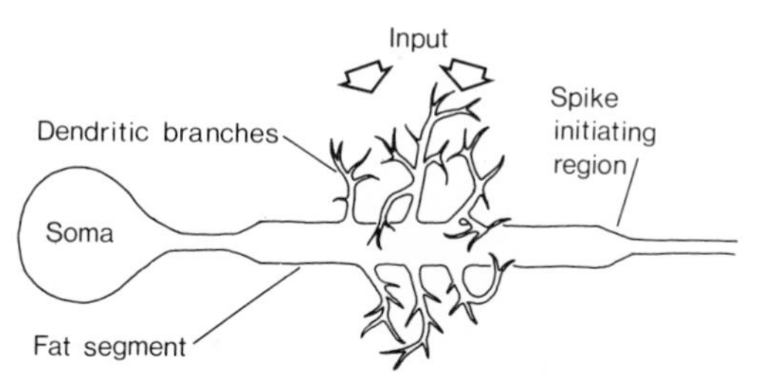

Fig. 1. A schematized motor axon at the somatic region. Information pooled from several sources shows that motor axons in some invertebrate groups, including decapod Crustacea and annelids, are composed of a trophic cell soma connected via a narrow diameter segment to a much broader region of the axon. At this region there are dendritic branches that synapse with various efferent sources and inputs. This is the integrating region (Sandeman, 1969). Distal to this the axon narrows again and the spike initiation area is reached. Potentials within the fat segment do not propagate, but are integrated. Dendrites closer to the axon are likely to be more effective at evoking spike potentials than those originating nearer the cell body.

thesizes that sensory discrimination is conferred by the spatial separation of inputs onto the various portions of the motor neuron.

Procion yellow reveals that motor neurons in abdominal ganglia of crayfish and lobsters as well as insects (Bentley, 1970) all have similar morphologies with fat proximal segments. The same is true of stomatogastric ganglion cells in lobsters (Maynard, 1971), and of leech motor neurons (Stuart, 1970).

3. *Innervation of Muscle*

While I do not propose to précis the great volume of literature of the last decade that has dealt with the innervation of muscle in Crustacea and other groups, it is, I feel, appropriate to point out one major discovery of recent years. This rather changes our viewpoint that muscles are innervated only by prolongations of nerves that end at neuromuscular junctions at some point on the muscle surface. In certain phyla this no longer holds true, at least for some muscles.

In echinoderms (notably *Echinus*) two alternative innervation patterns have been described (Cobb and Laverack, 1966, 1967; Cobb, 1967). In the case of the muscles of Aristotle's lantern, the neuromuscular organization is close to the usual pattern. Nerve fibers originate in hyponeural layers of the oral system and run to the muscles, ending low down on relatively long fibers. There is a neuromuscular junction enveloped in a small fold of the membrane of the muscle fiber, and which contains small vesicles. In the region of the ampullary muscles of the tube foot, however, a radically different innervation is shown.

The tube foot is narrowed at the neck where the ampulla joins the projecting part of the podium. At this neck there is a nerve ring in the wall of the podium. This nerve ring originates at the radial nerve cord, and there is one for each tube foot. There are no obvious dendritic extensions from this ring, or tracts of fibers, although there is an ampullary "seam," which has a nonmuscular but fibrous structure. The innervation of the musculature is not therefore by nerve fiber dendrites. Instead the opposite situation occurs. The *muscle* fibers possess long extensions, devoid of the normal contractile apparatus, and in electron micrographs look remarkably bereft of the cytoplasmic content typical of muscle. These long muscle "tails" run down the stalk of the ampulla to the neck and there intermingle with the nerve fibers. At this point some synaptic structures have been seen, and while there may be some axon–axon synapses, many are interpreted as axon–muscle junctions (NMJ's in fact). Muscle "tails" also occur in pedicellarial muscle (Cobb and Laverack, 1967).

Pentreath and Cottrell (1971) have published a description of "giant" neurons in *Ophiothrix*, that are thought to be motor in function. While the description is convincing, a valid alternative explanation may be that the giant neurons are in fact muscle tails. The structure is like that described in other echinoderms; the diameter is much greater than any other axon in the system, they run into the radial nerve cord, and specific NMJ's have not been described. Against this it can be stated that ophiuroids move rapidly and hence may be expected *a priori* to have larger motor nerves than other echinoderms; there are areas where apparently motor nerve and muscle extensions are intermingled, the cytoplasm of axons and "giant" neurons is similar, synapses are not described in the hyponeural or ectoneural collections of nerve fibers. The system perhaps needs more work before positive identification is achieved.

These muscle "tail" arrangements are not unique to echinoderms. They have been reported by Rosenbluth (1965) in *Ascaris*, and by Flood (1966, 1970) in *Amphioxus*. It seems possible that they will be described in other phyla also, especially among the smaller marine groups. Figures 2 and 3 show the situation as it is known for *Amphioxus* muscles. Muscles of the echinoderm tube foot and pedicellariae are shown in Fig. 4.

Some information is available on the electrical and mechanical responses of such muscles to electrical stimulation (Guthrie and Banks, 1970a,b) but we have no certain knowledge of the properties of the "tail." From its structure it must be assumed that it plays no part at all in contraction and is simply a conduction device. del Castillo *et al.* (1967) suggest that in *Ascaris* small abortive spikes in the muscle cell may be due to subthreshold stimulation of each of a number of terminal "fingers" ("tails" in present parlance). Summation from the several endings (3? in *Ascaris*) may give rise to an all-or-none discharge. The amplitude and time course of the small abortive transients is presumably related to the length and diameter of the tail. The safety factor for transmission toward the main muscle contractile region is small. Why this should be more satisfactory than the more usual arrangement of nerve and muscle junctions is not yet clear, but the interaction of motor nerve and muscle may be similar in both cases. In the classic case, as Sandeman (1969) and others have shown, integration of synaptic input onto the motor neuron is accomplished at the "fat" segment and the output of the motor nerve is a function of the integrated signals at this region. In the muscle "tail" situation perhaps this integration is accomplished at the tails. Summation of input from the nervous system takes place in the tails, but only gives rise to propagated potentials when it rises above threshold level. It may thus represent the "fat" integrating segment of the muscle. Neuromuscular systems of whatever kind therefore have intrinsic integrative capacity of their own regardless of the input derived from other sources.

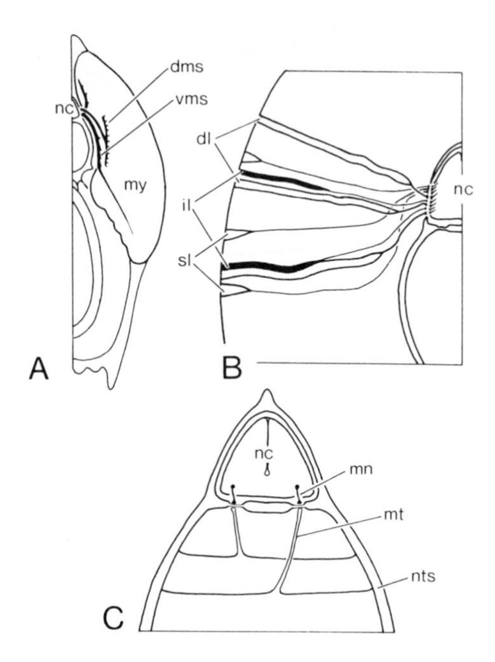

Fig. 2. Muscle tails are numerous in *Amphioxus* musculature. (A) A hemisection of the lancelet indicating the trunks that represent muscle tail collections (dorsal, dms; ventral, vms). These aggregations of muscle prolongations from the myotomal muscle (my) run back to the nerve cord (nc). (B) The myotomal muscle is comprised of differing layers of muscle fibers, with corresponding lengths of muscle tail. Superficial layers (sl) have long "tails"; intermediate layers (il) have medium lengths and deep layers (dl) have short tails. Each projects to the nerve cord (nc) centrally. (C) The notochord is composed of muscle (Flood, 1966), enclosed in a sheath (nts). The fibers of this muscle have muscle tail (mt) projections that extend dorsally to the ventral side of the nerve cord (nc). A thin layer of collagenous sheath separates these muscle tail terminations from nerve dendrites (mn) that end on the ventral floor of the nerve cord. These nerve endings carry synaptic vesicles (after Guthrie and Banks, 1970a; redrawn).

4. *Notochords*

While dealing with unusual forms of muscle innervation it should be mentioned that our views of the notochord as exhibited in some higher marine invertebrates have recently taken a surprising turn.

An anatomical study of the CNS and the notochord of *Amphioxus* by Flood (1966) showed that synapses and thinning of the bounding layer of sheath material occurs at the ends of nerve projections from the CNS, but the nerve does not project through the sheath in the region of the notochord that underlies the nerve cord. The existence of synapses above what has been long thought of as a skeletal structure was unexpected, but even more

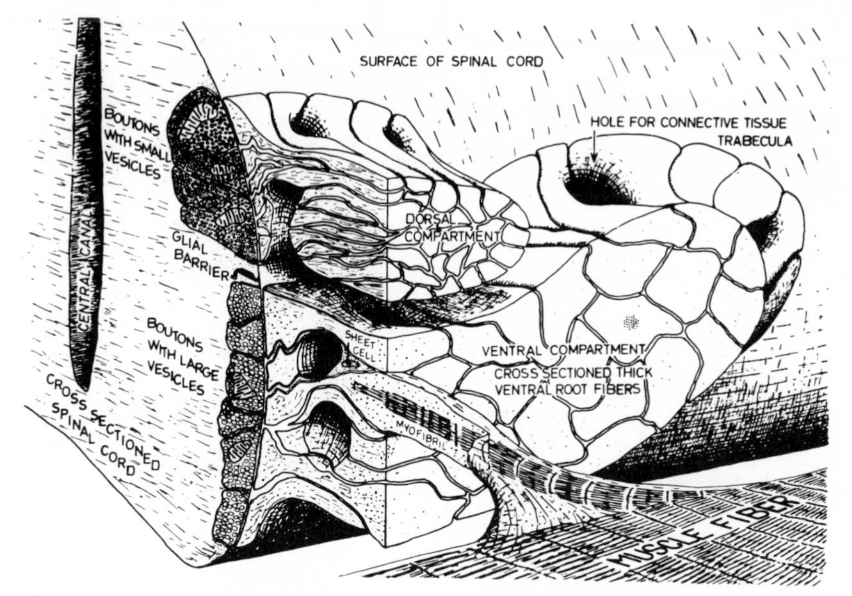

Fig. 3. Muscle "tails" of *Amphioxus* somatic muscles are collected into two bundles (dorsal and ventral) that extend toward the nerve cord. The muscle tails end bluntly upon the surface of the nerve cord where they make synaptic contact with motor neurons within the cord. A thin basement membrane separates the two cellular entities (Flood, 1966).

unexpected is the nature of the notochord. This is muscular. The individual muscle fibers have projections that reach dorsally toward the CNS, ending beneath the collagenous sheath (Fig. 2c).

The electrical and mechanical characteristics of this notochordal muscle are described by Guthrie and Banks (1970a). The muscle serves to increase the stiffness of the notochord, and there is an increase in internal pressures during contraction. It therefore acts as a form of hydrostatic skeleton. The physical nature of the muscle is like that of paramyosin muscle, but faster than those of mollusks. It is probable that the notochord makes its contribution to function during the swimming of the animal; each activation of the muscle being preceded by a giant fiber potential (but see Schramek, 1970, for a warning).

The notochord of other invertebrates is not of similar construction; in hemichordates for example it is composed of vacuolated epithelial cells interspersed with mucous cells (Welsch and Storch, 1970). While there may be homology between this and other chordates there seems to be no similarity with *Amphioxus*. Since all protochordate groups (Cephalochordata, Urochordata, Hemichordata) are marine this seems a fallow field that would be worth ploughing further.

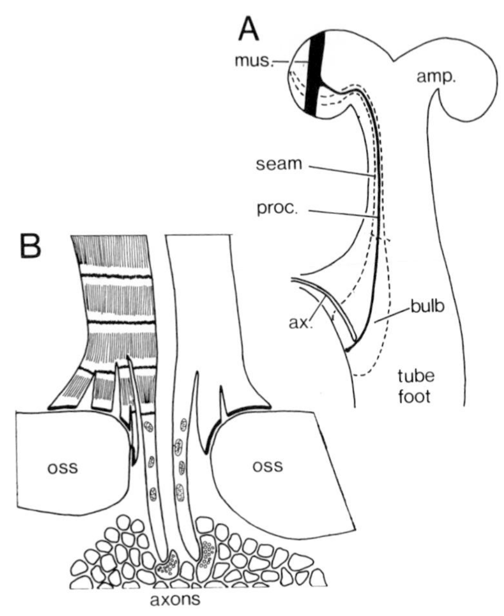

Fig. 4. In echinoderms muscle tails exist in at least two separate muscular systems, and may well be found in more. (A) The ampullary muscles of the tube foot of *Echinus* possess long extensions that are gathered into a fibrous "seam" that runs along one side and ends at the constricted "neck" of the tube foot. At this point synaptic connexions are seen between axons originating in the radial nerve cord and the muscle tails (after Cobb and Laverack, 1967; redrawn). (B) In the pedicellariae of *Echinus* both smooth and striped muscle fibers insert on the ossicles of the skeleton, but a muscle tail projects between the ossicles to synapse (NMJ) with axons forming a small neuropil at the base of the ossicles (after Cobb and Laverack, 1967; redrawn).

5. *Peripheral vs Central*

The old-fashioned division of the invertebrate nervous system into peripheral and central regions is beginning to be eroded. Although it may still be convenient to consider portions in separation it should also be realized that neurons can no longer be considered as (1) sensory and hence peripheral, (2) interneurons with central connections, and (3) motor fiber with perhaps central cell bodies and peripheral nerves.

This standpoint requires modification because recently it has become evident that not all sensory cells are at the periphery and not all motor cells necessarily originate in the CNS, nor do all interneurons exist only in the CNS.

The mantle nerve of bivalve mollusks runs around the edge of the mantle. It receives input from sensory tentacles projecting from the mantle, and is also connected to central ganglia via mantle nerves. It nonetheless acts as an independent reflex organizer (shown for *Pecten* by Wales, 1968, unpublished) which requires no interpolation of CNS influence to conduct its affairs. Olivo (1970a,b) has shown for *Ensis* that the digging movements of the animal when burrowing do not need the cerebral ganglion in the system. Peretz (1970) in *Aplysia* also indicates that peripheral reflexes not involving the CNS are quite common. Peripheral ganglia, or small collections of cells are also presumed integrative areas (Lawry, 1970 on *Aphrodite*).

Another manner in which change in status is occurring is shown by the number of cases in which sense cells subserving mechanoreception are now known to have fibers extending to the body wall and the epidermal surface, but which have distant cell somata located in central ganglia. The best documented of these cases is that of the leech *Hirudo* (Nicholls and Baylor in a series of papers, 1968–1971, not all referred to here). Mechanoreceptors responding to touch, pressure, and noxious stimuli are identifiable and constant in position in the ganglia of the ventral chain in this annelid. Similar cells are described in oligochaetes (Günther, 1970) and one might expect to find them in polychaetes.

Aplysia central and pedal ganglia also contain sensory cell bodies with peripheral endings (Castelluci *et al.*, 1970; Weevers, 1971); so also do the lampreys (Martin and Wickelgren, 1971). In the latter example the cells have been classified in the same way as for leeches, as T, P, and N cells. Martin and Wickelgren have given details as to the geography of the cell bodies, which lie dorsally in the nerve cord. They also give information about the receptive fields, and it is shown that each cell has some particular area of the epidermis that is their perquisite site of action (Fig. 5).

6. *Central Program vs Sensory Input as the Governor of Behavior in Invertebrates*

The attempt to characterize total input into an animal's nervous system is probably doomed to failure. It is almost impossible to envisage all the parameters of the environment that may be meaningful to the animal and affect behavior, and it is difficult to comprehend the mechanisms that allow analysis of the changing environment. The effects of pressure on various planktonic and shore dwelling species (Enright, 1962; Knight-Jones and Morgan, 1966) have led to various explanations (in fish swimbladders, Qutob, 1962; for copepods, Digby, 1964), that are insufficiently detailed to be considered final. Magnetism, barometric pressure, radiation, and so on all seem to have some influence on animals without having a background of sensory mechanism for explanation (see Laverack, 1968).

All the same, the input of information into any animal is enormous, even for those animals seemingly underequipped with sense organs. The adaptive

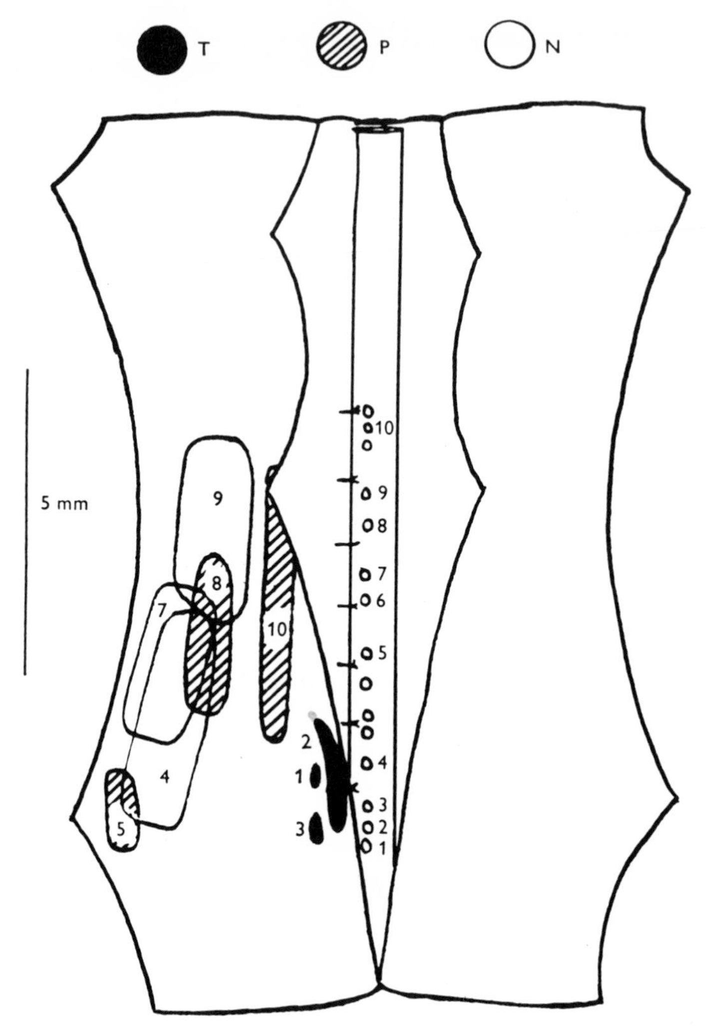

Fig. 5. Receptive fields of T, P, and N cells in the gill region of the lamprey spinal cord. The centrally placed dorsal sensory cells are shown on the left side of the cord. The numbers refer to cells from which recordings have been made and the receptive fields of these particular cells are numbered accordingly. The field of cell 6 could not be found (Martin and Wickelgren, 1971).

significance of sense organs is self evident; the information they provide is needed in order to maintain the animal in optimal conditions. Those optimal conditions are probably different for every species, and perhaps for every individual animal. The more information that can be obtained for determining behavioral variation, the better.

The basic argument of overall control by either exclusively peripheral or centrally governed behavior patterns is a nonstarter. It is apparent that under all conditions the behavior of an animal is a compromise between centrally generated activity within neurons that control the walking or respiration or excretion or feeding of the animal, and the sensory cues that stimulate the animal to modify the ongoing activity. An animal at rest is stimulated to movement or to alter respiratory rhythm by changes in the external environment such as the ingress of a predator, or a change in gaseous content. Subtle as well as overt and obvious changes are important. Once the trigger is applied, centrally generated patterns may be released within the CNS (Willows and Hoyle, 1968a,b; Dorsett *et al.*, 1969). These patterns must then be themselves the subject of modification by external influence (including the activity of proprioceptors, changes in hormone levels, blood chemistry, interneuron activity). Once turned on, the central program must also be turned off; either by a new signal or by the loss of the old, or by habituation and adaptation. The sensory input does not of itself provide behavior, the central mechanism is meaningless unless governed and modified by the sensory information (see Bullock, 1969).

Control must be exerted at all levels on the behavioral activities of animals so that movements, for example, are not only gross clumsy events, but may be capable of infinite delicate adjustment. Such adjustments are not built into the centrally generated pattern, but rather are the consequence of continually changing signals derived from the environment altering the output of the generator.

In order to understand all the probable forms of input into a central nervous system, and to determine their effect, a survey of all sources of information must be made. This has been the primary purpose of the work carried out by myself and colleagues in the last few years at the Gatty Marine Laboratory. We are still at the beginning of the task; much of the earlier work has been previously summarized (Laverack, 1968).

Typical of the questions raised by examples of behavior is what happens when a lobster walks, autotomizes legs, and feeds? These quite routine behavior patterns are exhibited regularly (though perhaps not daily in the case of autotomy!). Each of these events requires a barrage of sensory cues to trigger, maintain, and adjust. Innumerable sense organs operate to provide the information required. Frequently locomotion and other muscular activity is described as being a driven sequence with signals emanating from a central oscillator; but with no indication of how that oscillator may be modulated. Figure 6 (Burrows and Willows, 1969; Stein 1971) shows examples of this type of circuit thinking. I contend that each diagram should have another component (at least one, probably more) that is the sensory variation of central oscillators. The oscillators described by Mendelson (1971) are

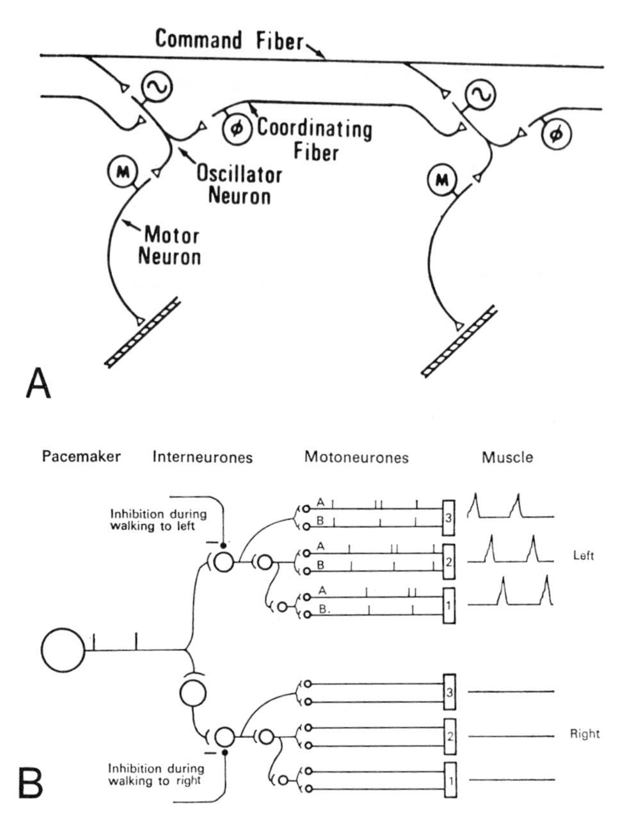

Fig. 6. (A) Schematic representation of connexions involved in control of swimmeret beating in the crayfish. This diagram shows two successive hemiganglia in the abdominal nerve cord containing command fibers that modulate activity in an oscillator neuron. This oscillator drives motor neurons to the appendage and at the same time is lined to the oscillators in other ganglia (Stein, 1971). (B) Schematic neuronal model to explain coordinated beating of maxilliped flagella in brachyuran and anomuran Crustacea. Both sides are driven by a common pacemaker (oscillator neuron?) that discharges regularly with various but constant synaptic delays interposed in the lines to each flagellum. The interneurons shown merely indicate that delay occurs at synapses and do not represent the actual number of interneurons involved. Coordination of beating with walking is accomplished by the inhibition of coupling interneurons (Burrows and Willows, 1969).

typical units that, like pacemakers, continually produce a patterned discharge, but one that is controlled extrinsically.

A lobster (or other arthropod) may be considered an ideal object for study of sensory systems. Since the animal has an exoskeleton, the places at which the external environment may act specifically on sense organs is limited. The general cuticle surface is unspecialized but is pierced in numerous places by projections of nerve fibers that end superficially in end organs.

The end organs may be campaniform organs (Fig. 7) (Shelton and Laverack, 1968), CAP organs (Fig. 8), *büschelorganen* (hair fans, Fig. 9), soft cuticle (Fig. 10) (Pabst and Kennedy, 1967), stout bristles and hairs (Fig. 11) and other types not yet described. Internally there are other organs, proprioceptors that are also monitoring the environment, but less directly.

Let us consider the activities mentioned above; stimulation to walking, autotomy, and feeding.

7. *Sensory Physiology of the Decapod Crustacea Especially the Lobster*

a. *Chemoreception.* The detection of food is an important part of the life of any animal, but active location and collection is the prerequisite of predators and scavengers, both of which terms are applicable to crabs and lobsters.

The distant perception of attractive (and also noxious) substances is highly significant in the location of food, mates, or enemies. Some of these aspects have recently been reviewed (Lindstedt, 1971; Laverack, 1974). For lobsters the mechanisms involved are several, and by no means yet fully determined.

Chemoreceptors are distributed widely over the body, and may be of several different modalities subserving different functions at these locations. Receptors in or on the gill covers may be functional in influencing respiratory rates, and are responsive to partial pressures of the gases dissolved in seawater (Larimer, 1964). Those on the walking legs are probably concerned with substances diffusing around the leg. Those on the subchelate portion of propodite and dactylopodite are much modified hairs, squat in shape and with a considerable sensory innervation (Fig. 11) (Shelton and Laverack, 1970). These are ideally placed as contact receptors for food particles that are grasped, passed forward, and then presented to the mouth for ingestion. Little is known of the responses of these receptors to sapient material. Chemoreceptors carried on the hairs of the maxillipeds are as yet also undescribed.

Distance chemoreception in the decapod Crustacea is certainly at least partly fulfilled by the aesthetasc hairs of the antennule (Copeland, 1923; Holmes and Homuth, 1910). These have been fully described elsewhere (Laverack, 1964; Laverack and Ardill, 1965) as thin walled, untanned, unpigmented, and heavily innervated projections of the cuticle of the outer ramus of the antennule (Fig. 12). They lie in rows of 12–15 in number, two per annulus. Methylene blue reveals about 15 neurons per hair, but the EM indicates that a figure of 300–350 is correct. These sensory cells have dendrites that project upward into the lumen of the hair, and carry ciliated endings (Laverack and Ardill, 1965; Ghiradella *et al.*, 1968a,b). These receptors cannot be completely characterized by physiological means since there

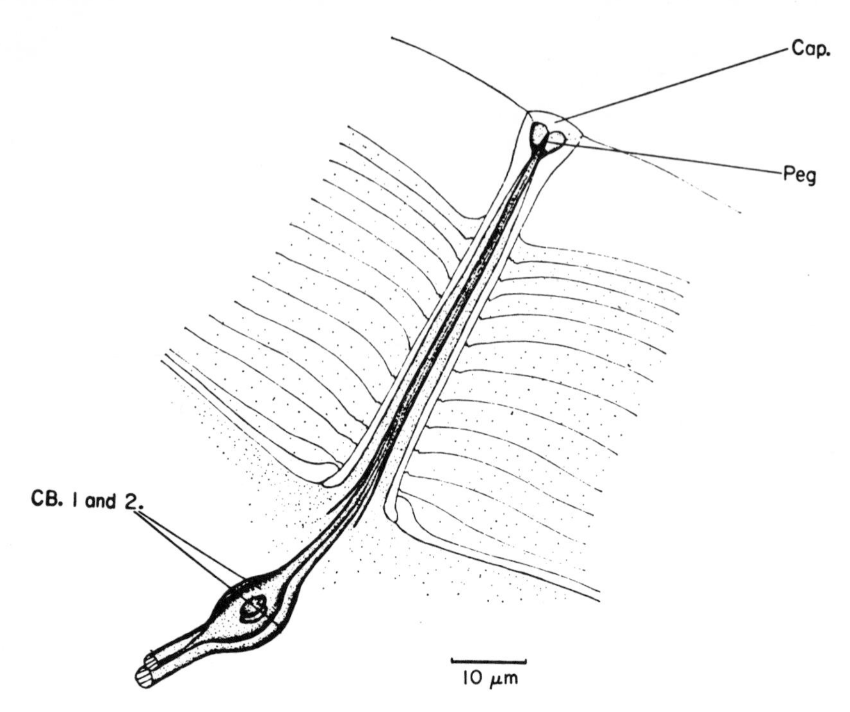

Fig. 7. Crustacean campaniform organ typical of the dactyl termination in walking legs of decapods. There is a distal cap (cap) and peg (peg) served by two sensory cells (Shelton and Laverack, 1968). This contrasts with the situation in insects where recent EM studies indicate a single innervation (Moran *et al.*, 1971).

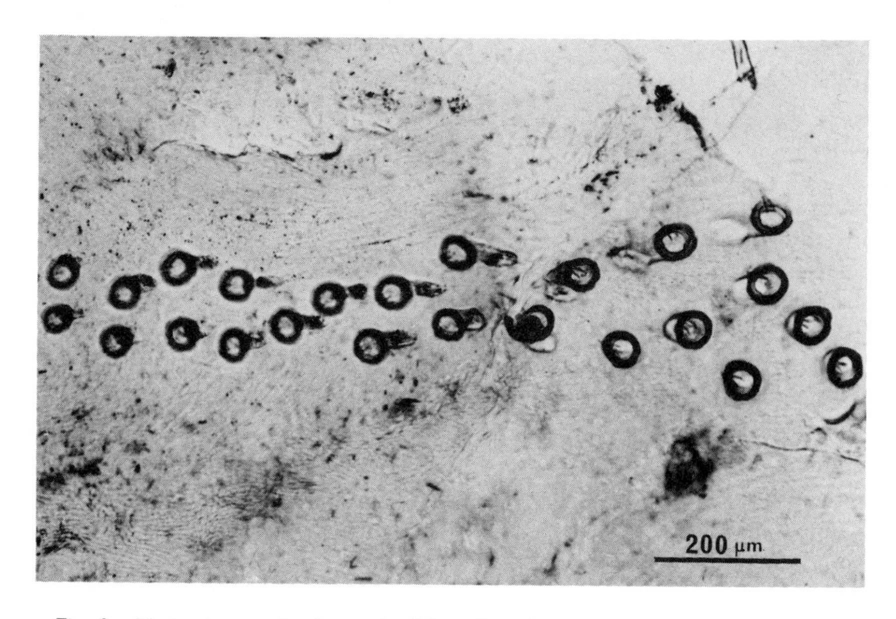

Fig. 8. Photomicrograph of unstained but cleared CAP sensilla from *Panulirus argus*. Note the canals through the cuticle with short spines mounted on a flexible membrane at the apex (Wales *et al.*, 1970).

Fig. 9. Scanning electron micrograph of a hair fan organ (*büschelorganen*). A small globular base is covered by branches arranged like petals on a flower bud; the whole organ lies in a shallow depression on the surface of the animal (Shelton and Laverack, 1970).

are very large numbers present, and they may have somewhat different properties. The situation is not amenable to very precise analysis on the style of insect preparations. Nonetheless Laverack (1964), Levandowsky and Hodgson (1965), van Weel and Christofferson (1966) have recorded successfully from antennules and shown that certain substances are stimulatory (see Fig. 13), although it is not certain what the behavioral conse-

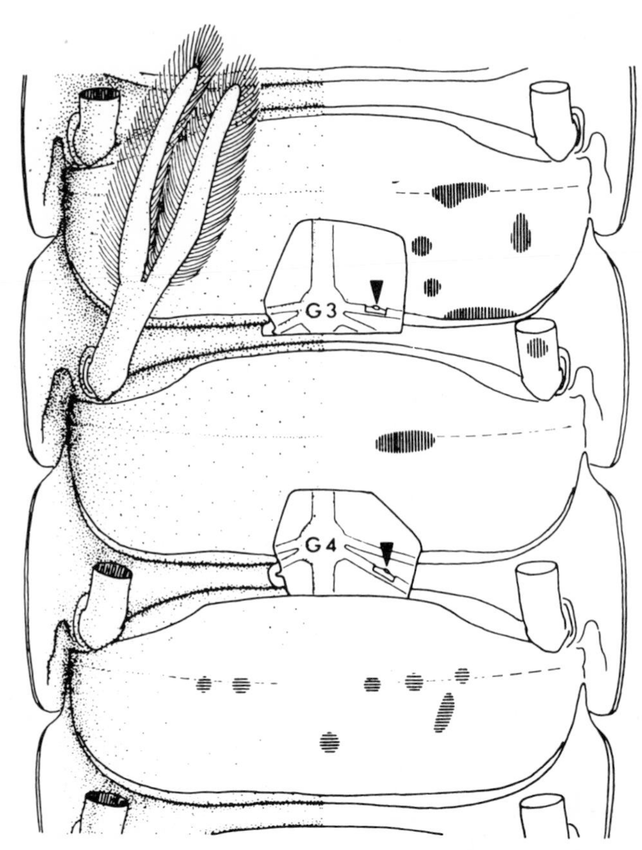

Fig. 10. Ventral view of a portion of the crayfish abdomen, showing ganglia 3 and 4 exposed with the first and second roots. The sensory fields of the two cells indicated by arrows in the first root of ganglion 3 and the second root of ganglion 4 are shown by the vertically and horizontally striped areas respectively. The dotted lines show the attachments of the superficial flexor muscles (Pabst and Kennedy, 1967).

quences may be. In the shrimps *Betaeus* certain commensal attributes with regard to echinoderm hosts may be explained by specific responses of antennular chemoreceptors (Ache and Case, 1969). Reproductive behavior may be mediated via this population of receptors (Ryan, 1966; Atema and Engstrom, 1971).

The manner in which ingoing signals are integrated and generate behavior patterns is a function of the medulla terminalis (Maynard and Yager, 1968) and this has been discussed by Maynard and Sallee (1970) (Fig. 14). The various inputs that originate in the periphery are coordinated in the medulla terminalis located in the eyestalk. The various activities of the animal

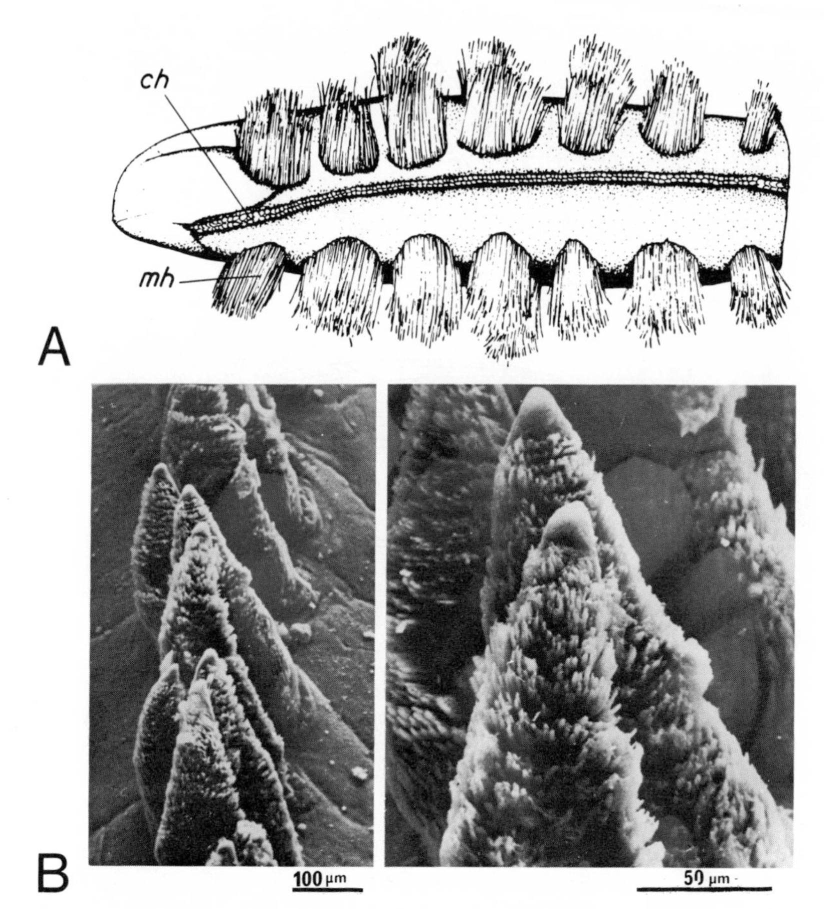

Fig. 11. (Top) Inner edge of the chela of the first pereiopod of *Homarus gammarus*; the first two pairs of pereiopods terminate in small chelae which bear prominent bunches of mechanically sensitive hairs (mh); the inner edge of the chela is formed by a row of squat branched hairs (ch) which are highly sensitive to chemical stimulation (Shelton and Laverack, 1970). (B) Scanning electron micrograph of the squat hairs of the chela edge.

reinforce and stimulate new pathways as the behavior reaches a new phase. Interference with the medulla terminalis acts as an effective block in the integration of chemical signals.

b. *Locomotion.* Adequate provision of stimulants in the environment release walking in the lobster (the basis of lobster fishing) and the animal walks toward the interesting material. When moving, and also when at rest

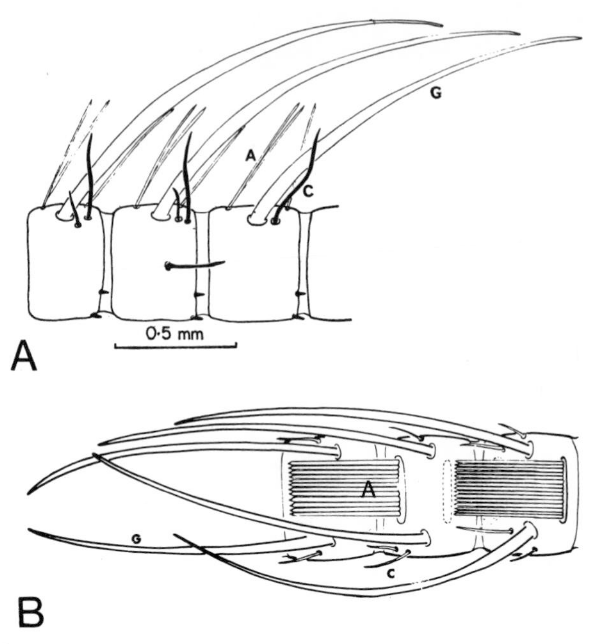

Fig. 12. The outer flagellum of the antennule of *Panulirus argus* carries a tuft of hair sensilla distally. These are of three main types, the guard hairs (G) which are long and stout; companion hairs (C) which are arranged at the base of the guard hairs; and aesthetasc hairs (A) which are long thin and unpigmented lying in two rows on each annulus of the flagellum. (A) represents the arrangement seen laterally; (B) as seen from ventrally (in the latter two rows of aesthetasc hairs have been omitted for clarity, the positions being shown by a broken line) (Laverack, 1964).

of course, a whole series of receptors are in action. Among these are the proprioceptors of the pereiopods. What is known about these has been summarized by Wales *et al.* (1970). The disposition of the receptors is shown in Fig. 15, together with the occurrence of CAP organs. CAP organs are small spines at the MC and CP joints and may represent a vestige of external hairs which have gradually been withdrawn during evolutionary history until they are now small abortive cuticular endings. The cell bodies are placed internally alongside the proprioceptors of the joints. There are no CAP organs at the other joints of the walking legs. At these positions it is postulated that withdrawal may be complete and the proprioceptors are entirely internal. The appropriate stimulus for CAP organs has not been determined. Wiersma (1959) claimed that response to mechanical stimulation could be recorded, but at a position far from the site of action. With so many highly sensitive mechanical units associated with other hairs in the area, the necessary precision for certain correlation of receptor with record

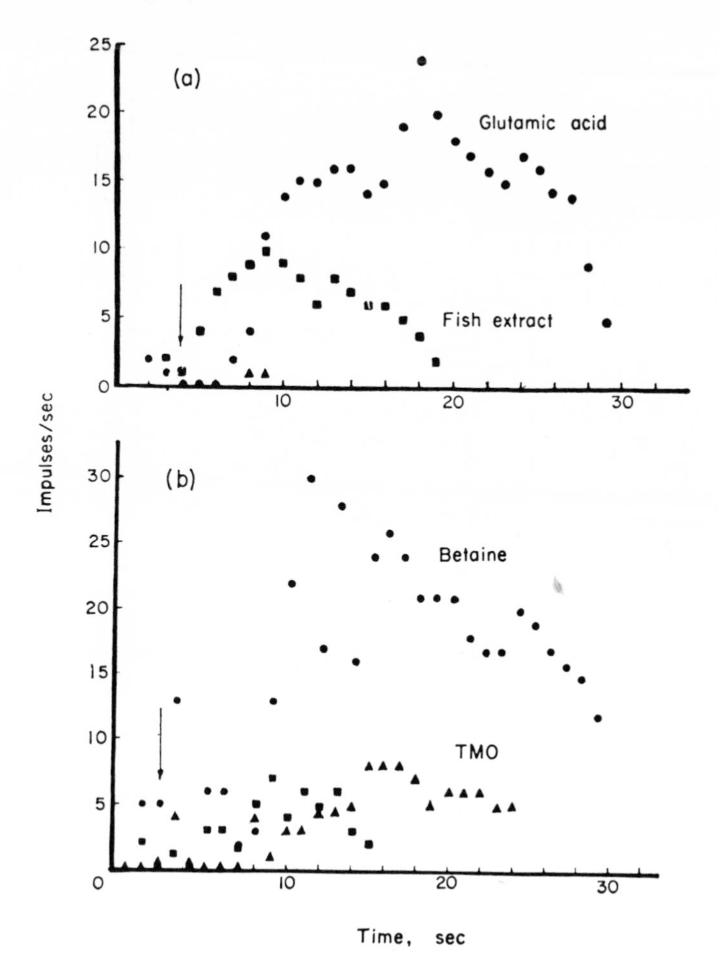

Fig. 13. The response of a chemoreceptor from the antennule of *Panulirus argus*. The addition of various stimulant substances to the preparation gave rise to varying responses by the sensor. (TMO = trimethylamine oxide.) To indicate effects of differing chemical stimulation on responses of a single nerve unit, the arrow indicates the time of addition. (a) ▲, Response to addition of seawater—only a negligible, possibly spontaneous, discharge. ■, Fish extract added to the preparation invoked activity that rose to a peak frequency after 5 sec, thereafter adapting back to zero at 15 sec. ●, After stimulation with saturated L-glutamic acid. A latency of 3 sec was followed by a rapid frequency rise, then a period of stable frequency terminating abruptly after 52 sec. (b) ■, Indicates a small response of this preparation to a repetition of the addition of seawater. ▲, A small, low-frequency, response after stimulation with 0.1 *M* TMO, only commencing following a latent period of some 7 sec. ●, A background discharge of about 5/sec continued almost unaltered for 6 sec after treatment with 0.1 *M* betaine but then a rapid rise in frequency to a peak of 50/sec was followed by a period of gradual adaptation toward the low spontaneous discharge rate (Laverack, 1964).

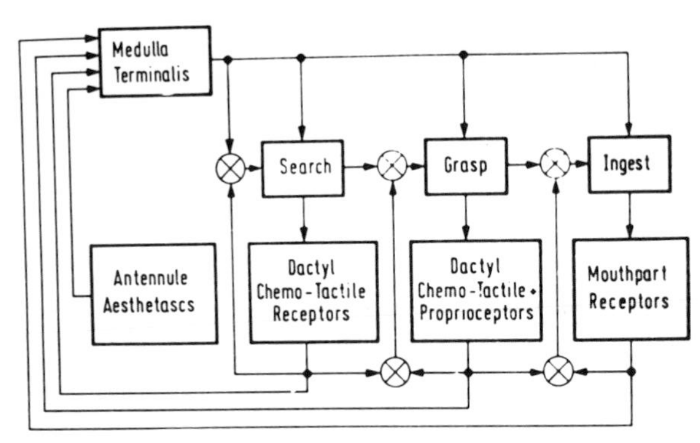

Fig. 14. In decapod Crustacea the three major feeding acts of search, grasp and hold, ingestion and mouthpart activity, are each presumed to be controlled by a separate center of nervous activity that can be turned on or off or modulated in intensity. The output of each center produces patterned motor activity and at the same time acts as a partial direct input to the next sequence. Peripheral motor activity initiated at each stage results in sensory feedback to the same center and feed forward to the next. Chemical, tactile and proprioceptive information is all provided. Some portion of the sensory input passes through to the medulla terminales in the eyestalks. Olfactory input from the antennules must pass through the medulla terminales if it is to turn on the feeding sequence. The diagram also suggests that output from the medulla terminalis may not only initiate the sequence through input to the first center, but may also set or tune the various generators to produce the appropriate responses to given inputs (Maynard and Sallee, 1970).

was difficult. Later work (Alexandrowicz, 1970) suggested that mechanical effects due to rolling of the articulating membrane across the surface during flexion and extension might be important, but in some cases the CAP organs were not covered during these movements. Wales *et al.* (1971) discovered that they could not stimulate these neurons mechanically at the surface with certainty. Their final opinion was that these may be a secondary group of proprioceptors and that the same stresses as stimulate their neighbors are important for CAP organs; that is, internal not external stimulation. Theories on the relative roles of central command and peripheral control may have to be revised in the light of recent work by Macmillan and Dando (1972). These authors have demonstrated the presence of unsuspected tension receptors located in the apodemes of the walking leg muscles of *Cancer*. These respond to tension without movement and hence provide another comparative input parallel to chordotonal organs and muscle receptors.

c. *Autotomy.* Under certain circumstances, the walking limbs and chelae of decapods may be autotomized; that is they are shed from the body, to be replaced by regrowth at a later time. This behavior is especially well shown by some land crabs such as *Gecarcinus* in which it forms part of a defensive

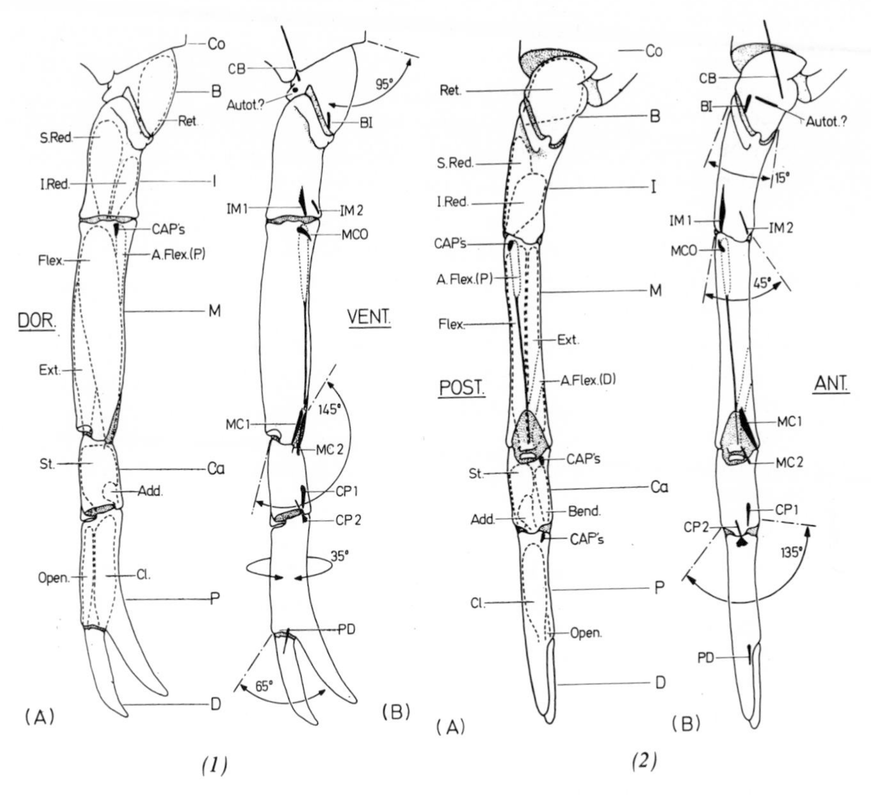

Fig. 15(1). Ventral views of a left walking leg of *Homarus gammarus* with the relevant parts of the internal anatomy shown in transparency. (A) To show the musculature and the position of the externally located cuticular articulated peg sensilla (CAP's). (B) To show the position of the internal chordotonal organs and the extent of movement possible at each of the joints. Segments: coxopodite (Co), basipodite (B), ischiopodite (I), meropodite (M), carpopodite (Ca), propodite (P), dactylopodite (D). Muscles: retractor (Ret.), superior reductor (S. Red.), inferior reductor (I. Red.), flexor (Flex.), extensor (Ext.), accessory flexor proximal head [A. Flex. (P.)] and distal head [A. Flex. (D)], stretcher (St.), bender (Bend.), additional muscle (Add.), closer (Cl.), opener (Open.). Receptors: The receptors are named by the initials of the limb segments which constitute the joint at which they are situated. An exception to this rule is the myochordotonal organ (MCO) situated close to the I-M joint. Autot. indicates the position of the possible autotomy receptor in the basipodite. For the orientation shown, posterior (POST.) and anterior (ANT.) relate to the position of the limb in its normal attitude.

Fig. 15(2). Posterior views of the left walking legs of *Homarus gammarus* with the relevant parts of their internal anatomy shown in transparency. (A) To show the musculature and the position of the externally located cuticular peg sensilla (CAP's). (B) To show the position of the internal chordotonal organs and the extent of movement possible at each of the joints. The lettering is the same as for Fig. 15(1).

attitude (Robinson *et al.*, 1970), but is also found in many others. In all cases the chela carries out functions in the normal life of the animal, and the causes of autotomy are not always clear.

Our concern in the present review is to consider what receptor mechanisms may be involved in determining whether or not autotomy occurs. Specific receptors have never been described and yet it seems likely that very precise stimulation must be required to invoke such drastic action. The familiar complement of hair sensilla, chemoreceptors, proprioceptors and muscle receptor organs may all play a part in the reflex, but are probably only generally implicated.

Possible contenders for the title of specific stress receptors that may monitor alterations in the stress of the cuticle have recently been discovered by Wales *et al.* (1971) and Clarac *et al.* (1971). These sense organs resemble proprioceptors in so far as they are collections of bipolar neurons that have dendrites running onto connective tissue sheets and strands. This connective tissue, however, is not arranged to cross any joint, nor is it associated in any way with muscles that move joints. Instead it inserts onto areas of thin flexible cuticle (visible externally) and which may be distorted under certain conditions. In the species examined; *Carcinus maenas, Homarus gammarus, Pagurus bernhardus, Galathea strigosa, Astacus leptodactylus,* and *Palinurus vulgaris* the situation is similar, differing in detail only.

There are two groups of receptors (CSD 1 and 2; cuticular stress detectors), in each chela and pereiopod. CSD 1 (Figs. 16 and 17) lies proximal to the breakage plane; CSD 2 distal to it in the ischiopodite. In *Palinurus vulgaris* a third set of receptors has been found on the membrane that seals the aperture when the limb is removed. Each of these receptors responds to pressure exerted on the flexible cuticular window upon which the connective tissue strand inserts. Tension in the anterior levator (autotomizer) muscle and depressor muscle tendons of the coxopodite–basipodite joint is also a potent stimulus. During autotomy both receptors, CSD 1 and 2, respond, especially the former. CSD 2 is, of course, lost with the leg and can play no further part in the behavior. General stressful situations in which the leg is trapped and held are probably appropriate stimuli, but even so, there are other factors that need to be considered since legs are autotomized under a variety of circumstances, which are not all relatable to severe mechanical strain.

The evidence provided is not so far conclusive that the usual role of CSD 1 and 2 is in the autotomy reflex. It does, however, seem probable that they are involved in some way. They may also have alternative functions, for example, in the reception of vibrations in the substrate. Horch and Salmon (1969) report that *Ocypode* responds to substrate-born vibrational stimuli, and Horch (1971) contends that this is a function of the MCO

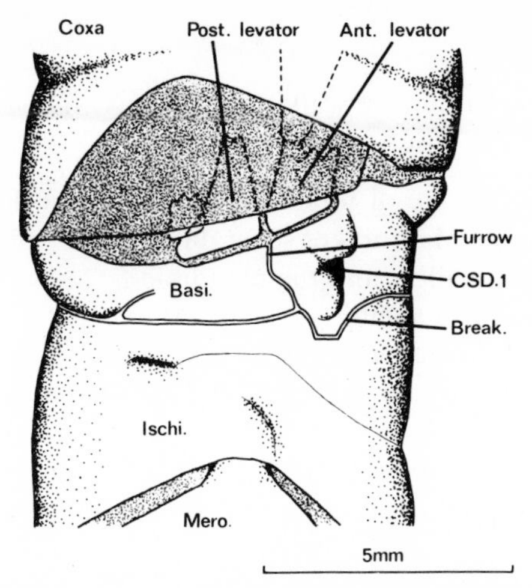

Fig. 16. A dorsal view of the basi-ischiopodite region of the second pereiopod of *Carcinus maenas* to show the position of the soft cuticle associated with CSD 1 (*CSD 1*) relative to the insertion of the posterior (*Post. levator*) and anterior levator muscle (*Ant. levator*) tendons, Paul's furrow (furrow) and the preformed breakage plane (*Break.*). Coxopodite (*Coxa.*), basipodite (*Basi.*), ischiopodite (*Ischi.*), meropodite (*Mero.*) (Wales *et al.*, 1971).

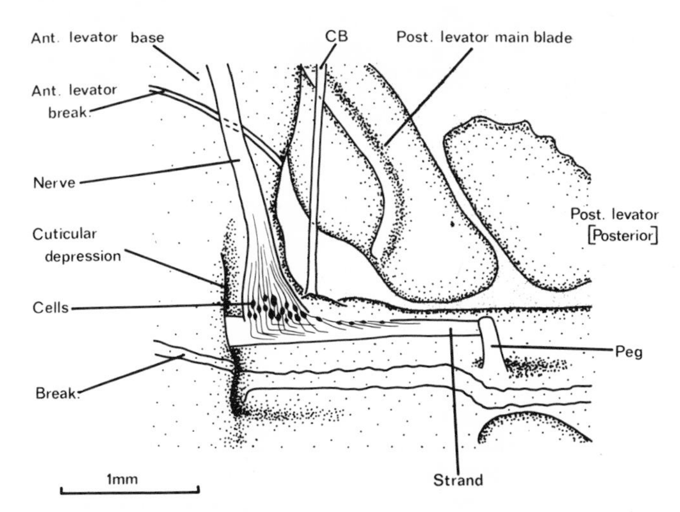

Fig. 17. The arrangement of CSD 1 in the right second pereiopod of *Carcinus maenas*. The top of the figure is proximal and the left side represents anterior in the walking leg. The receptor strand lies close to the breaking plane (*Break.*) and is attached posteriorly to a peg that projects internally. Anteriorly it is attached to an area of soft cuticle situated in a cuticular depression. The chordotonal organ of the coxopodite–basipodite (*CB*) inserts on a small protrusion close to CSD 1. The block of cuticle at the base of the anterior levator muscle tendon (*Ant. levator base*) is connected to the "basipodite" at a preformed breakage plane (*Ant. levator break*). The tendon of the posterior levator muscle is normally in two parts, the main blade, which is orientated perpendicular to the plane of the figure, and a smaller posterior portion [Post. levator (posterior)]. (Wales *et al.*, 1971).

(myochordotonal organ) of the walking legs. The CSD system seems another likely pathway for the input.

d. *Feeding*. The intake of food into the mouth of the lobster is accomplished by the combined efforts of several limbs, the second and third pairs of pereiopods, the maxillae, and the mandibles.

Little information is available regarding the receptors of any limb, other than the pereiopods. Wales *et al*. (1970) have given some details of the joint proprioceptors of the third maxilliped, and correlated these with the movements of the appendage. The notable difference between the maxilliped and the pereiopod lies in the reduction of number of receptors at the mero–carpopodite and carpopodite–propodite joints. The myochordotonal organ of the walking leg, in particular, is missing.

The second maxilliped has been examined by Pasztor (1969), who described a novel form of sense organ situated on a thin piece of cuticle crossed by a ridge (the oval organ). The innervation of the underlying connective tissue is accomplished by dendrites from two large nerve fibers. No cell bodies were seen; these may be located in the subesophageal ganglion. It now seems that this is an example of large diameter neurons similar to those mentioned elsewhere in this article. In view of the responses of these other organs it seems surprising that active propagated spikes are generated in the oral organ dendrites, rather than graded potentials that are conducted with little decrement to the CNS.

Some discrimination of food selected may take place as the food is picked up. Chemoreceptors exist on the dactylopodite of the limb (Laverack, 1963; Case, 1964).

The mandibles are heavily calcified and massively constructed for the purposes of crushing the food. They are slightly asymmetrical with the tooth

Fig. 18(A). Diagram of the right hemisection of the anterior cephalothorax of *Homarus gammarus*, to show musculature and innervation of mandible. The ventral portion of the anterior adductor muscle (Ml), muscle M2 and receptor muscle M3 are seen in transparency through the cuticle. M3 is stippled for clarity. The mandibular muscle receptor is innervated by a branch of the inner mandibular nerve (Inner mand. n) which arises close to the sub-oesophageal ganglion (sub. oes. gang.). The receptor branch is imn 4. This nerve also contains motor fibers to muscles, M3, M2, and a few to M1. Endophragmal skel. is endophragmal skeleton; M. part appendages, mouthpart appendages; ant, anterior; post, posterior; dors, dorsal; vent, ventral. Arrows indicate the direction of mandibular movement.

Fig. 18(B). Ventral portion of the mandibular muscle receptor organ. This is a posterior view of the muscle receptor in the right mandible. The receptor muscle M3 is closely apposed to muscle M2 throughout its length. The nerve supply is mixed motor and sensory to the area and the sensory cells are located at the distal insertion of the receptor muscle where it joins the hypodermis (Wales and Laverack, 1972).

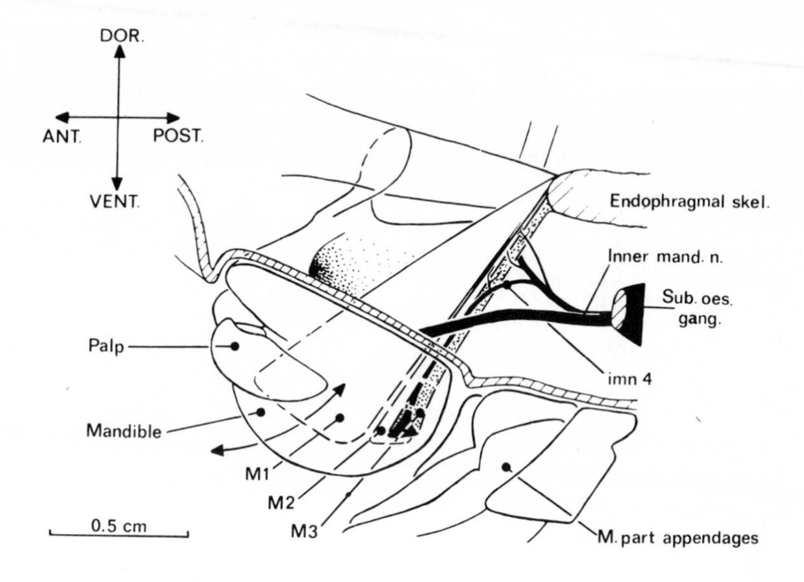
DOR.
ANT.
POST.
VENT.
Endophragmal skel.
Inner mand. n.
Sub. oes. gang.
imn 4
Palp
Mandible
M1
M2
M3
M. part appendages
A
0.5 cm

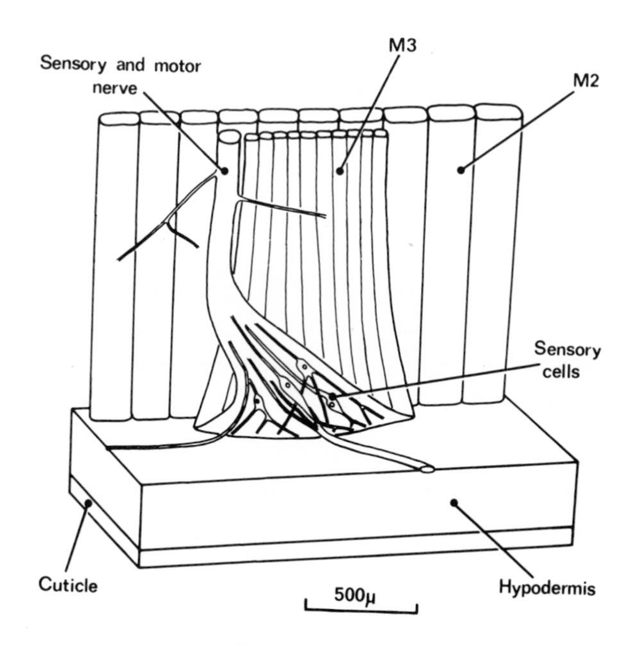
Sensory and motor nerve
M3
M2
Sensory cells
Cuticle
Hypodermis
500μ
B

Fig. 18.

portion being of different but complementary shapes on the right and left side so that they interlock with one another.

The appendage is hinged to swing down and out, thus opening the mouth. The labrum moves at the front border, the paragnaths to the rear, and these structures are also implicated in feeding. The mandible, however, is the major limb involved, together with the large palp at its anterior edge. The mandible is heavily muscularized with large adductor muscles serving to move the mandible about the hinge. There is a large adductor that fills the cavity of the mandible at the distal insertion, and attaches to the endophragmal skeleton proximally. It is innervated by motor neurons originating at the suboesophageal ganglion as the inner mandibular nerve (Fig. 18).

Alongside the large adductor muscle are two very much smaller slips of muscle that share the same insertions on the inner surface of the mandible and the endophragmal skeleton, lying slightly posterior and medial to the large adductor. The smaller of these muscles is heavily innervated at the distal insertion. A nerve trunk runs alongside the muscle strip giving off motor endings along the length to both larger and smaller muscles. As the nerve approaches the hypodermis, there is a divergence of fibers and a number of multipolar sensory cells are located just at the position at which the muscle and hypodermis meet. The dendrites of these cells run into the hypodermis underlying the muscle insertion. The cell bodies reach about 60–70 μm in diameter. Other supernumerary fibers spread out from this area and enter the hypodermis elsewhere.

This complex of sensory neurons is situated at a place where any tension exerted by the muscle slip is transmitted to the hypodermal attachment. It seems likely, therefore, that these may be muscle receptor organs, in that muscle contraction is the prime stimulation mechanism. This even though movement of the mandible allows shortening of the muscle and may also stimulate the sense cells (as is the case in the abdominal MRO).

Stimulation of the muscle electrically via the motor neurons brings about contraction and causes discharge of the sensory nerve (Fig. 19). The

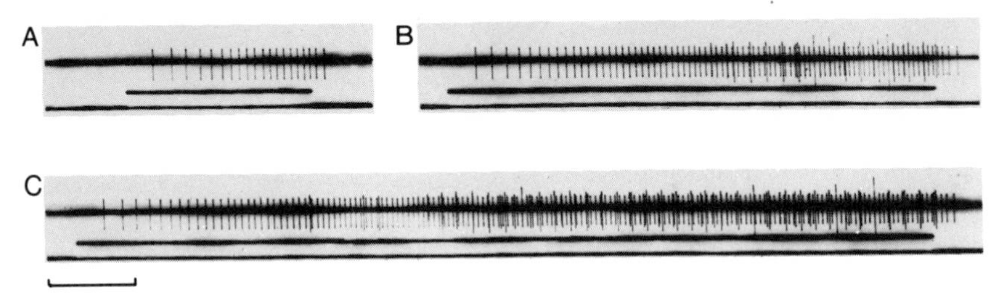

Fig. 19. (A–C) The response of some units of the mandibular MRO to contraction of the receptor muscle. A,B,C show the effect of increasing the period of stimulation, duration being monitored by the lower trace in each record. Muscle fibers were directly stimulated at a frequency of 100 Hz. (Scale mark 1 sec.)

muscle slip alone does not generate sufficient power to lift the mandible and hence alter the imposed stretch on the receptor. It is tension alone exerted at the muscle end that is the effective stimulant. Raising of the mandible during contraction of the main adductor muscle undoubtedly affects the response characteristics of the receptor by altering the passive stretch on the muscle slip, so there is a dynamic system in operation here. The response of the receptor will depend upon the tension imposed by contraction of the receptor muscle, but the discharge may vary according to the initial starting position due to the activity of the other musculature (Wales and Laverack, 1972).

The movements of the mouthparts culminate in the entrance of food into the gut. This is facilitated by dilation of the mouth and the alteration in position of the labrum and paragnatha. Changes in disposition of the soft flexible cuticle of the mouth region are monitored by a series of proprioceptors (termed mouthpart receptors, MPR 1, 2, 3 by Laverack and Dando, 1968; see also Moulins *et al.*, 1970). These receptors are associated with a strip of stout tendinous connective tissue that traverses the top of the mandible from an internal anterior thoracic spine laterally to the esophagus to a position close to the endophragmal skeleton posteriorly. This strip of connective tissue circumnavigates the mouth and is deformed by movement of the mouth and sundry mouthpart limbs. Differential responses are recorded during movements (Moulins *et al.*, 1970), and some peripheral analysis of position is achieved by these receptors. Members of the anteriormost group, MPR 1, are autogenic, discharging with constant frequency over long periods. The other groups, MPR 2 and 3, are phaso-tonic in response. Figure 20 shows an example of the type of nerve cell typical of MPR 2 at the rear of the esophagus in *Panulirus argus*.

Food entering the mouth is carried by the action of the esophagus to the gastric mill. Here it is ground and masticated by the action of the ossicles of the mill, the movement of which is occasioned by muscles innervated from the stomatogastric ganglion. This has been the subject of intensive work by Maynard over the last few years (Maynard, 1966, 1969; personal communication). From the sensory standpoint, however, the interesting feature lies in a cell described first by Orlov (1927) and later by Larimer and Kennedy (1966). This sense cell is of curious morphology being bifid both dendritically and axonally (Fig. 21). It is autogenic, discharging with great regularity when unstimulated, the frequency rising or falling during movement of the ossicles of the mill (Fig. 21). The grinding action of the mill is thus monitored by this system, but the input does not seem to make any synaptic contact with motor elements in the stomatogastric ganglion, passing through to some as yet undetermined termination.

The position of the component parts of the gastric mill is also monitored by other receptors, as described by Dando and Laverack (1969). A posterior

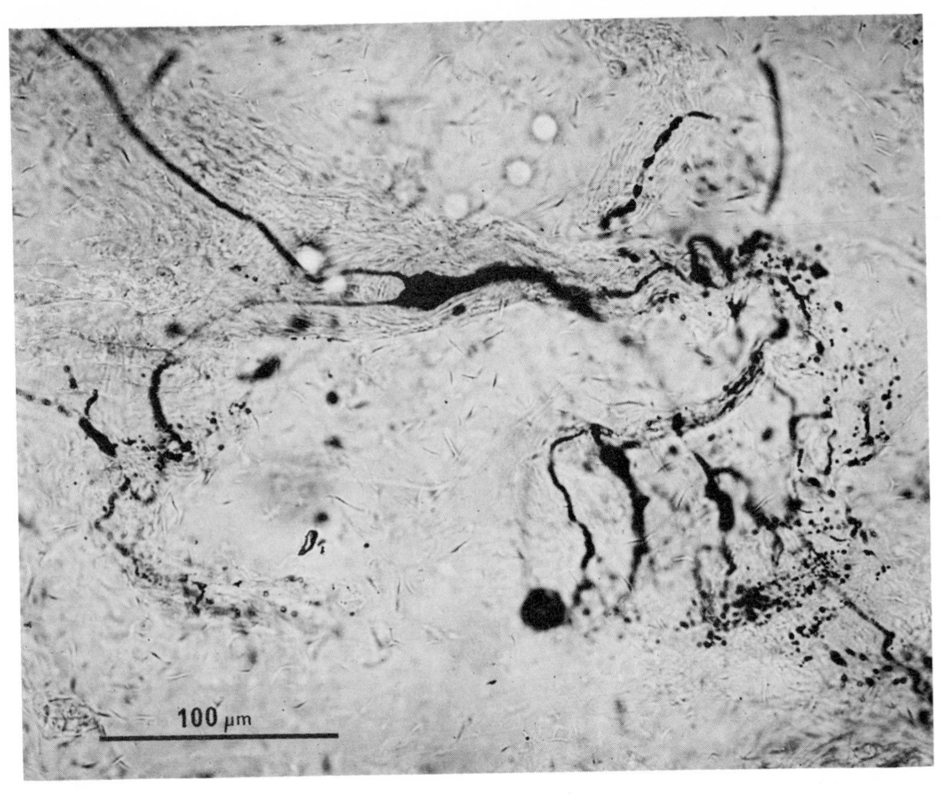

Fig. 20. A single neuron of MPR 2 in *Panulirus argus*. This group of mechanoreceptors exists in a thick sheet of connective tissue that extends across the lateral portions of the esophagus and upper mandible. There is a cluster of dendritic endings that in this case extend in two directions. The axon (A) leaves from another portion of the cell body. (Scale mark 100 μm.) (Moulins *et al*., 1970).

stomach nerve serves the posterior portion of the gastric mill and is a substantial tract of fibers containing about 105–125 fibers. All of these are mechanoreceptors with axons running to the subesophageal ganglion, having cell bodies that are located within the nerve trunk, and ending in long dendrites that are embedded in connective tissue on the dorsal side of the gastric mill (Fig. 22).

Changes in the position of the ossicles of the mill are indicated by discharges in the receptors (Fig. 23). The spike initiation site is located close to the termination of the dendrite as shown by recordings of potentials close to the distal extremity. Intracellular records indicate that the cell body plays no part in modifying the passage of the impulse traffic.

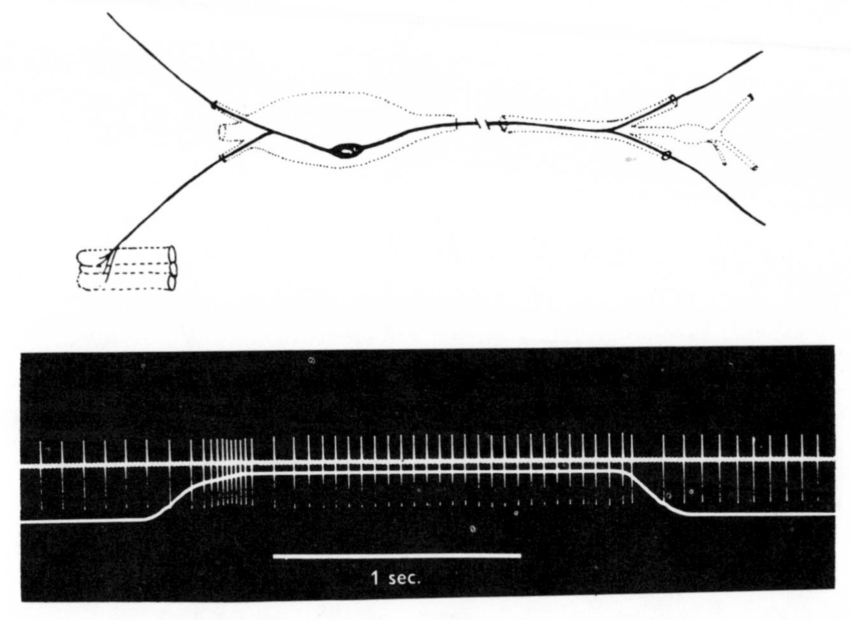

Fig. 21. (Top) The solitary bipolar mechanoreceptor cell from the stomatogastric ganglion of decapod Crustacea (Larimer and Kennedy, 1966). The cell has a bifid axon and a divided dendrite that ends on the tissues around the cardiac ossicle. (Bottom) The response recorded from the lateral nerve of the stomatogastric ganglion after depression of the cardiac ossicle. The receptor is autogenic with a regular discharge prior to stimulation that is accelerated during movement, and reaches a new frequency maintained during the new position, but decreasing again upon return to the original level. This kind of response is also demonstrated by MPR 1 near the mouth and esophagus (Dando and Laverack, 1969).

Other sensory organs also exist in this area to monitor passage of food through the gastrointestinal tract, as demonstrated by Dando (1974). The influence of those already described is indirectly transferred to the stomatogastric ganglion. Stimulation of the p.s.n.* leads to a modification of the output of the stomatogastric ganglion.

Food that has passed along the whole length of the alimentary canal is voided through the anus by the action of the musculature of the rectum and the anal area (Larimer and Kennedy, 1969; Winlow and Laverack, 1970, 1972). The activity of the anus dilation and constriction is governed by nervous output from the sixth abdominal ganglion initiated by command fibers that link more anterior centers with the abdomen. These almost certainly originate in the cerebral ganglion. Why the anus should be opened due to stimulation from the brain remains obscure at present.

*p.s.n. = Posterior Stomach Nerve.

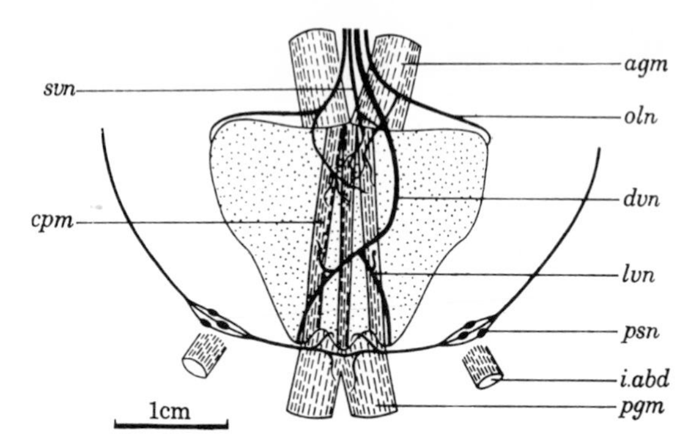

Fig. 22. The arrangement of the posterior part of the stomatogastric nervous system on the dorsal surface of gastric mill of *Cancer pagurus*. On each side of the mill the posterior stomach nerve (*psn*) contains a group of cell bodies at a position close to the internal abductor muscle (*iabd*). The psn's then run onto the gut near the insertion of the posterior gastric muscles. The psn's join up behind the propyloric ossicle and fibers from one side probably often cross to the other. *dvn*, dorsal ventricular nerve; *cpm*, cardio–pyloric muscles; *svn*, short ventricular nerve; *oln*, outer lateral nerves; *agm*, anterior gastric muscles; *pgm*, posterior gastric muscles (Dando and Laverack, 1969).

Fig. 23. Intracellular recordings from a unit in the psn of *Homarus* showing the response to forward movements of the urocardiac ossicle. (A) Response to a movement of 3 mm. (Time mark 1 sec.) (Dando and Laverack, 1969).

Information that may be relevant can be obtained from the anal region. The cuticle of the area is soft and not heavily calcified. It is hence quite pliable. Distortion of the anal circumference is monitored by mechanoreceptors that respond to both opening and closing movements. The cell bodies of those concerned with opening have been observed lying within the nerves that innervate the area, the dendrite protruding onto a sheet of connective tissue surrounding the anus. The closing monitors have not been identified (Winlow and Laverack, 1972).

8. *General Aspects of Nerve Cell Responses Illustrated from a Variety of Sources*

Certain marine animals have become important because their neurophysiological attributes allow particular experiments to be conducted. The squid giant axon is perhaps the most famous of these. We can attempt a summary here of what phenomena have been reported. The examples do not all come from the same animals, or even groups, but such interactions *may* be common to nervous systems in general.

a. *Cell to Cell Transmission.* Numerous examples are now at hand that indicate the passage of information from one nerve cell to another without the interpolation of a chemical transmitter as messenger. This activity may take place alone, or in combination with chemical transmission.

i. *Electrical transmission.* Nerve cells may be joined electrically with one another due to some morphological peculiarity such as the existence of tight junctions or nexuses associated with specific electrical properties. Such junctions have short latencies, with transmission of information rapidly across from one cell to the other. In *Hirudo* (the medicinal leech) stimulation of the touch sensitive cells (T) leads to depolarization of motor neurons (M) innervating longitudinal muscles (two in number, acting together). This junction is rectifying and excitation spreads only in one direction from T to motor cell (M) while hyperpolarizing current passes only in the opposite direction, M to T. P cells (pressure) stimulate M via an electrical synapse, but also have a chemical synaptic site (longer latency); the electrical synapse rectifies as T. Noxious stimuli receptors (N) transmit only via chemical synapses (Nicholls and Purves, 1970).

In the motor system of *Hirudo* the large motor neurons are functionally linked and act as a unit. This coordination is donated by an electrical synapse that allows transmission with little attenuation from one to the other (Stuart, 1970), so both are brought into synchrony (Fig. 24 summarizes these events).

Electrical interaction between cells in this manner in marine animals has been reported for Crustacea (lobster giant axon: motor synapse) and more particularly the mollusks, *Aplysia*, and *Navanax*. The bag cells that surround in a cluster the right and left abdominal connectives close to the abdominal ganglion of *Aplysia* number about 300–400. They have been shown by Kupfermann and Kandel (1970) to be linked electrically to each other in a way that permits synchrony among the population (Fig. 25). *Navanax* has a group of 10 readily identifiable cells in the buccal ganglion that may be recorded from two at a time (Levitan *et al.*, 1970). These cells also are electrically coupled, with the signal between them attenuated as a function of the relative size of the cells involved, and independent of direction of current flow. Spontaneous postsynaptic potentials are noted in the cells due to a common (chemical) synaptic input.

ii. *Chemical transmission.* The history of the study of chemical transmission is a very long one and it is now well established that nerve cells interact with others and with muscles via chemicals liberated at their endings. In the present context there is no need to dwell on this, but two lines of research demand some comment.

Squid giant axons have been consistently productive of results in the hands

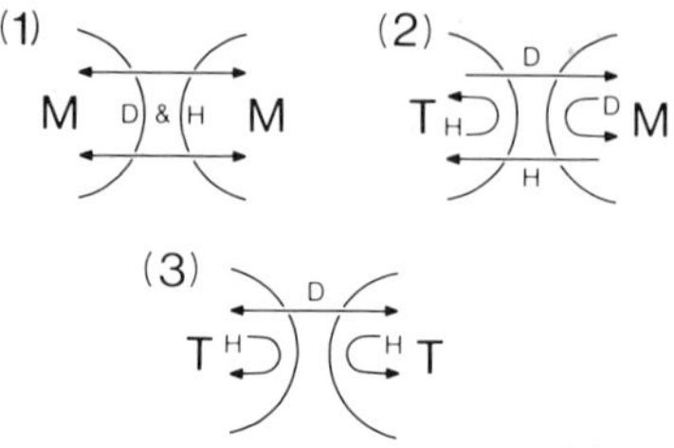

Fig. 24. The membrane properties of nerve cells are not rigid in type and character. In cases in which electrical interaction has been reported without the interpolation of chemical transmitters the situation is not a simple one. The examples here are taken from the leech though similar examples can be drawn from the mollusks and the arthropods. (1) The large motor axons of the leech segmental ganglion are directly linked by electrical synapses that are bidirectional, i.e., depolarizing and hyperpolarizing current passes readily in both directions, influencing the postsynaptic cell by either raising or lowering the cellular potential. (2) The interaction of the tactile (T) cells within a ganglion upon the motor cell is directional in character; depolarizing current passing from T to M, and hyperpolarizing current from M to T. The T cell membrane rectifies hyperpolarizing current and the M cell rectifies depolarizing current so that there is no interaction between cells as indicated. (3) The T cells within the ganglion are interlinked and cross depolarizing currents pass in both directions to affect the second cell. Hyperpolarizing currents are rectified in both cells. (Drawn from data in Nicholls and Purves, 1970.)

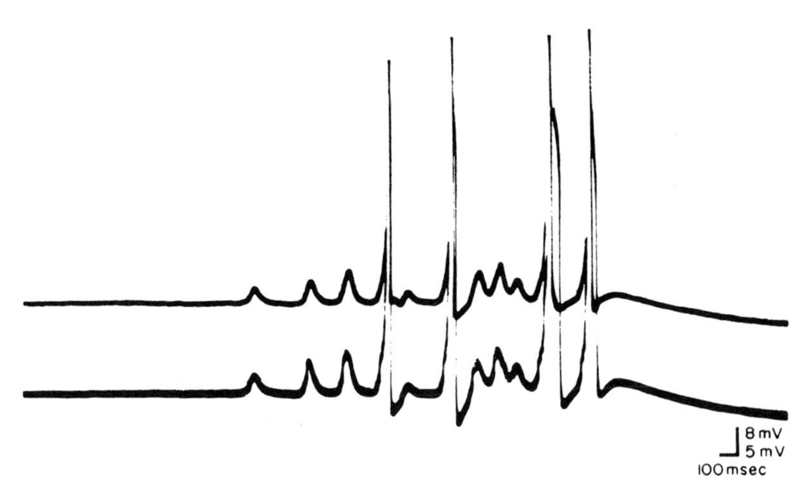

Fig. 25. Bag cells are neurosecretory neurons arranged around the abdominal connectives of *Aplysia*. Intracellular recording of potentials reveals that the responses of cells in a cluster on one connective are synchronized. In the present case both spikes and graded potentials are common to both cells. Bag cells arranged contralaterally are not so closely linked, but it seems that one side may act as pacemaker for the second side (Kupfermann and Kandel, 1970).

of high class research teams and provide the basis of much of our knowledge of ionic events during electrical propagation of signals in nerve fibers. These events in the giant axons are, however, followed by transmission to secondary giant axons across a synapse. This synapse is now being studied with micromethods because it is suitable for both pre- and postsynaptic elements to be impaled with microelectrodes close to the synapse. It is therefore relatively easy to observe synaptic phenomena at the junction (Fig. 26). Miledi

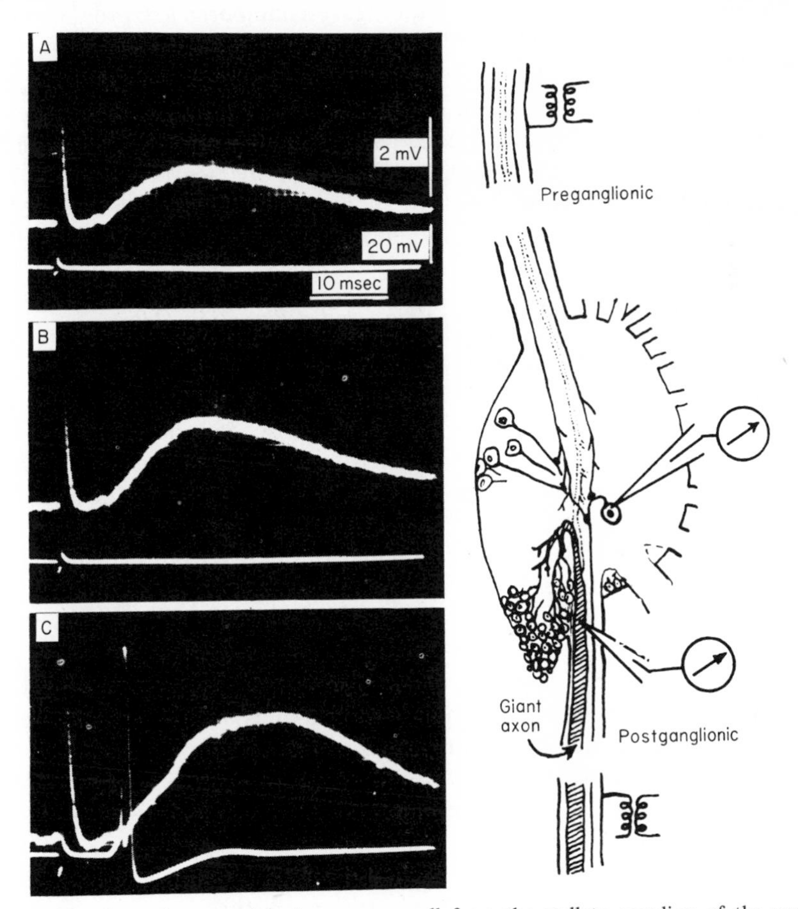

Fig. 26. Synaptic potentials in a nerve cell from the stellate ganglion of the squid *Loligo vulgaris*. To the right is shown the arrangement of the experiment. One microelectrode is inside a nerve cell of which the axon runs in the last stellar nerve; a second electrode is inside the giant axon in the last stellar nerve. On the left is shown the electrical response both pre- and postsynaptically. In A–C the preganglionic nerve was stimulated with increasing intensity and in C one of the fibers making synaptic contact with the giant axon was stimulated and evoked a postsynaptic potential and spike in the giant axon (Miledi, 1966).

(1966, 1967) and Katz and Miledi (1966) have reported on the transmission properties in squids, and suggested L-glutamate as the transmitter. There is a small electrical coupling also.

The other comment worth making is that *Aplysia* also provides ideal material for the study of synaptic events. Recordings from cell bodies show certain aspects of synaptic transmission, and Kandel and his colleagues have conclusively demonstrated that one nerve cell, synthesizing one transmitter may have contrasting effects upon the postsynaptic follower cells with which it makes contact. The interneuron numbered L10 provides input to some 20–30 cells in three different regions of the abdominal ganglion. This input has been characterized to 14 follower cells, and found to effect (a) excitation in one group, (b) inhibition in a second, (c) combined excitation–inhibition in the third group. The driver cell releases one transmitter, but the postsynaptic cells may have more than one acceptor for the same transmitter, and their response is to allow different ionic exchanges across the membrane (Kandel and Kupfermann, 1970; Kandel *et al.*, 1967; Wachtel and Kandel, 1967).

9. *Neurons Demonstrating Graded Potentials Only*

Recent advances in neurophysiology have demonstrated that the commonly accepted story of depolarization followed by action potential and restoration of resting potential is too simplistic. It is now apparent that the activities of nerve cells are (infinitely?) variable. The levels of ions (see for example Kerkut and Meech, 1966), the resting potentials, the responses to external stimulation, graded depolarizations, rectification, and other membrane responses are obviously reflections of myriad processes that we can thus far only surmise. Each neuron has to be examined as an entity in its own right that may well be different to contiguous cells. It may show considerable differences in response to similar stimuli, or to the same transmitter; it may have a different resting potential, a different ionic content, contain a different synaptic transmitter, as well as being of different morphology, dendritic field and function, to any other cell.

It still comes as a surprise, however, to discover examples of neurons with markedly unusual or unexpected properties, although they are becoming more commonplace.

One such unusual nerve cell type is the nonpropagating neuron. Two varieties have so far been described in Crustacea, and there are no doubt others yet to be found.

a. *Sensory Systems.* In sensory systems the classic picture of firing in primary neurons is that of a burst of spike potentials when the stimulus is applied or when it is removed. This response is due to depolarization of the receptor ending with graded decrementing potentials recordable within the cell body, but giving rise to fully fledged propagated potentials at some site in the cell. This may be in the dendrite [as in hair mechanoreceptors (Mellon and Kennedy, 1964)], and in some internal proprioceptors (Dando and Laverack, 1969) or central to the cell body as in the MRO of the abdomen (Edwards and Ottoson, 1958).

The situation is different for some sense organs. Three crustacean examples are best known.

i. *The coxal muscle receptors of decapods* (described by Alexandrowicz and Whitear, 1957; Whitear, 1965). Two large sensory neurons innervate the receptor system. They are both 2–5 mm long and 50 μm in diameter; they originate at the thoracic ganglion (where the cell somata are located) and branch just before reaching the muscle. One fiber (T) has many branches that innervate the connective tissue at the proximal end of the receptor muscle. The other fiber (S) has two major portions that run to connective tissue strands on either side of the receptor muscle (Fig. 27).

ii. *The barnacle eye* (Gwilliam, 1965; Brown *et al.*, 1971; Shaw, 1972). The lateral ocellus is located beneath the opening of the shell valves of the barnacle and is sensitive to light falling upon it. They are situated at

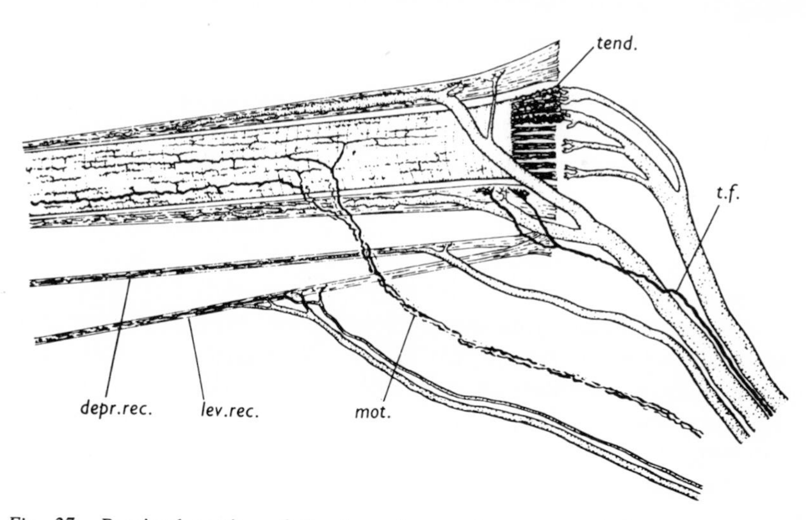

Fig. 27. Proximal portion of the thoracico–coxal receptors of *Carcinus maenas*. This semidiagrammatic drawing applies to the fourth to eighth thoracic segments of the left side. A receptor muscle is flanked by two strands of connective tissue into which penetrate branches of a thick nerve fiber. *tend*, tendon of the receptor muscle; *t.f.*, thin fiber ending on ventral strand; *mot.*, motor nerves of the receptor muscle; *lev.rec.*, *dep.rec.*, levator and depressor receptors with their innervation (Alexandrowicz and Whitear, 1957).

the end of long nerves that pass to the cerebral ganglion. No action potentials have been recorded in these nerves under the influence of depolarizing conditions.

iii. *The ventral photoreceptors of Limulus* (Clark *et al.*, 1969; Millechia and Mauro, 1969). These are discrete units situated on certain ventral nerves. They are connected to the CNS by large diameter axons.

These receptors all respond to stimulation by graded depolarization (receptor potential) which do not usually generate propagated impulses. The receptor potentials spread passively to the CNS, where they then excite motor neurons that form part of a reflex.

In the case of the TC organs Bush and Roberts (1968) and Roberts and Bush (1971) have shown that there are dynamic and static components to the response (Fig. 28). A variable impulse discharge is sometimes seen in response to very fast stimulation via the receptor muscle. This latter is abolished by tetrodotoxin and is presumably sodium dependent; the resting potential of the cell is potassium dependent. Similar dynamic and static responses may occur in the other examples quoted.

b. *Motor Systems.* In motor systems Pasztor (1968) described rhythmic activity in neurons of the subesophageal ganglion of *Panulirus argus*. These are motor units that control the ventilatory appendages, the scaphognathites (second maxilla). The motor activity is very regular and long lasting in isolated preparations. It is possible to record intracellularly from units involved in this rhythmic activity within the ganglion. The assumption is that there is some oscillatory mechanism that governs such rhythmic activity and that this might be recordable from the cell bodies. Mendelson (1971) shows that certain cells exhibit oscillating resting potentials that are in phase with the bursting potentials of motor units (Fig. 29). There is no sign of spiking within the oscillatory neurons. That this activity is meaningful in terms of control of output was demonstrated by depolarizing and hyperpolarizing experiments conducted upon them. Records were made from motor axons in two branches of the second maxilla root showing rhythms that were in opposing phases. Depolarization of the oscillator led to slowing of one group, and discharge of the second; hyperpolarization led to the opposite result. Mendelson concludes from his work that the penetrated cells are indeed the control command oscillators for the ventilatory rhythm of the lobster. They are interneurons.

From these examples, and others in vertebrates it is apparent that some short neurons that have overall lengths comparable to their space constant (space constant = distance over which potential declines to $1/e$ or about 37% of its original magnitude) do not give rise to action potentials. The passive propagation of electronic potentials is sufficent to transmit information

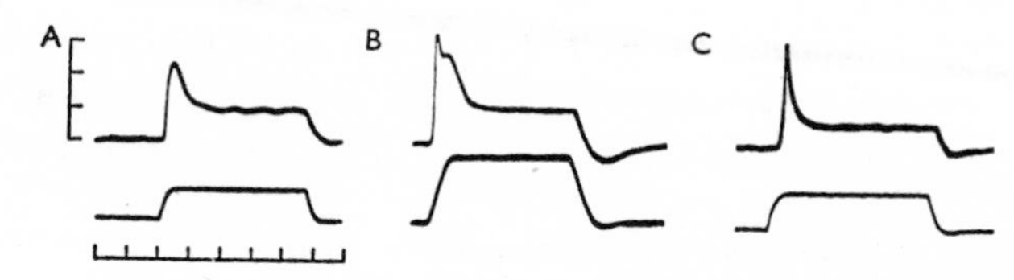

Fig. 28. Examples of T fiber receptor potentials in response to fast "step" pulls on the receptor muscle and showing different degrees of spiking in the dynamic component. The stimulus is indicated in the lower beam, increased stretch moving this up. Calibrations 20mV and 20 ms/division (Roberts and Bush, 1971).

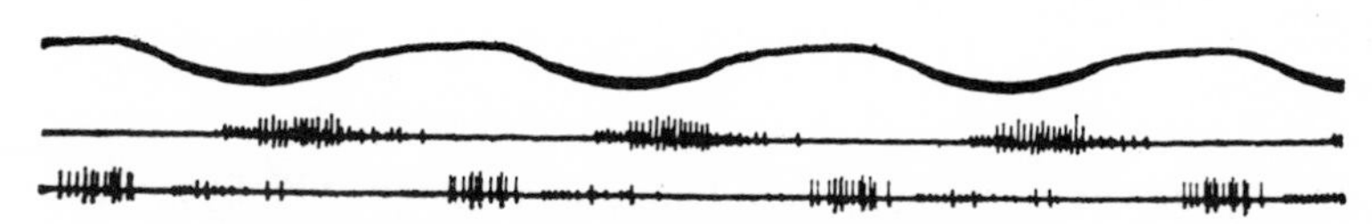

Fig. 29. The oscillator neuron. Recordings made by Mendelson (1971) in the subesophageal ganglion of *Homarus*. The upper trace shows the intracellular potential of the oscillator neuron; the middle and lower traces are from motor root branches. The potential goes from about −15 mV to −25 mV. Depolarization via the microelectrode slows motor neurons in the upper trace (proximal root), and at the same time elicits prolonged firing in the lower group (distal root). In the recording shown it can be seen that the membrane potential moves up and down in exact phase with the timing of the axon bursts.

from the source to the CNS. In the case of TC receptors (Bush and Roberts, 1968) a motor reflex with motor fiber discharge has been shown that results from nonspiking stimulation of the final motor pathway.

Several examples of rhythmic activity that may be governed in similar ways by oscillators within the nervous system have been described. In Crustacea two cases are worth mentioning.

The respiratory rhythm used by Mendelson has a correlated system alongside it which is also concerned with assisting in ventilation of the gill cavity. This is due to the rhythmic beat of the flagellum carried on the exopodite of the first, second and third maxillipeds. Each of these is moved by a single muscle, contraction of which draws the flagellum upward and outward; return to the rest position is by passive means due to the elasticity of the cuticle. The beat of the flagella of both right and left sides is coordinated so that the right side ceases to beat when movement is to the right and the left side inhibited when walking is to the left.

The simplest circuitry of nervous pathways that may account for the range of activity exhibited is shown in Fig. 6B (Burrows and Willows, 1969). The diagram holds for both Brachyura and Anomura although there are differences in detail between them. The critical aspect here is that Burrows and Willows postulate a pacemaker system that fires at a constant frequency to drive the motor neurons via an interposed series of interneurons. It seems possible that as for the respiratory neurons, the central driving cell could be an oscillator with a constantly varying cellular potential, but not carrying action potentials.

A circuit model has also been proposed to account for the rhythmicity of beat of the swimmerets (abdominal appendages) in decapods (mainly shown for crayfish, *Procambarus* and lobster, *Homarus americanus*) (Fig. 6A). Stein (1971) illustrates the present thinking along these lines with a set of motor neurons controlled, in a coordinated fashion via interneurons, from an oscillator within the central nervous system.

These oscillators, of which there may be one in each hemiganglion, are not known anatomically.

10. *Why Marine Invertebrates as a Source of Comparative Neurophysiological Material?*

In 1967 a book edited by Wiersma and entitled "Invertebrate Nervous Systems" and subtitled "Their Significance for Mammalian Neurophysiology" was published by the University of Chicago Press. This volume contains 27 chapters, an introduction and a concluding résumé. It might therefore be suspected that the last word had been spoken about the value of such investigations. And yet upon reading the various contributions one is lead to the conclusion that most of the authors have not truly shown the value of their work in comparison with work being carried out on vertebrate systems. In fact perhaps the most pertinent statement in the volume may be that quoted from G. H. Parker in whose honor the book was dedicated. Parker posed the question "How did the simplest animal reflexes that could be called nervous arise, and how, out of these primitive activities, did that enormously complex body of responses that we look on as evidence of mentality in higher creatures like ourselves originate?"

Physiologists working on invertebrate preparations, and among them the various examples taken from the sea, are all basically interested in the same question. The analysis of behavior in reductionist terms, i.e., how the component parts react and respond to changing conditions and how might they be linked together to give rise to behavior in the animal, preoccupies all those who take a scalpel to an animal (see Kennedy 1969). Whether or not human or vertebrate behavior is explicable in similar terms is not yet clear, though it is an underlying assumption.

We might briefly consider here why invertebrates are favored by some, and possibly a growing number of physiologists.

a. *The Problem of Scale.* Vertebrates are complex animals, frequently of great size. The number of neurons may be enormous [Young, 1967 gives figures of between 1 and 2×10^{10} cells in the human brain and 1 and 2×10^8 in *Octopus*]. On the grounds of numbers then, humans have rather more nerve cells in the CNS than the most developed invertebrates (though *Octopus* scores elsewhere by having large numbers of nerve cells in the arms). But *Octopus* may be taken as the zenith of invertebrate nervous complexity; when we consider other examples we find the absolute number of neurons considerably less. I hesitate to say the systems are necessarily simpler, but they are smaller.

b. *The Disposition of Units.* It is both the strength of the system, and also a weakness for investigatory purposes, that vertebrate nervous systems are highly concentrated with most cell bodies close to the CNS or in it. Advancing up the evolutionary scale leads to greater and greater concentration of units in the brain. Among invertebrates the overall numbers are smaller, and they are subdivided into smaller packets; thus serially repeated ganglia in metameric animals may have a restricted number of nerve cells each of which can be treated separately from its neighbor.

c. *The Size of Units.* With spatial separation may come also ease of experimentation because of the accessibility of units. It helps of course if the units are large, as in giant fibers of squid or polychaetes, or the cell bodies of *Tritonia* or *Aplysia*, but small collections are frequently more readily dealt with than large ones. This is of importance because intracellular individual studies are facilitated by size and accessibility. It is easier to gain entrance to a single cell when that is near the surface of a small ganglion than when it is embedded deep among hundreds of other similar units.

d. *Simplicity of Preparations.* One of the crushing liabilities of vertebrates is the necessity for maintenance of a high pressure blood flow, particularly to the brain which otherwise rapidly fails. In studies on invertebrate brains blood flow is imperative especially in well developed animals such as cephalopods and decapod Crustacea, but in other mollusks, in annelids and other ganglia of Crustacea it may be unnecessary and an impediment to easy working. A copious supply of seawater, or saline, preferably cooled, is all that is required for at least short-term experiments. Protein may be missing but otherwise the bathing medium is closely similar to the normal circulating fluids of the vascular system.

e. *Possibility of Precision in Studies on Activity of Single Cells in Normal Activities of the Animal.* While it is possible to record brain waves and other gross potential changes in vertebrates, and it is also possible to insert probes at various levels of the central mass of nervous tissue, it is not yet possible to deal with individual identified units. Willows and Hoyle (1968a) have shown that for selected invertebrates it is possible to insert electrodes into single cells and then observe consequent experimentally induced behavior changes of the animal. This is not quite so refined a technique as one would like since the animal is restricted in operational field and is not "normal," but it is an enormous stride forward.

f. *Length of Nerves.* In small animals of course no nerve can be long in absolute terms, but in many invertebrates peripheral nerves are relatively accessible and can be used for recording purposes *en passant*. Most nerves, however, suffer from the drawback that they are not purely sensory or motor, but are rather a mixture of these modalities. The EM frequently reveals in detail that what appears a simple motor nerve in fact is a complex sensory–motor mixture of fibers, although physiological function is not, of course, ascribed by histological study alone.

g. *Possibility of New Preparations.* The variety of phyla and species living in the sea allows the expectation that more preparations will be developed that enable special questions to be asked and answered. The development may take a hypothetical course of (a) anatomical, (b), histological, (c) histochemical, (d) fine structural, (e) extracellular, (f) intracellular, and (g) intact animal studies. We do not yet by any means know *all* the valuable possibilities that exist. *Myxicola* giant axons have started to be useful in biophysics in competition with squid axons (Binstock and Goldman, 1969); sea anemones show phenomena of cell to cell conduction in a particularly amenable way (McFarlane, 1969); some barnacles have very large muscle fibers (Hoyle and Smyth, 1963), and many others yet await description.

h. *Variety of Form and Specific Systems.* The enormous range of types of marine organisms allows comparative investigations to be made on organs of all degrees of complexity. Among Crustacea, for example, the compound eye may be studied at all levels from the simplest types of a few facets through intermediates such as lobsters and crabs to enormous collections of facets as in the amphipod *Cystisoma*.

IV. The Future

There can be little doubt that the study of the physiology of animals is due to undergo a further violent upheaval.

Prior to this time it has been the usual procedure to study systems in a logical progression of morphology on a gross scale, then histology with the light microscope, followed by physiological and pharmacological investigations on the isolated organ system, studies with the electron microscope, and probably redescription of morphology from a functional rather than anatomical point of view.

In neurophysiology we have reached this stage and most work continues along these lines, although the ability to analyse systems becomes greater and the organs more complex as isolates. This approach has enabled us to begin to reassemble the animal as an article manufactured from many pieces, each of which needs fitting into place to understand the whole. Nonetheless, the pieces are as yet rather small and we have yet to discover whether the whole is greater than the sum of its parts.

The present trend toward miniaturization enables one to predict part of the next revolution. The suction electrode, insulated wire, radio emitting capsules, printed circuit, and so on will allow of the implantation of recording circuitry and its carriage by relatively small animals in the watery environment. This means that while analysis of small isolated fragments will still be enormously valuable it is now becoming possible to record activity in legs, muscles, nerves, central nervous system, sensory receptors, etc. while the animal remains intact, virtually unloaded and perhaps unimpeded by wires and harnesses.

This advance will allow students of behavior and physiology to investigate complex interactions of organs with a minimum of interference and a maximum of information available. The technique will allow the animal to move freely, to indulge in its normal behavior such as walking, swimming, feeding, breeding, and so on. The interaction of muscle groups and sense organs will be as the animal uses them, not as the experimenter sees fit to impose in an arbitrary way. Something along these lines has already been attempted, notably in studies of insect flight, but in marine organisms it is a field which has been generally unexplored.

That this may have considerable importance in fields other than those directly associated with physiology is indicated by observations made by the team of the undersea habitat, Tektite. *Panulirus argus* is a large spiny lobster of uncertain habits. Many large decapods are well known commercially and culinarily but not physiologically, behaviorally, or ecologically; and *P. argus* falls into this group. Scuba divers are able to follow individual animals for short distances but the use of small electronic devices (sonar pingers) allows this distance to be much increased, multiplied to a larger number, and to be carried out in the dark. The Tektite team carried out simple observations on mobile and unhindered lobsters and have added interesting facts to knowledge of behavior, migration, and habitats (e.g., Herrnkind, this volume). Carry this one stage further by implantation of small transmitters within the animals, rather than on the carapace, and the possibilities become almost innumerable (Clifton *et al.*, 1970). Bottom-living fish have been tracked at sea from a boat on the surface by means of acoustic signals (also see Adler, 1971).

The next step is telemetry *à les astronauts* to give details of heartbeat, nerve impulses, respiration, and so on (Walker *et al.*, 1971).

A. TECHNIQUES

The advance of physiology, as with other experimental sciences, relies not only on the ingenuity of the investigator but also upon his scientific colleagues. It is not possible to progress without methods being available and for this the efforts of chemists, physicists, biochemists, electronics experts (especially these), as well as fellow biologists are necessary. The rapid strides of the past 20 years have been due to (a) the availability of trained manpower, (b) the development of sophisticated tools, and (c) the present progress toward miniaturization that allows more channels in the same size package.

As methods become more precise, selective, and sensitive, so the possibilities open to the experimenter become greater. With better defined histochemistry, the staining of individual cells to demonstrate transmitters becomes likely. Better biochemistry allows characterization of enzymes at single cell level (Giller and Schwartz, 1971a,b); microchromatography permits analysis of single cell contents (Osborne *et al.*, 1971a); autoradiography gives an introduction to the more obscure pathways of synthesis of active substances in the interior of cells. All taken together would provide a firm platform for the study of particular examples.

Among marine invertebrates such examples abound. Giant cells, giant fibers, isolatable preparations of simple muscle fibers of great size (*Balanus smythii*, Hoyle and Smyth, 1963), Ca^{++} specific proteins that luminesce (isolated from the jellyfish *Aequorea*), Mauthner cells in lower vertebrates, etc., are but a few. Further comparative studies on many phyla will enrich the possibilities. Reference to Bullock and Horridge (1965) will show how many opportunities exist. Marine animals provide the largest reservoir of types that live in any habitat. Some unknown (at present) will surely provide the material of tomorrow.

B. USE OF WHOLE ANIMAL PREPARATIONS

One of the drawbacks of a great deal of neurophysiology, or indeed any physiology, is that it relies upon analysis of portions of the whole. Whether or not the intact animal is greater than the sum of its component parts, it is quite obvious that the investigation of some aspect of isolated organs is only likely to tell part of the story because as soon as the organ is removed from the body the tissue is no longer subjected to the "normal" environment. The removal of blood proteins, hormones, trace ions, carbohydrates, and even formed elements, which all contribute in some way to the welfare of the whole, must be detrimental to performance. If further subdivision is attempted and portions of the nervous system are removed from others (if, say, the cerebral ganglion is studied without the subesophageal ganglion in many invertebrates) then interactions between those parts are circumvented.

This may be of great consequence and largely unsuspected. The activity of sense organs is not only a function of the individual cells when stimulated, but also a question of the interaction between adjacent cells (as in lateral inhibition in *Limulus* eye), or of biasing and modulation from the CNS (via perhaps centrifugal fibers as in *Octopus* eye, Lund, 1966), Thus to study a sense organ can only give part of the story, and it may not be easy in some cases to obtain information about the whole.

Similarly the responses and reactions of nerve cells, central or peripheral, are in many cases mere reflections of the conditions imposed upon them by the experimenter. There may be valid reasons for the conditions and the results may be very illuminating, but at the same time they give false impressions as the organ did not evolve to answer experimental problems but rather everyday requirements. If a large cell in the totally isolated ganglionic mass of *Tritonia* is stimulated electrically it may discharge with a characteristic burst of impulses (Dorsett *et al.*, 1969). This burst is taken as evidence of centrally generated rhythmic activity of significance in normal behavior, and indicating that locomotion (and perhaps by analogy other aspects of behavior also) is governed by genetic coded "tapes" located in specific cells (motor units). The sensory input is relegated only to a trigger mechanism: a suggestion that is claimed to be reinforced by similar observations in intact preparations in which the stimulus was a drop of salt solution applied to the body wall. The sensory input stimulated the motor elements to fire in bursts, as for the isolated ganglion. Proprioceptive feedback is eliminated in the isolated preparation so there is no feedback of sensory stimulation. There are very slight differences in the figures reproduced in the paper by Dorsett *et al.* (1969), and synaptic events from chemoreceptors over the body surface may be continually stimulated and provide input for some while after application and spread to other parts. It is also notable that the cell was active before stimulation by this means.

It is very difficult to be sure of the precision with which stimulation is accomplished in this situation, and consequently it may be unwise to be too dogmatic about the results.

The difficulty of the "bits and pieces" approach, even on this sophisticated scale, is shown by work carried out on giant fibers in the crayfish, by Schramek (1970). It has often been presumed that giant fibers in the crustacean abdomen are "command fibers" mediating swimming especially by the tail flip reflex. In this way they are analagous to giant fibers in other animals such as oligochaetes. Schramek's question therefore is "Do the giant fibers act in swimming or do they not? If they do not, what do they do?" The question perhaps is an old one. The experiment, however, is not. Realizing that isolated fibers may tell what the experimenter imposes upon the material Schramek worked upon the whole walking animal. She was able to attach a suction electrode to the nerve cord from the dorsal side, and to record electrical activity within the giant fibers during walking and other forms of locomotion. Giant fibers are *not* active during single tail flips or swims. Occasional lateral giant fiber activity was seen, usually in later cycles than the first. She concludes that giant fibers provide only small input into the mechanism that controls swimming. She suggests that the giant fibers may help to synchronize flexor motor neurons, but are not an essential part of the generating cycle of impulses that ultimately drive the flexor muscles.

This is the type of work that will surely increase in volume in the next few years. It is necessary to know the fundamentals of biophysics, membrane properties, motor patterns, sensory input, and integrative functions of interneurons. It is also equally important that we should know how these various parameters are distributed in the animal and how they are operative during normal activities. Preconceived notions, and hypotheses erected after experimentation on isolated pieces must be subjected to test by observation on intact animals.

This analysis will come about in more amenable fashion in the future. One problem in the study of marine animals has always been the effect of sea-water surrounding the animal and the recording electrodes. Now that well insulated flexible electrode materials are available this is no longer a limitation and whole animals can carry their own recording or broadcasting gear.

The miniaturizing revolution, which has allowed electronic components to become increasingly smaller, allows us to use animals to carry the whole amplification arrangements for monitoring activity both in the laboratory and field. Animals in aquaria can be watched and used under close scrutiny, and can carry equipment with a harness, feeding into recording equipment close at hand.

Another line of approach is to attach small transmitting gear to the animal and then return it to the sea for monitoring from a floating platform. Thus at the moment sonic tags giving forth an oscillator signal and detectable on board a boat or by a diver, are being used to plot the movement of fish and lobsters (e.g. Herrnkind & McLean, 1971). The power available is limited, and hence the range is comparatively small, but no doubt this will improve with better battery sources. The use of this tactic is valuable for knowledge of migratory swimming, daily rhythms, feeding habits, shoaling, and net avoidance behavior. It is only a step further for the first implanted electrode to be placed in the musculature or nervous system in a fish, or a large decapod crustacean returned to the sea and then monitored during life in its own environment under the usual conditions of that environment. In a way something along these lines has been done by Hodgson and his colleagues (1967) in the study of electroencephalograms during chemical stimulation of sharks. These animals were, however, captive, and may therefore still be subject to the experimenter's whim. The nervous system, like all other bodily organ systems, has evolved to meet certain stringent criteria in a particular habitat. How does it function in that situation? As but one example, *Homarus americanus* increases stomach activity after feeding as shown by chronic implanted electrodes in intact lobsters (Morris and Maynard, 1970).

C. DEVELOPMENT AND GROWTH

There have long been many unresolved problems relating to the growth of nerve fibers and the establishment of specific connexions. It is perhaps a somewhat "chicken and egg" problem. Which happened first, development of a specific site of interaction, or the growth of the nerve fiber toward a preordained ending? Which guides the other? Does the nerve fiber induce changes in the postsynaptic membrane, or does the postsynaptic cell induce the establishment of the connexion by a randomly wandering nerve fiber? That there is specificity there can be no doubt, but how much is another question.

The work now in progress on the nature of membranes (see de Robertis, 1971; Miledi *et al.*, 1971) shows that these structures are of very specialized types, and that they exist probably as a mosaic of molecules and sites adapted to accomplish various functions. The detailed demonstration of cellular anatomy revealed by intracellular injection must one day be repeated on a biochemical scale for the surface of the individual cell. We will be able to draw a nerve cell as a patchwork of areas each "colored" differently (like a political map of the world) according to the molecules present at that place. This will show which portions are purely structural and binding, which are enzymatic, which are acceptor sites for external agents, which are concerned with nutrient transfer, which with transmitters, which with hormones, which are electrically inexcitable and which excitable (and why), and all the other activities of the cell. The work quoted above relies heavily on marine material, the electric organ of *Electrophorus* (electric ray), and it seems obvious that other marine examples will offer the experimentalist further suitable preparations for this work.

Nerve growth and specific connectivity patterns are already showing signs of becoming well worked fields as molecular biologists move into the discipline ready to apply their methods to the question of function in the nervous system.

Eventually the knottiest problem of all may yield its secrets; the structure and function of the neuropil. It is not unreasonable at the moment to say that we have little idea of what actually goes on in the neuropil of the invertebrate ganglion. The information from procion yellow now can tell us where the dendrites of individual cells run. Chromatography of individual cells shows us what chemicals are within such cells and electron microscopy tells us what such cells look like internally, but still we cannot say anything much about the interaction of each cell with all others. The tangle of fibers within the ganglion makes precision difficult, but all the newly founded lines of research will show profit in this direction before long.

The question of connectivity yields its secrets to painstaking and tedious analysis as shown by Horridge and Meinertzhagen (1970) on their investigation of the connexion of primary receptors to optic lamina in a variety of insects (the complexity of even this restricted system with its interwoven fibers corkscrewing round one another and then transferring from one ommatidium to make contact with those from neighboring facets, makes the final analysis of even the optic lamina a massive undertaking). Translated to a large ganglion the size of that of *Octopus*, it seems that the serial sectioning of one brain and the tracing of all fibers would take a lifetime. Given also the fact that the light microscope, even at its best, cannot resolve the finest detail so that fine dendrites may go unsuspected, and it begins to appear almost beyond contemplation.

Biochemical and transmitter analysis of single cells from separate regions of a ganglion reveals that the closer the cells are together, the more alike they are. Kandel and Kupfermann (1970b) report that one driver interneuron may induce 3 different effects in three different groups of nerve cells. These groups are compact aggregations and seem to have similar properties. No doubt this shows they have a common origin in the development of the animal from the embryonic stages. Late divergence of cells in development probably confers more common properties among the group, than earlier divergence which leads to radically different properties. The position of cells within the ganglion indicates morphogenetic processes that arrange cells in particular patterns, and these processes are under the influence of biochemical gradients that ensure common biochemical properties of contiguous cells. It seems likely that any one ganglion may have nerve cells that grade gradually across the complex from various extremes at the periphery of the ganglion to other extremes elsewhere via a series of cells of intermediate characteristics between them.

The consistency with which cells take up position within a ganglion has been remarked in numerous examples where the individual cells are large and easy to demonstrate. They are in special places that are remarkably invariant; in *Ascaris* where the total cell number is constant this might be expected, but in other more obviously variable animals it is less so. Nonetheless recognizable cells turn up again and again in the same relative positions within the ganglion thus demonstrating the precision of morphogenetic processes, the necessity for specific placement of neurons in order that growth of nerve fibers enables contact to be made at the right place and with the right fibers, and the similarities of origin of contiguous cells. It is this factor that contributes so much to the value of giant cells in some groups, but it is to be expected that other animals with much larger numbers of small cells also have precise and regular conformations, but this is less readily understood because of the microscale involved.

V. Last Comments

This has been an attempt to indicate some of the lines upon which work has proceeded in the recent past, and may perhaps reasonably be expected to pursue in the immediate future.

The more work that accumulates in neurophysiology, the more it becomes apparent that the nervous system is dynamic, variable, and very incompletely known. Even on the level of single neurons we are still learning much; short term changes may by now be fairly well known, but we know little of long term changes, and yet these may be of the greatest consequence for consideration of the same cell over long periods, and of behavior of the animal in the environment (see Tauc, 1969). So although unit studies are well advanced we have by no means reached the end.

The interaction of small groups of cells is now receiving more and more attention, and the various methods by which cells communicate with one another are becoming more obvious. How such communications are translated into behavior patterns is still something for the future.

References

Ache, B., and Case, J. F. (1969). An analysis of antennular chemoreception in two commensal shrimps of the genus *Betaeus*. *Physiol. Zool.* **42**, 361–371.

Adler, H. (ed.) (1971). Orientation: sensory basis. *Ann. N.Y. Acad. Sci.* **188**, 1–408.

Alexandrowicz, J. S. (1970). Report of Council 1969–70. *J. Mar. Biol. Ass. U.K.* **50**.

Alexandrowicz, J. S., and Whitear, M. (1957). Receptor elements in the coxal region of Decapoda Crustacea. *J. Mar. Biol. Ass. U.K.* **36**, 603–628.

Anderson, M. (1968). Electrophysiological studies on initiation and reversal of the heart beat in *Ciona intestinalis*. *J. Exp. Biol.* **49**, 363–386.

Anderson, M., and Smith, D. S. (1971). Electrophysiological and structural studies on the heart muscle of the lobster *Homarus americanus*. *Tissue and Cell* **3**, 191–205.

Atema, J., and Engstrom, D. G. (1971). Sex pheromone in the lobster, *Homarus americanus*. *Nature (London)* **232**, 261–263.

Bentley, D. R. (1970). A Topological map of the locust flight system motor neurons. *J. Insect Physiol.* **16**, 905–918.

Binstock, L., and Goldman, L. (1969). Current- and voltage-clamped studies on *Myxicola* giant axons. *J. Gen. Physiol.* **54**, 730–740.

Björklund, A., Falck, B., and Owman, C. (1972). Fluorescence microscopic and microspectrofluorometric techniques for the cellular localisation and characterization of biogenic amines. *In* "Methods of Investigative and Diagnostic Endocrinology" (J. Kopin, ed.). North-Holland Publ. Amsterdam.

Brown, H. M., Hagiwara, S., Koike, H., and Meech, R. W. (1971). Electrical characteristics of a barnacle photoreceptor. *Fed. Proc.* **30**, 69–78.

Bullock, T. H. (1969). *In* "The Interneurone" (M. Brazier, ed.), pp. 31–34. Univ. of California Press, Berkeley.

Bullock, T. H., and Horridge, G. A. (1965). *In* "Structure and Function in the Nervous Systems of Invertebrates," 2 vols. Freeman, San Francisco, California.

Burrows, M., and Willows, A. O. D. (1969). Neuronal coordination of rhythmic maxilliped beating in brachyuran and anomuran Crustacea. *Comp. Biochem. Physiol.* **31**, 121–135.

Bush, B. M. H., and Roberts, A. (1968). Resistance reflexes from a crab muscle receptor without impulses. *Nature (London)* **218**, 1171–1173.

Case, J. F. (1964). Properties of the dactyl chemoreceptors of *Cancer antennarious* Stimpson and *C. productus* Randall *Biol. Bull.* **127**, 428–446.

Castelluci, V., Pinsker, H., Kupfermann, I., and Kandel, E. R. (1970). Neuronal mechanisms of habituation and dishabituation of the gill-withdrawal reflex in *Aplysia*. *Science* **167**, 1745–1748.

Clarac, F., Wales, W., and Laverack, M. S. (1971). Stress detection at the anatomy plane in the Decapod Crustacea. II The function of receptors associated with the cuticle of the basi-ischiopodite. *Z. Vergl. Physiol.* **73**, 383–407.

Clark, A. W., Millechia, R., and Mauro, A. (1969). The ventral photoreceptor cells of *Limulus*. I. The microanatomy. *J. Gen. Phsyiol.* **54**, 289–309.

Clifton, H. E., Mahnken, C. V. W., van Derwalker, J. C., and Walker, R. J. (1970). Tektite 1, Man-in-the-Sea-Project: Marine Science Program. *Science* **168**, 659–663.

Cobb, J. L. S. (1967). The lantern of *Echinus esculentus* (L.) III The fine structure of the lantern retractor muscle and its innervation. *Proc. Roy. Soc. B* **168**, 624–640.

Cobb, J. L. S., and Laverack, M. S. (1966). The lantern of *Echinus esculentus* (L.) III. *Proc. Roy. Soc. B* **164**, 651–658.

Cobb, J. L. S., and Laverack, M. S. (1967). Neuromuscular systems in echinoderms. *Symp. Zool. Soc. London.* **20**, 22–51.

Copeland, M. (1923). The chemical senses of *Palaemonetes vulgaris*. *Anat. Res.* **24**, 394.

Cottrell, G. A. and Laverack, M. S. (1968). Invertebrate Pharmacology. *Ann. Rev. Pharmacol.* **8**, 273–298.

Dando, M. R., and Laverack, M. S. (1969). The anatomy and physiology of the posterior stomach nerve (p.s.n.) in some decapod Crustacea. *Proc. Roy. Soc. B* **171**, 465–482.

Davis, W. J. (1970). Motoneuron morphology and synaptic contacts: Determination by intracellular dye injection. *Science* **168**, 1358–1360.

Davis, W. J., and friends (1970). Notes from the Neurophysiology Underground No. I.

Debaisieux, P. (1949). Les foils sensoriels d'arthropodes et l'histologie nerveuse. 1. *Praunus flexuosus* Mull et *Crangon crangon* L. *Cellule* **52**, 311–360.

del Castillo, J., de Mello, W. C., and Morales, T. (1967). The initiation of action potentials in the somatic musculature of *Ascaris lumbricoidea*. *J. Exp. Biol.* **46**, 263–280.

de Robertis, E. P. (1971). Molecular biology of synaptic receptors. *Science* **171**, 963–971.

Digby, P. S. E. (1964). Semi-conduction and electrode processes in biological material. I. Crustacea and certain softbodied forms. *Proc. Roy. Soc. B.* **161**, 504–525.

Dorsett, D. A. (1964). The sensory and motor innervation of *Nereis*. *Proc. Roy. Soc. B.* **159**, 652–667.

Dorsett, D. A., Willows, A. O. D., and Hoyle, G. (1969). Centrally generated nerve impulse sequences determining swimming behaviour in *Tritonia*. *Nature (London)* **224**, 711–712.

Edwards, C., and Ottoson, D. (1958). The site of impulse initiation in a nerve cell of a crustacean stretch receptor. *J. Physiol.* **143**, 138–148.

Enright, J. T. (1962). Responses of an amphipod to pressure changes. *Comp. Biochem. Physiol.* **7**, 131–145.

Falck, B. (1962). Observations on the possibilities of the cellular localization of monoamines by a fluorescence method. *Acta Physiol. Scand. Suppl. 197* **56**, 1–25.

Falck, B., and Owman, C. (1965). A detailed description of the fluorescence method for the cellular localization of biogenic monoamines. *Acta Univ. Lund.* Sect. II, No. 7, 1–23.

Flood, P. R. (1966). A peculiar mode of muscular innervation in *Amphioxus*. Light and electron microscope studies of the so-called ventral root. *J. Comp. Neurol.* **126**, 181–217.

Flood, P. R. (1970). The connection between spinal cord and notochord in *Amphioxus* (*Branchiostoma lanceolatum*) *Z. Zellforsch.* **103**, 115–128.

Ghiradella, H., Case, J. F., and Cronshaw, J. (1968a). Structure of aesthetascs in selected marine and terrestrial decapods; chemoreceptor morphology and environment. *Amer. Zool.* **8**, 603–621.

Ghiradella, H., Cronshaw, J., and Case, J. F. (1968b). Fine structure of the aesthetasc hairs of *Pagurus hirsutiusculus* Dana. *Protoplasma* **66**, 1–20.

Giller, E., Jr., and Schwartz, J. H. (1971a). Choline acetyltransferase in identified neurons of abdominal ganglion of *Aplysia californica*. *J. Neurophysiol.* **34**, 93–107.

Giller, E., Jr., and Schwartz, J. H. (1971b). Choline acetyltransferase in identified neurons of abdominal ganglion of *Aplysia californica*. *J. Neurophysiol.* **34**, 108–115.

Graham, J., and Gerard, R. W. (1946). Membrane potentials and excitation of impaled single muscle fibres. *J. Cell. Comp. Physiol.* **28**, 99–117.

Günther, J. (1970). On the organisation of the exteroceptive afferences in the body segments of the earthworm. *Vehr. Deutsch. Zool. Gesell.* **64**, 261–265.

Guthrie, D. M., and Banks, J. R. (1970a). Observations on the function and physiological properties of a fast paramyosin muscle the notochord of *Amphioxus* (*Branchiostoma lanceolatum*). *J. Exp. Biol.* **52**, 125–138.

Guthrie, D. M., and Banks, J. R. (1970b). Observations on the electrical and mechanical properties of the myotomes of the lancelet (*Branchiostoma lanceolatum*). *J. Exp. Biol.* **52**, 401–417.

Gwilliam, G. F. (1965). The mechanism of the shadow reflex in Cirripedia. II Photoreceptor cell response, secured order responses, and motor cell output. *Biol. Bull. Mar. Biol. Lab. Woods Hole* **129**, 244–256.

Heimer, L. (1971). Pathways in the Brain. *Sci. Amer.* **225**, 48–60.

Herrnkind, W. F., and McLean, R. (1971). Field studies of homing, mass emigration, and orientation in the spiny lobster, *Panulirus argus*. *Ann. N.Y. Acad. Sci.* **188**, 359–377.

Hodgson, E. S., Mathewson, R. F., and Gilbert, P. W. (1967). Electroencephalographic studies of chemoreception in sharks. *In* "Sharks, Skates and Rays" (P. W. Gilbert, R. F. Mathewson, and D. P. Rall, eds.), pp. 491–501. Johns Hopkins Univ. Press, Baltimore, Maryland.

Holmes, S. J., and Homuth, E. S. (1910). The seat of smell in the crayfish. *Biol. Bull.* **18**, 155–160.

Horch, K. W. (1971). An organ for hearing and vibration sense in the ghost crab *Ocypode*. *Z. Vergl. Physiol.* **73**, 1–21.

Horch, K. W., and Salmon, M. (1969). production, perception and reception of accessive stimulation by terrestrial crab. (*Genus Ocypode, Uca, Family Ocypodiatae*). *Forma Functio* **1**, 1–25.

Horridge, G. A. (1963). Proprioceptors, bristle receptors, efferent sensory impulses, neurofibrils and number of axons in the parapodial nerve of the polychaete. *Proc. Roy. Soc. B* **157**, 199–222.

Horridge, G. A. (1965a). Relations between nerves and cilia in ctenophores. *Amer. Zool.* **5**, 357–375.

Horridge, G. A. (1965b). Relations between nerves and cilia in ctenophores. *Nature (London)* **205**, 602.

Horridge, G. A. (1968). Intracellular action potentials associated with the beating of the cilia in ctenophore comb plate cells. *In* "Structure and Function of Nervous Tissue" (G. H. Bourne, ed.), pp. 1–29. Academic Press, New York.

Horridge, G. A., and Meinertzhagen, I. A. (1970). The accuracy of the patterns of connexions of the first- and second-order neurons of the visual system of *Calliphora*. *Proc. Roy. Soc. B* **175**, 69–82.

Hoyle, G., and Smyth, T., Jr. (1963). Neuromuscular physiology of giant muscle fibres of a barnacle *Balanus nubilis* Darwin. *Comp. Biochem. Physiol.* **10**, 291–314.

Kandel, E. R., and Kupfermann, I. (1970). The functional organisation of invertebrate ganglia. *Ann. Rev. Physiol.* **32**, 198–258.

Kandel, E. R., Frazier, W. T., Waziri, R., and Coggeshall, R. E. (1967). Direct and common connections among identified neurons in *Aplysia*. *J. Neurophysiol.* **30**, 1352–1376.

Katz, B., and Miledi, R. (1966). Input–Output relation of a single synapse. *Nature (London)* **212**, 1242–1245.

Kennedy, D. (1969). The interface between organismal and populational biology–I. *Amer. Zool.* **9**, 253–259.

Kennedy, D. M., Evoy, W. H., and Fields, H. L. (1966). The unit basis of some crustacean reflexes. *Symp. Soc. Exp. Biol.* **20**, 75–109.

Kerkut, G. A., and Meech, R. W. (1966). The internal chloride concentration of H and D cells in the snail brain. *Comp. Biochem. Physiol.* **19**, 819–832.

Kerkut, G. A., and Walker, R. J. (1962). Marking individual nerve cells through electrophoresis of ferrocyanide from a microelectrode. *Stain Tech.* **37**, 217–219.

Knight-Jones, E. W., and Morgan, E. (1966). Responses of marine animals to changes in hydrostatic pressure. *Oceanogr. Mar. Biol. Ann. Rev.* **4**, 267–299.

Kriebel, M. E. (1968a). Electrical characteristics of tunicate heart cell membranes, nexuses. *J. Gen. Physiol.* **52**, 46–59.

Kriebel, M. E. (1968b). Studies on cardiovascular physiology of tunicates. *Biol. Bull.* **134**, 434–455.

Kupfermann, J., and Kandel, E. R. (1970). Electrophysiological properties and functional interconnection of two symmetrical neurosecretory clusters (bag cells) in abdominal ganglion of *Aplysia*. *J. Neurophysiol.* **33**, 865–876.

Langdon, F. E. (1895). The sense organs of *Lumbricus agricola* Hoffner *J. Morphol.* **11**, 193–234.

Langdon, F. E. (1900). The sense organs of *Nereis vireus* *J. Comp. Neurol.* **10**, 1–77.

Larimer, J. L. (1964). Sensory-induced modifications of ventilation and heart-rate in crayfish. *Comp. Biochem. Physiol.* **12**, 25–36.

Larimer, J. L., and Kennedy, D. (1966). Visceral afferent signals in the crayfish stomatogastric ganglion. *J. Exp. Biol.* **44**, 345–354.

Larimer, J. L., and Kennedy, D. (1969). Innervation patterns of fast and slow muscle in the uropods of crayfish. *J. Exp. Biol.* **51**, 119–133.

Laverack, M. S. (1963). Responses of cuticular sense organs of the lobster *Homarus vulgaris* (Crustacea)–III Activity invoked in sense organs of the carapace. *Comp. Biochem. Physiol.* **10**, 261–272.

Laverack, M. S. (1963). Aspects of chemoreception in Crustacea. *Comp. Biochem. Physiol.* **8**, 141–151.

Laverack, M. S. (1964). The antennular sense organs of *Panulirus argus* *Comp. Biochem. Physiol.* **13**, 301–321.

Laverack, M. S. (1968). On the receptors of marine invertebrates. *Oceanogr. Mar. Biol. Ann. Rev.* **6**, 249–324.

Laverack, M. S. (1974). The structure and function of chemoreceptor cells. *In* "Chemical Sensitivity in Aquatic Organisms" (P. Grant and A. Mackie, eds.). Academic Press, New York. (in press).

Laverack, M. S., and Ardill, D. J. (1965). The innervation of the aesthetasc hairs of *Panulirus argus*. *Quart. J. Microsc. Sci.* **106**, 45–60.

Laverack, M. S., and Dando, M. R. (1968). The anatomy and physiology of mouthpart receptors in the lobster, *Homarus vulgaris*. *Z. Vergl. Physiol.* **61**, 176–195.

Lawry, J. V. (1970). Mechanisms of locomotion in the polychaete, *Harmothoë*. *Comp. Biochem. Physiol.* **37**, 167–179.

Levandowsky, M., and Hodgson, E. S. (1965). Amino acids and amine receptors of lobsters. *Comp. Biochem. Physiol.* **16**, 159–161.

Levitan, H., Tauc, L., and Segundo, J. P. (1970). Electrical transmission among neurons in the buccal ganglion of a mollusc, *Navanax inermis*. *J. Gen. Physiol.* **55**, 484–496.

Lindstedt, K. J. (1971). Chemical control of feeding behaviour. *Comp. Biochem. Physiol.* **A39**, 553–581.

Ling, G., and Gerard, R. W. (1949). The normal membrane potential of frog sartorius fibre. *J. Cell. Comp. Physiol.* **34**, 383–396.

Livett, B. G., Geffen, L. B., and Rush, R. A. (1971a). Immunochemical methods for demonstrating macromolecules in sympathetic neurons. *Phil. Trans. Roy. Soc. London B*, **261**, 359–361.

Livett, B. G., Uttenthal, L. O., and Hope, D. B. (1971b). Localisation of neurophysic II in the Hypothalamo-neurophysiological system of the pig by immunofluorescence histochemistry. *Phil. Trans. Roy. Soc. London B* **261**, 371–378.

Lund, R. D. (1966). Centrifugal fibres to the retina of *Octopus vulgaris* *Exp. Neurol.* **15**, 100–112.

Mackie, G. O. (1970). Neuroid conduction and the evolution of conducting tissues, *Quart. Rev. Biol.* **45**, 319–332.

Martin, A. R., and Wickelgren, W. O. (1971). Sensory cells in the spinal cord of the sea lamprey. *J. Physiol.* **212**, 65–83.

Maynard, D. M. (1966). Integration via crustacean ganglia. *Symp. Soc. Exp. Biol.* **20**, 111–149.

Maynard, D. M. (1969). *In* "The Interneurone" (M. Brazier, ed.), pp. 56–68. Univ. of California Press, Berkeley.

Maynard, D. M., and Sallee, A. (1970). Disturbance of feeding behaviour in the spiny lobster, *Panulirus argus*, following bilateral ablation of the medulla terminalis. *Z. Vergl. Physiol.* **66**, 123–140.

Maynard, D. M., and Yager, J. G. (1968). Function of an eyestalk ganglion, the medulla terminalis in olfactory integration in the lobster *Panulirus argus*. *Z. Vergl. Physiol.* **59**, 241–249.

Maynard, E. A. (1971). Electron microscopy of stomatogastric ganglion in the lobster, *Homarus americanus*. *Tissue and Cell* **3**, 137–160.

McFarlane, I. D. (1969). Two slow conduction systems in the sea anemone *Calliactis parasitica*. *J. Exp. Biol.* **51**, 377–385.

MacMillan, D. L., and Dando, M. R. (1972). Tension receptors on the apodemes of muscles in the walking legs of the crab *Cancer magister*. *Mar. Behav. Physiol.* **1**, 185–208.

Mellon, M. de F. (1963). Electrical responses from dually innervated tactile receptors on the thorax of the crayfish. *J. Exp. Biol.* **40**, 137–148.

Mellon, M. de F., and Kennedy, D. (1964). Impulse origin and propagation in a bi-polar sensory neuron. *J. Gen. Physiol.* **47**, 487–499.

Mendelson, M. (1971). Oscillator neurons in crustacean ganglia. *Science* **171**, 1170–1173.

Mertens, L. (1970). "In-water photography, theory and practice." Wiley, New York.

Miledi, R. (1966). Miniature synaptic potentials in squid nerve cells. *Nature* (*London*) **212**, 1240–1242.

Miledi, R. (1967). Spontaneous synaptic potentials and quantal release of transmitter in the stellate ganglion of the squid. *J. Physiol.* **192**, 379–406.

Miledi, R., Molinoff, P., and Potter, L. T. (1971). Isolation of the cholinergic receptor protein of *Torpedo* electric tissue. *Nature* (*London*), **229**, 554–557.

Mill, P. J., and Knapp, M. (1967). Efferent sensory impulses and the innervation of tactile receptors in *Allolobophora longa* Ude and *Lumbricus terrestris Linn*. *Comp. Biochem. Physiol.* **23**, 263–276.

Millechia, R., and Mauro, A. (1969). The ventral photoreceptor cells of *Limulus*. II. The basic photoresponse. *J. Gen. Physiol.* **54**, 310–330.

Moores, S. R. (1970). A preliminary investigation into the fine structure of the planktonic larva of the sea urchin *Echinocardium cordatum* (Pennant). B.Sc. honours thesis, Univ. of St. Andrews.

Moran, D. T., Chapman, K. M., and Ellis, R. A. (1971). The fine structure of cockroach campaniform sensilla. *J. Cell. Biol.* **48**, 155–173.

Morris, J., and Maynard, D. M. (1970). Recordings from the stomatogastric nervous system in intact lobsters. *Comp. Biochem. Physiol.* **33**, 969–974.

Moulins, M., Dando, M. R., and Laverack, M. S. (1970). Further studies on mouthpart receptors in Decapoda Crustacea. *Z. Vergl. Physiol.* **69**, 225–248.

Mulloney, B. (1970). The structure of the giant fibers of Earthworm. *Science* **168**, 994–996.

Nicholls, J. G., and Baylor, D. A. (1968). Specific modalities and receptive fields of sensory neurons in the CNS of the leech. *J. Neurophysiol.* **31**, 740–756.

Nicholls, J. G., and Purves, D. (1970). Monosynaptic chemical and electrical connexions between sensory and motor cells in the central nervous system of the leech. *J. Physiol.* **209**, 647–667.

Ogawa, F. (1934). The number of ganglion cells and nerve fibres in the nervous system of the earthworm *Pheetima communissima*. *Sci. Rep. Tohoku Univ.* **8**, 345–368.

Olivo, R. F. (1970a). Motor aspects of reflex foot withdrawal in the razor clam. *Comp. Biochem. Physiol.* **35**, 787–808.

Olivo, R. F. (1970b). Central pathways involved in reflex foot withdrawal in the razor clam. *Comp. Biochem. Phsyiol.* **35**, 809–826.

Orlov, J. (1927). Die Magenganglion des Flusskrebses. Ein Beitrag zur vergleichenden Histologie des sympathischen Nervensystems. *Z. Mikrosk. Anat. Forsch.* **8**, 73–96.

Osborne, N. N., Ansong, R., and Neuhoff, V. (1971a). Micro-Disc Electrophoretic separation of soluble proteins from nervous and other tissues of *Helix* (Pulmonate Mollusca). *Int. J. Neurosci.* **1**, 259–264.

Osborne, N. N., and Dando, M. R. (1970). Monoamines in the stomatogastric ganglion of the lobster *Homarus vulgaris*. *Comp. Biochem.* Physiol. **32**, 327–331.

Osborne, N. N., Briel, G., and Neuhoff, V. (1971b). Distribution of Gaba and other amino acids in different tissues of the gastropod mollusc *Helix pomatia*, including *in vitro* experiments, with 14C glutamic acid. *Int. J. Neurosci.* **1**, 265–272.

Otsuka, M., Kravitz, E. A., and Potter, D. D. (1967). Physiological and chemical architecture of a lobster ganglion with particular reference to gamma aminobutyrate and glutamate. *J. Neurophysiol.* **30**, 725–752.

Pabst, H., and Kennedy, D. M. (1967). Cutaneous mechanoreceptors influencing motor output in the crayfish abdomen. *Z. Vergl. Physiol.* **57**, 190–208.

Pasztor, V. M. (1968). The neurophysiology of respiration in decapod Crustacea. I. The motor system. *Can. J. Zool.* **46**, 585–596.

Pasztor, V. M. (1969). The neurophysiology of respiration in decapod Crustacea. II. The sensory system. *Can. J. Zool.* **47**, 435–441.

Pentreath, V. M., and Cottrell, G. A. (1971). 'Giant' neurons and neurosecretion in the hyponeural tissue of *Ophiothrix fragilis* Abildgaard. *J. Exp. Mar. Biol. Ecol.* **6**, 249–264.

Peretz, B. (1970). Habituation and dishabituation in the absence of a central nervous system. *Science* **169**, 379–381.

Qutob, Z. (1962). The swimbladder of fishes as a pressure receptor. *Arch. Neerl. Zool.* **15**, 1–67.

Roberts, A., and Bush, B. M. H. (1971). Coxal muscle receptors in the crab: The receptor current and some properties of the receptor nerve fibers. *J. Exp. Biol.* **54**, 515–524.

Robinson, M. H., Abele, L. G., and Robinson, B. (1970). Attack autotomy: A defence against predators. *Science* **169**, 300–301.

Rosenbluth, J. (1965). Ultrastructure of somatic muscle cells in *Ascaris lumbricoides*. *J. Cell. Biol.* **26**, 579–591.

Rude, S., Coggeshall, R. E., and van Orden, L. F. (1969). Chemical and ultrastructural identification of 5-hydroxytryptamine. *J. Cell. Biol.* **41**, 832–854.

Ryan, E. P. (1966). Pheromone: evidence in a decapod crustacean. *Science.* **151**, 340–341.

Sandeman, D. C. (1969). Integrative properties of a reflex motoneuron in the brain of the crab *Carcinus maenas*. *Z. Vergl. Physiol.* **64**, 450–464.

Schramek, J. (1970). Crayfish swimming: Alternating motor output and giant fiber activity. *Science* **169**, 698–700.

Selverston, A. I., and Kennedy, D. M. (1969). Structure and function of identified nerve cells in the crayfish. *Endeavour*, **28**, 107–113.

Shaw, S. (1972). Decremental conduction of the visual signal in the barnacle lateral eye. *J. Physiol.*, **220**, 145–175.

Shelton, R. G. J., and Laverack, M. S. (1968). Observations on a redescribed crustacean cuticular sense organ. *Comp. Biochem. Physiol.* **25**, 1049–1059.

Shelton, R. G. J., and Laverack, M. S. (1970). Receptor hair structure and function in the lobster *Homarus gammarus*. *J. Exp. Mar. Biol. Ecol.* **4**, 201–210.

Smallwood, W. M. (1926). The peripheral nervous system of the common earthworm, *Lumbricus terrestris*. *J. Comp. Neurol.* **42**, 35–55.

Stein, P. S. G. (1971). Intersegmental coordination of swimmeret motoneuron activity in crayfish. *J. Neurophysiol.* **34**, 310–318.

Steinbrecht, R. A. (1969). On the question of nervous syncytia: lack of axon fusion in two insect sensory nerves. *J. Cell. Sci.* **4**, 39–53.

Strathmann, W. (1971). The feeding behaviour of planktotrophic echinoderm larvae: mechanisms, regulation, and rates of suspension-feeding. *J. Exp. Mar. Biol. Ecol.* **6**, 109–160.

Strausfeld, N. J. (1970). Golgi studies on insects. II. The optic lobes of Diptera. *Phil. Trans. Roy. Soc. London B* **258**, 135–223.

Strausfeld, N. J., and Blest, A. D. (1970). Golgi studies on insects. I. The optic lobes of Lepidoptera. *Phil. Trans. Roy. Soc. London B* **258**, 81–134.

Stretton, A. O. W., and Kravitz, E. A. (1968). Neuronal geometry: determination with a technique of intracellular dye injection. *Science* **162**, 132–134.

Stuart, A. E. (1970). Physiological and morphological properties of motoneurons in the central nervous system of the leech. *J. Physiol.* **209**, 627–646.

Tauc, L. (1969). Excitatory—inhibitory processes. *In* "The Interneurone" (M. Brazier, ed.). Univ. of California Press, Berkeley.

Thomas, W. J. (1970). The setae of *Austropotamobius pallipes* (Crustacea: Astacidae). *J. Zool. London.* **160**, 91–142.

Tupper, J., Saunders, J. W., Jr., and Edwards, C. (1970). The onset of electrical communication between cells in the developing starfish embryo. *J. Cell Biol.* **46**, 187–191.

van Well, P. B., and Christofferson, T. P. (1966). Electrophysiological studies on perception in the antennulae of certain crabs. *Physiol. Zool.* **39**, 317–325.

Wachtel, H., and Kandel, E. R. (1967). A direct synaptic connection mediating both excitation and inhibition. *Science* **158**. 1206–1208.

Wales, W. (1968). A investigation of the peripheral part of the mantle of *Pecten maximus* (Linnaeus). B. Sc. honours thesis, Univ. of St. Andrews.

Wales, W. and Laverack, M. S. (1972). Sensory activity of the mandibular muscle receptor organ of *Homarus vulgaris*. I. Response to receptor muscle stretch. *Mar. Behav. Physiol.*, **1**, 239–256.

Wales, W., Clarac, F., and Laverack, M. S. (1971). Stress detection at the autotomy plane in the decapod Crustacea. I. Comparative anatomy of the receptors of the Basi-ischiopodite region. *Z. Vergl. Physiol.* **73**, 357–382.

Wales, W., Clarac, F., Dando, M. R., and Laverack, M. S. (1970). Innervation of the receptors present at the various joints of the pereiopods and third maxilleped of *Homarus gammarus* and the macruran decapods (Crustacea). *Z. Vergl. Physiol.* **68**, 345–384.

Walker, M. G., Mitson, R. B., and Storeton—West, T. (1971). Trials with a transponding acoustic fish tag tracked with an electronic sector scanning sonar. *Nature* (*London*) **229**, 196–198.

Weevers, R. de G. (1971). A preparation of *Aplysia fasciata* for intrasomatic recording and stimulation of single neurones during locomotor movements. *J. Exp. Biol.* **54**, 659–676.

Welsch, U., and Storch, V. (1970). The fine structure of the stomochord of the enteropneusts *Harrimania kupfferi* and *Ptychodera flava*. *Z. Zellforsch.* **107**, 234–239.

Whitear, M. (1965). The fine structure of crustacean proprioceptors 2. The thoracic-coxal organs in *Carcinus*, *Pagurus* and *Astacus*. *Phil. Trans. Roy. Soc. B* **248**, 437–456.

Wiersma, C. A. G. (1959). Movement receptors in decapod Crustacea. *J. Mar. Biol. Ass. U.K.* **38**, 143–152.

Willows, A. O. D. (1967). Behavioural acts elicited by stimulation of single identifiable brain cells. *Science* **157**, 570–574.

Willows, A. O. D., and Hoyle, G. (1968a). Neuronal network triggering a fixed action pattern. *Science*, *166*, 1549–1551.

Willows, A. O. D., and Hoyle, G. (1968b). Correlation of behaviour with the activity of single identifiable neurons in the brain of *Tritonia*. *In* "Neurobiology of Invertebrates" (J. Salanki, ed.), pp. 443. Plenum Press. New York.

Winlow, W., and Laverack, M. S. (1970). The occurrence of an anal proprioceptor in the decapod Crustacea *Homarus gammarus* (L). (Syn. *H. vulgaris* M. Ed.) and *Nephrops norvegicus* (Leach) *Life Sci.* **9**, 93–97.

Winlow, W., and Laverack, M. S. (1972). The control of hind-gut motility in the lobster, *Homarus gammarus* L. *Mar. Behav. Physiol.* **1**, 1–48.

Wolpert, L., and Mercer, E. (1963). An electron microscope study of the development of the blastula of the sea urchin embryo and its radial polarity. *Exp. Cell. Res.* **30**, 280–300.

Young, J. Z. (1939). Fused neurones and synaptic contacts in the giant nerve fibres of cephalopods. *Phil. Trans. Roy. Soc. London B* **229**, 465–503.

Young, J. Z. (1967). Some comparisons between the nervous systems of cephalopods and mammals. *In* "Invertebrate Nervous Systems" (C. A. G. Wiersma, ed.), pp. 352–362. Univ. of Chicago Press, Chicago, Illinois.

Chapter 5

Comparative Endocrinology

Milton Fingerman

I. Introduction

Comparative endocrinology is one of the areas of experimental biology where new information is accumulating at a rapid pace. Although the endocrine mechanisms of only a small percentage of the marine animals have been investigated, considerable information about the subject is now available. The classic method of establishing whether or not a structure is endocrine consists of the following steps. The suspected tissue or gland is ablated and the effects of the hormone deficiency are observed. Then the tissue or gland is implanted or an extract of the tissue or gland is injected to determine if the deficiency effects can be ameliorated or corrected. The endocrinologist then seeks to isolate and identify the active substance in the tissue or gland.

The product of an endocrine gland or cell is a chemical messenger, a hormone. Hormones have been classically considered as blood-borne chemical mediators. However, this view is too restrictive in view of more recent discoveries of endocrine mechanisms in some of the lower invertebrates which do not have a circulatory system. Thus, the newer definition of a hormone provided by Zarrow *et al.* (1964) as "a physiological organic substance that is secreted by living cells in relatively restricted areas of the organism and that diffuses or is transported to a site in the same organism, where it brings about an adjustment that tends to integrate the component parts and actions of the organism" is in keeping with our current knowledge of the origin and action of hormones. Classically, two systems have been looked on as having a coordinating function, the nervous and endocrine systems. The former has been described as being involved in rapid coordination between different parts of the body by means of transient nerve impulses, whereas the endocrine system functions over protracted periods of time, thereby having long-term effects. However, the discovery of the process of neurosecretion whereby certain neurons are themselves capable of secreting hormones has shown that these two systems are in fact functionally highly intertwined with one another.

Hormones have been divided into three categories on the basis of their actions by Carlisle and Knowles (1959), morphogenetic, metabolic, and kinetic. The hormones involved in morphogenesis are those which regulate growth, differentiation, and maturation of the organism. Metabolic hormones are those which affect metabolism by acting upon blood sugar, salt and water balance, and other metabolic processes. Kinetic hormones are those which act immediately upon effector organs such as the heart and pigment cells involved in color changes or even on other endocrine glands causing them to release their own hormone. Because some hormones have multiple effects, they could fall into more than one category.

Endocrinologists distinguish between the action of a hormone (what it does) and its mechanism of action (how it does it at the molecular level). The mechanisms of hormone action are probably the least understood area of endocrinology. Karlson (1963) has described the three mechanisms of action that are now commonly visualized for hormones. One mechanism that has been proposed for some hormones is that they directly control the activity of some enzymes. A second possible mechanism is that some hormones act to control cell permeability. The third theory is that some hormones produce their effects by activating or repressing the action of particular genes.

Endocrine mechanisms have been described in a wide variety of invertebrates as well as vertebrates. The invertebrate phyla, which will be discussed, consist of the coelenterates (cnidarians), platyhelminths, nemerteans, nematodes, echinoderms, annelids, mollusks, and arthropods. The discussion of invertebrates will emphasize crustaceans, particularly the decapods, although the highlights of the endocrinology of these other invertebrate groups will be treated. In addition, although among the marine vertebrates we find fishes, reptiles, birds, and mammals, this chapter will be restricted, so far as vertebrates are concerned, to the fishes.

II. Endocrine Systems

This section will be devoted to a discussion of the morphology of the endocrine systems of invertebrates and fishes. The functions of the hormones will be discussed in Section III. Among the invertebrates epithelial (nonneural) endocrine glands are scarce compared with the number found in vertebrates. Hormones produced by neurosecretory cells, on the other hand, have great importance among the invertebrates.

Because of the significant role of neurosecretory cells in the endocrine mechanisms of vertebrates, especially invertebrates, the structure of such cells will be discussed before the endocrine systems in the various groups of animals are described. Neurosecretory cells can be distinguished from ordinary neurons by their large content of membrane-bound granules which usually have a high electron density (Scharrer, 1969). These granules usually have a diameter of 500–3000 Å. Proteinaceous neurosecretory material appears to be synthesized by the rough-surfaced endoplasmic reticulum in the perikaryon. The packaging of the granule occurs in the Golgi apparatus which provides it with the membrane covering. The "classic" neurohormones are polypeptides or low molecular weight proteins. However, some neurosecretory neurons have a hormonal product which is a nonpeptide such as a biogenic amine. We can, therefore, distinguish between "peptidergic" and "aminergic" neurosecretory fibers. Generally the granules in the former are larger (1000–3000 Å) than in the latter (500–1400 Å). After the neurosecretory granules have been manufactured in the perikaryon they are transported distally along the axon to the axonal terminal where the hormone is released. These terminals are often bulbous, allowing accumulation of the granules. Release may also occur in some instances along the axon or even from the perikaryon itself. The neurohormones may be released in three different ways. In the earthworm, *Lumbricus terrestris*, it has been suggested that the granules pass across the cell membrane as intact entities (Röhlich *et al.*, 1962). A second method is exocytosis which involves fusion of the granule's membrane with that of the cell, followed by discharge of the granule's contents (Normann, 1965). The third method is simple diffusion of the hormone out of the granule into the exoplasm and then out of the cell (Gerschenfeld *et al.*, 1960). The axonal terminals of several neurosecretory cells sometimes form a cluster in or next to the walls of blood vessels or blood sinuses. Such an aggregation of neurosecretory terminals is known as a neurohemal organ.

A. COELENTERATES, PLATYHELMINTHS, NEMERTEANS, NEMATODES, AND ECHINODERMS

Neurosecretory cells containing many membrane-bound neurosecretory granules have been reported in three species of *Hydra* by Lentz and Barrnett (1965) and Davis *et al.* (1968). Most of these cells are found in the hypostome and tentacle bases. The terminations of these neurosecretory cells are found on or near other cell types and in intercellular spaces. Lentz and Barrnett suggested that in the absence of a circulatory system in coelenterates these intercellular spaces correspond to the circulatory systems of higher animals. Because coelenterates seem to be the most primitive organisms to have a nervous system, i.e., their nerve net, neurosecretory cells may be phylogenetically as old as the nervous system itself.

The nervous system of all but the most primitive platyhelminths consists of a brain and a number of longitudinal nerve cords. Platyhelminths appear to be the most primitive animals in which a brain is found. Although most of the work on neurosecretion in this phylum has utilized species which are not marine, the evidence reveals that neurosecretory cells are found in the brains of freeliving (marine and freshwater) and parasitic platyhelminths (Turner, 1946; Scharrer and Scharrer, 1954; Lender and Klein, 1961; Ude, 1962; Oosaki and Ishii, 1965; Morita and Best, 1965; Davey and Breckenridge, 1967; Lentz, 1967). Axon terminals of neurosecretory cells have been observed in the neuropil of the brain. Furthermore, neurosecretory cells have been found in the longitudinal nerve cords as well as the brain of the planarian, *Procotyla fluviatilis*, by Lentz (1967).

Neurosecretory cells also occur in the brain of nemerteans (Lechenault, 1962; Bianchi, 1969). Most species of this phylum are marine. Nemerteans are the only acoelomates which possess a circulating system. In *Cerebratulus marginatus* a marine species, neurosecretory cells in the dorsal lobes of the brain have axons which run toward the neuropil where, in what is presumed to be a neurohemal area, the axon terminals are in close contact with a blood vessel (Bianchi, 1969).

With respect to nematodes, neurosecretory cells have been reported in *Ascaris lumbricoides* by Gersch (1957), Gersch and Scheffel (1958), and Davey (1964, 1966), and in *Phocanema decipiens* by Davey (1966). *Ascaris lumbricoides* parasitizes humans and pigs. Davey's description of the distribution of neurosecretory cells in this nematode differs from that of Gersch and Scheffel. Whereas Davey reported that they are located in the lateral ganglia which are associated with the circumentric nerve ring which surrounds the pharynx, Gersch and Scheffel reported them in the ganglia of the major lateral papillary nerves with their axons running toward the circumentric ring. Davey found that some of the neurosecretory axons ran toward the cuticle. *Phocanema decipiens* is a nematode whose last stage larva is in the muscles of the cod, but the adult is found in the intestine of seals. This worm has neurosecretory cells in the dorsal and ventral ganglia associated with the circumenteric ring.

The neurosecretory cells of echinoderms have been described by Unger (1960, 1962) utilizing the sea star, *Asterias glacialis*. These cells were found in both the circumoral nerve ring of the ectoneural or oral part of the nervous system and the radial nerves which arise from this nerve ring and run the length of each arm. The neurosecretion-bearing axons are oriented in such a direction that it appears their product is released into the hemal system from which the neurohormone could be transferred to the water vascular system.

B. ANNELIDS

Neurosecretory cells have been described for the polychaetes, oligochaetes, and leeches. After the initial report of neurosecretory cells in the brain of the polychaete, *Nereis virens*, by Scharrer (1936) a number of other investigators have described the types and distribution of such cells in several other annelids. Golding (1967a) and Dhainaut-Courtois (1968a) have published detailed electron microscopic studies of such cells in nereids. Cells that are probably neurosecretory have been found in the ventral ganglionic nerve cords of *Nereis pelagica* as well as in its brain (Dhainaut-Courtois and Warembourg, 1967). An interesting epithelial structure is attached to the brain capsule on the ventral surface of the posterior part of the brain.

Dhainaut-Courtois (1966a,b) described it in *Nereis pelagica* and with the electron miscroscope found endings of neurosecretory axons both adjacent to the capsule and passing through it from the brain to contact these epithelial cells. Golding *et al.* (1968) later described this cerebrovascular complex in four other species of nereids and named this epithelioneurosecretory complex "the infracerebral gland." The latter authors and Dhainaut-Courtois (1968b) suggested that the endocrine role assigned to the brain may in reality be due to activity of the infracerebral gland whose activity would be regulated by the neurosecretory axons. However, Baskin (1970) more recently concluded that it is unlikely that the epithelial cells of the infracerebral gland in *Nereis limnicola* are themselves the source of any of the known annelid hormones.

C. MOLLUSKS

Mollusks are an extremely diverse group of organisms. The gastropod and bivalve mollusks have cells which stain as neurosecretory cells in practically all of the ganglia of their nervous systems. Scharrer (1935) was the first investigator to report such cells in a mollusk. Using opisthobranch mollusks, she found cells filled with secretory droplets in the cerebral and visceral ganglia of *Aplysia* and in the cerebral ganglia of *Pleurobranchaea*. Neurosecretory cells have now been reported in all orders of gastropods and all orders of pelecypods except the septibranchs. An investigator has to be very cautious in declaring nerve cells of mollusks "neurosecretory" because they often contain pigment granules, glycogen granules, or cell organelles such as lysosomes which can be stained with the commonly used neurosecretory stains and can hence lead to erroneous findings (Simpson *et al.*, 1966). However, the electron miscroscope has been successfully used with some mollusks to establish the presence in their ganglionic neurons of granules that one would call "neurosecretory" (Rosenbluth, 1963; Vicente, 1969).

Frazier *et al.* (1967) found two types of neurosecretory cells in the abdominal ganglion of *Aplysia californica*. One type, called the bag cells, occurs in clusters at the junctions of the left and right pleuroabdominal connectives with the abdominal ganglion. The bag cells occur in a pocket, hence their name, in the connective tissue sheath of the ganglion. Their granule-filled axons terminate in the sheath where the neurosecretion is presumed to be discharged. The sheath is vascularized, allowing the neurosecretion to reach the blood. The second type of neurosecretory cell consists of the white cells which are found in the outer cortex of the ganglion and have an opalescent apperance when observed with transmitted light. Each white cell has a stout axon which leaves the ganglion via the branchial nerve (the site of release of the neurosecretion from these axons was not determined) and, in addition, granule-containing processes which intermingle in the vascular sheath with those of the bag cells.

A number of different structures have been implicated in mollusks as possible neurohemal regions or even endocrine glands (Joose, 1964; Nolte, 1965; Simpson *et al*., 1966; Vicente, 1969). However, experimental proof that hormones are actually stored in or produced by them is lacking. For example, in some gastropods dorsal bodies (paired mediodorsal and laterodorsal bodies) have been found in close association with the cerebral ganglia. These bodies appear to consist primarily of glial cells (supporting cells of the central nervous system) and the perineurium of the cerebral ganglia. The mediodorsal bodies have received most of the attention. Granule-filled axons on or near the mediodorsal bodies (Simpson *et al*., 1964) suggest a functional relationship between the cerebral ganglia and the mediodorsal bodies. Joose (1964) considers the mediodorsal bodies of *Lymnaea stagnalis* to be endocrine glands and not neurohemal organs, but their exact nature needs to be clarified.

In addition to the neurosecretory cells in the major ganglia of mollusks, cells which appear to be neurosecretory have been found in the cephalic tentacles of gastropod mollusks where they form the "tentacular neuroglandular complex" and are thought to be the source of a tentacular hormone (Pelluet and Lane, 1961). In the slug, *Arion*, these cells, named collar cells, surround the optic and tentacular nerves and the blood vessels in the cephalic tentacles. Lane (1964) has investigated the ultrastructure of the cells in the cephalic tentacles of the common terrestrial snail, *Helix aspersa*, and found that the collar cells contained electron-dense granules which appeared to be neurosecretory granules whose size was within the range for such granules.

Although various structures in cephalopods (for example, the epistellar bodies which are probably rudimentary photoreceptors) have at one time or another been considered as possible neuroendocrine glands, the only one for which there is satisfactory evidence that neurosecretory cells are actually involved is the neurosecretory system of the vena cava which was first found with use of the light microscope in an octopus, *Eledone cirrosa*, by Alexandrowicz (1964). He later (Alexandrowicz, 1965) found comparable structures in the cuttlefish, *Sepia officinalis*, and in *Octopus vulgaris*. The cell bodies of the cells which he presumed to be neurosecretory on the basis of their staining behavior with paraldehyde-fuchsin, which is commonly used to stain neurosecretory cells, occur in the outer layer of the visceral lobe of the brain and in the ganglionic trunks continuous with this layer near the roots of the nerves to the vena cava. The neurosecretory axons terminate in the wall of the vena cava. Berry and Cottrell (1970) examined the vena cava of *Eledone cirrosa* with the electron microscope and confirmed Alexandrowicz's assumption and observations. They found neurosecretory terminals which were filled with electron-dense granules similar to those seen in neurosecretory tissues of other animals.

The optic glands of cephalopods are endocrine glands whose ultrastructure suggests they are nervous in origin but nonneurosecretory (Björkman, 1963). They are found on the optic stalks in all cephalopods which have been examined except *Nautilus*.

The branchial glands which are closely associated with the gills in cephalopods may also be epithelial endocrine glands (Taki, 1964). These are large glands which cannot be removed without death occurring within about 2 days. Replacement therapy has very little effect; injected animals do not survive longer than do the controls. Therefore, the endocrine status of these glands is still doubtful. The recent observations of Schipp *et al.* (1973) suggest a hemocyanin–synthesizing function for the branchial glands. Sereni (1930) has suggested that color changes in cephalopods may be regulated in part by a hormone from the posterior salivary gland, but this view has been challenged by Taki (1964).

D. CRUSTACEANS

There has been very little investigation of the endocrine systems of the smaller species of crustaceans, the entomostracans. On the other hand, the larger species, the malacostracans, have received considerable attention, particularly the order Decapoda which contains the more familiar crustaceans such as the crabs, shrimps, and lobsters. Crustaceans have an endocrine system that rivals that of the vertebrates in complexity and the number of different functions that it regulates. The hormones are the products both of neurosecretory cells and epithelial endocrine glands. Three well-developed neurohemal organs occur in crustaceans: the sinus glands, postcommissural organs, and pericardial organs.

1. *Sinus Glands*

These glands, discovered by Hanström (1931, 1933), are composed primarily of axonal terminals of neurosecretory cells. In freshly dissected tissues the sinus gland appears as a bluish-white, opalescent structure. In the vast majority of crustaceans having eyestalks the sinus glands are located in the eyestalks, whereas in a few stalk-eyed species and in those without eyestalks the sinus glands lie close to the brain. The structure of the sinus gland can vary from simply a thickened disk of the epineurium enveloping the nervous elements in the eyestalk in the Mysidacea to a cupshaped structure in the eyestalks of natantians (e.g., prawns) and a branched structure in the eyestalks of astacurans (e.g., freshwater crayfishes) lining a blood sinus. The origins of the aponal terminals present in sinus glands that lie in the eyestalks have been well investigated. These axonal terminals are derived principally from a large cluster of neurosecretory cells in a portion of one of the optic ganglia, the medulla terminalis (Bliss, 1951; Enami, 1951; Passano, 1951; Bliss and Welsh, 1952). These particular cells are known as the medulla terminalis X-organ. In addition, these sinus glands receive axons from neurosecretory cells in other areas of the optic ganglia in the eyestalks, from the brain, and probably from the ventral ganglia. Sinus glands also contain some glial cells whose function is not clear. These cells do not produce the elementary neurosecretory granules, but they may influence the release process (Hodge and Chapman, 1958), bring about *in situ* changes in the neurosecretory products (Gabe, 1966), or release a secretory product of their own (Adiyodi, 1969).

Electron microscopic examination of the sinus glands in different crustaceans, *Gecarcinus lateralis* (Hodge and Chapman, 1958; Weitzman, 1969), *Cambarellus shufeldti* (Fingerman and Aoto, 1959), *Procambarus clarki* (Bunt and Ashby, 1967), *Orconectes nais* (Shivers, 1967, 1969); *Carcinus maenas* (Meusy, 1968), and *Callinectes sapidus* (Andrews *et al*., 1971) revealed the presence of 2–5 different types of neurosecretory granules (Fig. 1). The granules differ from each other in electron density, dimensions, shape, and response to stains used in electron microscopy. Each type is confined to different axonal terminals. The differences in the number of different granule types observed by the various investigators may represent species differences, different neurohormones, stages in the release of the contained hormone, or the effects of different methods of fixation and staining.

2. *Postcommissural Organs*

The postcommissural organs are located in the body proper, in the region of the esophagus. They have so far been found only in decapod and stomatopod crustaceans (Knowles, 1953; Maynard, 1961; Fingerman, 1966). Postcommissural organs consist of the terminations of axons from the postcommissural nerves. These nerves run from the tritocerebral commissure which itself runs just posterior to the esophagus from one circumesophageal connective to the other. In prawns and shrimps the postcommissural organs which lie adjacent to a blood sinus consist of paired swellings (lamellae), sometimes joined by a horizontal lamella, in which the axonal terminals form fine branches. The cells bodies belonging to at least some of the axons in the postcommissural organs appear to lie in the brain. Although some of the cells in the tritocerebral commissure itself appear to be secretory, where their axons terminate has not been determined. In crabs each postcommissural organ consists of a network of terminations of one of the branches from each of the paired postcommissural nerves and ending near a dorsoventral blood sinus.

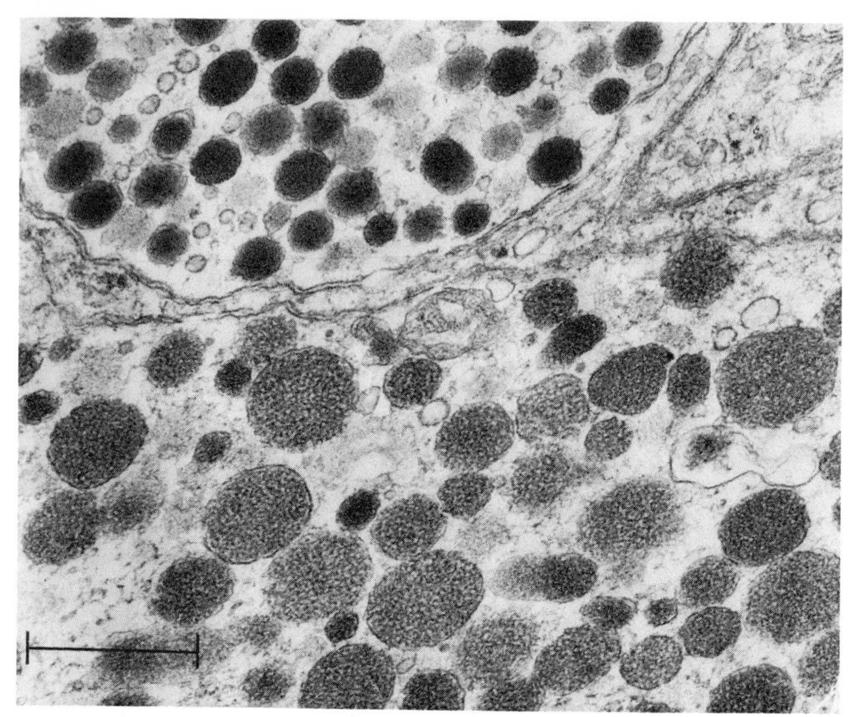

Fig. 1. Two types of neurosecretory granules in the sinus gland of the blue crab, *Callinectes sapidus*. (The scale marker represents 0.5 micron.) (From Andrews *et al.*, 1971.)

3. *Pericardial Organs*

The pericardial organs, discovered by Alexandrowicz (1952), consist of branching networks of nerve trunks and associated axonal terminals, with some connective tissue, which are suspended in the pericardial cavity. Among crabs these axons originate in the thoracic gonglia and quite likely from intrinsic neurosecretory cells also (Maynard, 1961). In an electron microscopic study of the pericardial organs of the mantis shrimp, *Squilla mantis*, Knowles (1962) found two types of neurosecretory fibers which differed in the size (1200 vs 1500 Å in diameter) and structure of the granules they contained and in the relationship of their terminals to the surface of the organs. Unlike the large granules, the smaller ones had a clear space between their contents and bounding membrane. The fibers containing the smaller granules terminated at the surface of the pericardial organs, whereas the fibers with the larger granules always were in contact with a basement membrane that separated them from direct contact with the blood. Maynard and Maynard (1962), also using an electron microscope, examined the pericardial organs of three species of crabs and found three types of terminals in the pericardial organs of *Carcinus maenas*, two containing neurosecretory granules, one with 1400 Å and the other with 1700 Å granules, and a third type which contains 300–500 Å vesicles. On the other hand, *Cancer irroratus* had only two types of terminals, one containing 300–500 Å vesicles and the other 1500 Å granules, whereas in *Libinia emarginata* the terminals had either only one kind of granule, 900 Å, or two kinds of granules, 600 Å and 1500 Å, within a single terminal.

4. *Y-Organs*

According to Gabe (1953) Y-organs are characteristic of higher crustaceans. He correctly hypothesized that these glands, which in crabs (brachyurans) are located near the excretory organs, are involved in their molting. The Y-organs are epithelial endocrine structures. Knowles (1965) published some preliminary observations of the ultrastructure of the Y-organ which indicated that electron-dense inclusions pass out of the nucleus into the surrounding cytoplasm. By administration of the molting hormone, 20-hydroxyecdysone (ecdysterone, crustecdysone, β-ecdysone), Miyawaki and Taketomi (1970) induced the formation in Y-organs of numerous granules (500–700 Å) rich in ribonucleic acid. Couch (1971) reported that the ultrastructure of the Y-organs parallels that of insect prothoracic glands in many respects. Changes in the mitochondrial structure, such as their becoming less dense and assuming complex shapes, were observed as *Procambarus simulans*, a freshwater crayfish, prepared for shedding its old exoskeleton. These mitochondrial changes may reflect increased rates of activity of enzymes necessary for synthesis of molting hormone(s) or a prohormone. However, Sochasky *et al.* (1972) found no obvious structure in the lobster, *Homarus americanus*, or the crayfishes, *Orconectes virilis* and *Orconectes propinquus*, comparable to the Y-organs of brachyurans and suggested that investigators who had been working with what they supposed were the Y-organs of macrurans may in fact have been studying the mandibular organs instead. The latter are probably not molting glands. Even more recently, Carlisle and Connick (1973) provided evidence in support of their conclusion that a gland in the antennary segment of *Orconectes propinquus* is the chief or only site of 20-hydroxyecdysone synthesis in this crayfish.

5. *Androgenic Glands*

An androgenic gland, first described by Charniaux-Cotton (1954) in the amphipod, *Orchestia gammarella*, is usually attached to the ejaculatory portion of each sperm duct, but sometimes lies closer to the testis. These glands are also epithelial. As will be seen below, they control the differentiation of the testes and the male secondary sexual characteristics, Although the chemical nature of the hormone from the androgenic glands has not yet been determined, the ultrastructure of this gland is more suggestive of tissue that secretes proteinaceous compounds rather than steroids King, 1964; Meusy, 1965). The morphology of the androgenic glands of the crayfish, *Orconectes nais*, at different stages of the reproductive cycle has been described by Carpenter and deRoos (1970).

6. *Ovaries*

The ovaries of crustaceans are the source of a substance that controls differentiation of secondary sexual characteristics in females. The chemical nature of the active material is not known nor have the cells that produce it been identified. In crustaceans the testes are not known to be a source of any endocrine substance.

E. FISHES

1. *Pituitary Gland*

The anatomy of the pituitary gland in fishes has been reviewed in great detail a number of times (e.g., Pickford and Atz, 1957; Wingstrand, 1966; Ball and Baker, 1969). The vertebrate pituitary consists of two portions, the adenohypophysis and the neurohypophysis (Fig. 2). The former develops

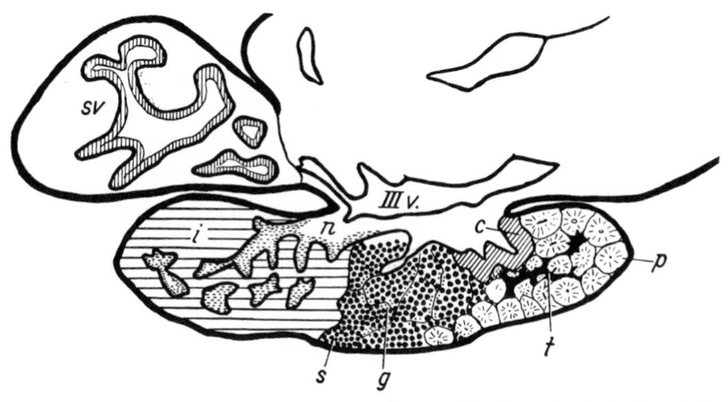

Fig. 2. The pituitary gland of the eel, *Anguilla anguilla*. Anterior is to the right: c, adrencorticotropic hormone cells; g, gonadotrops; i, intermediate lobe; n, neurohypophysis; p, follicles of prolactin cells; s, growth hormone cells (somatotrops); sv, saccus vasculosus; t, thyroid-stimulating hormone cells; IIIv, third ventricle. The saccus vasculosus is a thin-walled, folded sac growing out from the third ventricle and has no functional connection with the pituitary gland. (From Olivereau, 1967.)

from an ectodermal upgrowth (Rathke's pouch) from the roof of the embryonic buccal cavity. The neurohypophysis, on the other hand, develops from an outgrowth from the floor of the diencephalon. The adenohypophysis is the site of synthesis and release of most of the hormones from the pituitary gland. In fishes it consists of the pars intermedia and the pars distalis. The neurohypophysis consists of axons and axonal terminals of neurosecretory cells originating in the hypothalamus plus some intrinsic ependymal and glial cells collectively called "pituicytes." The hormones of the neurohypophysis are consequently synthesized in the hypothalamus and transported to the neurohypophysis for subsequent storage and release. It is a neurohemal organ. The pars intermedia is the posterior portion of the adenohypophysis. In fishes it typically receives long fingerlike projections from the neurohypophysis, whereas the pars distalis receives much smaller ones. The pars intermedia is most likely the source of intermedin (MSH, melanocyte-stimulating hormone) as it is in tetrapods, whereas the pars distalis synthesizes the other adenohypophysial hormones.

The adenohypophysis of teleosts receives two types (Type A and Type B) of neurosecretory fibers from the hypothalamus (Knowles, 1965; Knowles and Vollrath, 1966a,b). Type A predominates in the pars intermedia and Type B in the pars distalis. Type A fibers contain secretion granules more than 1000 Å in diameter, Type B less than 1000 Å (Knowles, 1965). Knowles suggested that the former contain peptide neurosecretion, the latter nonpeptide neurosecretion. These fibers may control the synthesis and release of hormones from the teleost adenohypophysis.

In addition to intermedin, the teleost adenohypophysis has been found to secrete prolactin, growth hormone, gonadotropins, thyroid-stimulating hormone, and adrenocorticotropic hormone. The particular cells in the teleost adenohypophysis which produce each of these hormones have been identified by varous experimental techniques and by the use of appropriate stains at the light microscope level (Olivereau, 1967; Ball and Baker, 1969).

2. *Thyroid Gland*

The basic unit of the thyroid gland in all vertebrates is the follicle which is a single-layered, fluid-filled sphere of cells. In cyclostomes and almost all teleosts the follicles instead of forming a single gland as in the rest of the fishes are scattered in connective tissue near the pharynx (Bern and Nandi, 1964; Gorbman, 1969), and, in addition, some teleosts have follicles that have migrated from the pharyngeal region into the kidneys and to other places such as the eye and heart (Baker-Cohen, 1959). The usually unencapsulated nature of the thyroid in teleosts may account for this dispersal. The thyroid hormones of fishes are the same as in higher vertebrates, thyroxine and triiodothyronine.

3. *Pancreas*

All fishes have endocrine and exocrine pancreatic tissue. In some fishes the two types of tissue are completely separate, whereas in others they are highly intermingled. Cyclostomes have no discrete pancreas. Furthermore, there is complete separation of their endocrine and exocrine tissues. The endocrine cells (islets of Langerhans) are located at the point where the bile duct enters the intestine and the exocrine cells are embedded in the submucosa of the intestine. In other fishes the pancreas is either a more or less compact structure in which the islets are scattered among the exocrine tissue, or as in some teleosts the pancreatic tissue is scattered throughout the body cavity and in some organs (liver, spleen, and ovary) with a partial separation of the exocrine and endocrine components, or as in some teleosts, the endocrine portion is split off from a compact pancreas forming a Brockmann body (Epple, 1969; Thomas, 1970).

Except for cyclostomes, all fishes have A-cells in their islets of Langerhans which produce glucagon and B-cells which produce insulin. Cyclostomes have B-cells. B-cells contain granules which are soluble in alcohol, whereas the granules in A-cells are insoluble in alcohol. A-cells and glucagon have not been found in cyclostomes although they were looked for (Falkmer and Winbladh, 1964; Falkmer, 1966). In conformity with these earlier reports, Van Noorden *et al.* (1972) recently found insulinlike immunoreactivity in the pancreatic islets of larval cyclostomes, *Lampetra fluviatilis* and *Lampetra ploneri*, but found no cross-reactivity with antiglucagon.

4. *Adrenal Tissues*

The two adrenal tissues are known by this name because of their close association with the kidneys. One component develops from the mesonephric blastema (mesodermal) and is capable of synthesizing steroid hormones; the other develops from neurectoderm and produces epinephrine (adrenaline) and norepinephrine (noradrenaline). Because of the fact that the adrenal glands of mammals consist of the internal medulla surrounded by the cortex, the homologous tissues in other vertebrates are often referred to as "medullary" and "cortical" tissues. However, because the medullary cells stain especially well by such stains as chromic acid and ferric chloride, these cells are often called "chromaffin cells." Also, because the cortical tissue does not envelope the medullary tissue in nonmammalian vertebrates, cortical tissue is often referred to as "interrenal tissue."

In the cyclostomes clusters of both chromaffin and presumptive interrenal cells are scattered in the walls of the cardinal veins (Gaskell, 1912; Sterba, 1955; Chester Jones and Bellamy, 1964). Clear proof that the latter cells are a source of hormone is still to be obtained (Chester Jones *et al.*, 1969). In elasmobranchs the chromaffin and interrenal cells are completely separated from each other, unlike the usual pattern in the other jawed vertebrates. The chromaffin tissue is segmentally distributed along the medial borders of the kidneys, whereas the interrenal tissue is located in one or more compact masses on or between the posterior lobes of the kidneys (Dittus, 1940). The chromaffin tissue of teleosts is confined to the anterior ends of the kidneys typically forming a sheath of cells around the posterior cardinal veins and their branches. The chromaffin cells usually intermingle with the interrenal cells although they are sometimes separate as in the rock sole, *Lepidopsetta bilineata* (Nandi, 1962; Chavin, 1966).

5. *Ultimobranchial Glands*

Among fishes, ultimobranchial glands are found only in species having jaws. These glands develop from the last pair of gill pouches. In the elasmobranch, *Squalus acanthias*, the gland occurs on the left side only between the pericardium and ventral surface of the pharynx just anterior to its junction with the esophagus (Camp, 1917). In teleosts the gland is either single along the midline or bilateral and located in the transverse septum between the abdominal cavity and sinus venosus just ventral to the esophagus (Krawarik, 1936). The gland is follicular in elasmobranchs but in teleosts it consists of cords of polygonal cells (Copp, 1969). As will be seen below, the substance (calcitonin) produced by the ultimobranchial glands may be involved in calcium regulation in some, but not all, jawed fishes. The parathyroid gland, another endocrine gland involved in calcium regulation in tetrapods, appears to be lacking in fishes.

6. *The Caudal Neurosecretory System and Urophysis*

The caudal neurosecretory system is characteristic of elasmobranchs and teleosts (Bern, 1969). It consists of neurosecretory cell bodies in the posterior spinal cord. Dahlgren (1914) first described these large cells but he supposed they were motor neurons involved in the control of electric organs. However, in 1919 Speidel came to the proper conclusion regarding the nature of these cells and correctly considered them to be endocrine cells. Speidel's paper has historical significance because it was the first description of neurosecretion in any animal. Enami (1959) showed that the neurosecretory axons of this system in teleosts terminated in the urophysis, a neurohemal organ, which is associated with a capillary network and located on the ventrolateral surface of the cord in the last vertebral element or urostyle. In contrast, elasmobranchs lack a urophysis, but there is a neurohemal region on the ventral surface of the caudal spinal cord; where the neurosecretory axons terminate there is a capillary plexus (Fridberg, 1962).

7. *Gonads*

Fish ovaries and testes are the sources of steroid gonadal hormones. Gonadectomy results in the failure of the development of secondary sexual characteristics in both sexes. Some investigators have concluded that the interstitial cells of Leydig are the source of the testicular hormone, but others have been unable to find such cells in other species of fishes (Hoar, 1957a). In the latter fishes it is conceivable that either the Sertoli cells which are ordinarily considered to be supportive, nutritive cells or the spermatogenic epithelium itself is the source of the hormone.

The ovaries contain numerous ovarian follicles which develop from the germinal epithelium (Hoar, 1957b). Estrogen is found in fishes, possibly produced by the granulosa, one of the layers of follicular cells surrounding the developing egg. Some of the follicles do not produce mature eggs but instead form a glandular structure called the "preovulatory corpus luteum." But most corpora lutea develop from the follicular cells of ruptured follicles after ovulation (postovulatory corpora lutea). However, it is not certain that corpora lutea in fishes are the source of a hormone which they are in mammals, although progesterone occurs in some fishes.

8. *Pineal Organ*

The pineal organ lies against the roof of the skull or projects through a foramen to occupy a subdermal position. It develops as a dorsal evagination of the roof of the diencephalon (Hoffman, 1970). There is some evidence that this organ is an endocrine gland in fishes but relatively few investigations of the function of this gland have been carried out with these organisms. It has been suggested that this organ functions by antagonizing the release of hormones from the pituitary gland in fishes (Grunewald-Lowenstein, 1956; Pflugfelder, 1956). It also appears to be involved in the control of their color changes (Eddy and Strahan, 1968). There is some evidence, as will be presented below, that the active substance of the pineal organ in fishes is melatonin. Investigators have two views of what the hormone(s) may be in mammalian pineal glands: (a) melatonin or related indoleamine derivatives and (b) polypeptide(s) or protein(s) (Quay, 1970).

9. *Corpuscles of Stannius*

The corpuscles of Stannius in teleosts lie on the peritoneal surface of the kidney. They develop from the posterior region of the opisthonephros and appear to be endocrine glands involved in serum electrolyte regulation.

III. Functions of the Hormones

A. Salt and Water Balance

Hormones regulating salt and water movements have been found in four phyla, Annelida, Mollusca, Arthropoda, and Chordata. However, the investigation with an annelid was done with an earthworm (Kamemoto, 1964). Only with the latter three phyla is there evidence for hormonal control in marine species.

1. *Mollusks*

Lever *et al.* (1961) obtained evidence for a diuretic hormone in the right pleural ganglion of the freshwater snail, *Lymnaea stagnalis*. Removal of this ganglion results in swelling because of the accumulation of water. More recently Vicente (1969) obtained evidence for an antidiuretic substance in the marine gastropod, *Aplysia rosea*. Removal of various ganglia, resulted in a loss of body weight, presumably due to water loss. The greatest loss of weight occurred when the pleural ganglia were removed. Implantation of any of the ganglia slowed this loss. Vicente concluded that the pleural ganglia were the major source of an antidiuretic hormone which normally prevents water loss and maintains the normal osmotic balance of the blood. It is presumably a product of neurosecretory cells in the ganglia.

2. *Crustaceans*

There is evidence for diuretic and antidiuretic hormones in the land crab, *Gecarcinus lateralis*, (Bliss *et al.*, 1966). The antidiuretic hormone which promotes retention of water during the premolt period when there is active preparation for shedding the old exoskeleton may be one of the ecdysones or a different molecule secreted in response to or in conjunction with one of the ecdysones. The diuretic hormone, on the other hand, is produced by the central nervous system and causes output of the accumulated water. In contrast to Bliss and her associates, Kamemoto *et al.* (1966) and Kato and Kamemoto (1969) have directed their attention to the hormonal control of salt and water balance during the intermolt period, a time when there is no active preparation for the next shedding of the exoskeleton. Using the grapsid crab, *Metopograpsus messor*, which is plentiful along the coast of Oahu, Hawaii, they found that eyestalk removal or ligation caused a rapid decrease in the blood osmotic concentration when crabs were placed in 25% seawater, but an increase when placed in 110% seawater. This crab regulates both hyposmotically and hyperosmotically in seawater concentrations ranging from 25 to 125%. The drop observed when the eyestalkless crabs were placed in 25% seawater could be prevented in part by injection of eyestalk extract.

Injection of extracts of thoracic ganglia or implantation of thoracic ganglia caused a decrease in the osmotic concentration of the blood in animals exposed to a hyposmotic medium but caused an increase in crabs in a hyperosmotic medium. The factor in the thoracic ganglia appears then to increase the permeability of the body surfaces to water, whereas that in the eyestalk decreases it. The eyestalk factor may act directly on the body surfaces causing a decrease in their permeability or indirectly by either inhibiting the production and/or release of the thoracic ganglion factor or by antagonizing the action of the latter factor. Ramamurthi and Scheer (1967) showed that extracts of the cephalothorax of the prawn, *Pandalus jordani*, decrease the outflux of sodium ions from the shore crab, *Hemigrapsus nudus*. Mantel (1968) showed that extracts of the thoracic ganglionic mass of *Gecarcinus lateralis* increased the salt and water permeability of the foregut. Tullis and Kamemoto (1971) found two substances in the brain and thoracic ganglia of a euryhaline crab, *Thalamita crenata*, that affected water movement when assayed on a freshwater crayfish, *Procambarus clarkii*. One which is acetone-soluble increased the influx of T_2O, whereas the second which is acetone-insoluble decreased the T_2O influx.

3. *Fishes*

Maintenance of salt and water balance in fishes appears to involve several different endocrine glands and hormones. The pituitary is one of the glands so involved. The role of the neurohypophysis will be treated first. Arginine vasotocin (8-arginine oxytocin) appears to be present in the neurohypophysis of all fishes (Perks, 1969). Teleost fishes have in addition 4 serine, 8-isoleucine oxytocin (ichthyotocin, isotocin). Elasmobranchs, particularly skates, contain, in addition to arginine vasotocin, glumitocin (4-serine, 8-glutamine oxytocin). The data of Sawyer *et al.* (1970) showed that the pituitary gland of the elasmobranch, *Squalus acanthias*, had to contain at least one new substance resembling oxytocin. More recently Acher *et al.* (1972) found not one but two new active neurohypophysial peptides, 8-valine oxytocin (valitocin) and 4-asparagine oxytocin (aspartocin), in the same species, *Squalus acanthias*. Cyclostomes contain arginine vasotocin, but while some species may contain an as yet unidentified second peptide, other cyclostomes may be the only vertebrates which possess only one neurohypophysial peptide hormone.

The functions of these neurohypophysial hormones in fishes have not been well worked out as yet. The role of the hagfish neurohypophysis in water and electrolyte homeostasis has not been studied sufficiently. However, what has been done revealed that hagfish pituitary extracts cause a rise in blood sodium of fish kept in 70% seawater, but a mammalian preparation (Pituitrin) injected into fish in 165% seawater caused a fall in the blood sodium level

(Chester Jones *et al.*, 1962). Adam (1963) found that another mammalian preparation (Pituifral) caused a rise in total body water in the same species, *Myxine glutinosa*. The hagfish is an osmoadjuster. Perhaps the neurohypophysial hormone(s) act to slow the changes in the salt concentration of the body fluids that occur when there are slight changes in the concentration of its marine environment.

There is no evidence that the neurohypophysis is involved in the control of salt and water balance in elasmobranchs (Heller and Bentley, 1965; Perks, 1969). In teleost fishes it seems that arginine vasotocin is normally involved in the maintenance of salt and water balance, whereas 4-serine, 8-isoleucine oxytocin is involved in reproduction, eliciting egg laying, and mating behavior. Oxytocin increased sodium outflux against the concentration gradient in the case of the flounder, *Platichthys flesus* (Motais and Maetz, 1964). Marine teleosts are hyposmotic to their environment and must conserve water and exclude salt. The neurohypophysial hormones may function to increase the rate of sodium excretion in seawater.

Some euryhaline teleosts, mostly cyprinodonts, are unable to survive for more than a few days in freshwater after hypophysectomy (Ball, 1969a). Pickford and Phillips (1959) found that prolactin injections can keep such fishes alive. The primary osmoregulatory function of prolactin is to reduce sodium loss through the gills. Pickford *et al.* (1970a) later showed that prolactin permits survival of hypophysectomized killifish, *Fundulus heteroclitus*, in freshwater by decreasing the activity of the sodium transport enzyme Na^+, K^+-ATPase in the gills, thereby reducing sodium loss from the gills and increases the activity of this enzyme in the kidney, thereby increasing sodium retention there. Teleost prolactin is not the typical tetrapod prolactin, it lacks some of the biological activities of the latter. For example, teleost prolactin has only a minimal pigeon crop-stimulating action, and is consequently sometimes called paralactin (Ball, 1969b). Olivereau (1971) found a prolactin release-inhibiting factor of hypothalamic origin in the eel, *Anguilla anguilla*.

The interrenal tissue is the source of steroid hormones which are also involved in the process of salt and water balance. Corticosterone and cortisol have been found in the blood of agnathans. Elasmobranchs also have these two hormones but their principal interrenal steroid is 1α-hydroxycorticosterone. The principal interrenal hormones found in teleost plasma are corticosterone, cortisol, and cortisone (Chester Jones *et al.*, 1969). The output of steroid hormones by the interrenal tissue is presumably under the control of the hypothalamus and adrenocorticotropic hormone from the pituitary gland as in higher vertebrates. Interrenal hormones appear to act on the gills, kidneys, and intestine of teleosts, the three organs which are the principal sites of salt and water exchange between the environment and their body fluids. Cortisol, for example, is a salt-excreting hormone for fishes living in hyperosmotic media (Maetz, 1969). More recently Pickford *et al.* (1970b) using hypophysectomized *Fundulus heteroclitus* in seawater found that the Na^+, K^+-ATPase activity in the gills increased in cortisol-treated specimens, just the reverse of the effect of prolactin on the gills.

The function of the caudal neurosecretory system and urophysis is not certain at the present time. But an osmoregulatory role is the one most often suggested by investigators and is supported by at least some of the experimental data (Yagi and Bern, 1965; Bern, 1969). Extracts of goldfish urophysis cause a net increase in the sodium absorption across the gills, diuresis, and a reduction in renal sodium excretion (Maetz *et al.*, 1964). These changes result in an increase in the sodium concentration of the fish. Yagi and Bern (1965) showed that changes in sodium concentration altered the spontaneous firing rate of caudal neurosecretory cells in *Tilapia mossambica*. The urophysis may provide a fine control of the body fluids which supplements the control exerted by the other hormones discussed in this section. More recently Lederis (1970a) showed that the teleost urophysis contains a factor that causes contraction of the urinary bladder of the rainbow trout. This substance appears to be a polypeptide and is not any of the following biologically active substances: acetylcholine, serotonin, histamine, epinephrine, norepinephrine, or neurohypophysial peptides (Lederis, 1970b). Release of this teleost bladder-contracting factor can be obtained by *in vitro* depolarization of urophyses from the goby, *Gillichthys mirabilis*, by use of a high concentration of potassium ions (Berlind, 1972a). This release in response to depolarization is dependent upon the presence of calcium ions in the medium.

Calcitonin, the hormone produced by cells of ultimobranchial origin, is capable of causing a rapid reduction in the level of blood calcium and phosphate in mammals, but the role of calcitonin in fishes is not well defined (Copp, 1969). Calcitonin from salmon is a peptide consisting of 32 amino acids and has a molecular weight of 3427 (O'Dor *et al.*, 1969). Pang and Pickford (1967) and Pang (1971a) found no change in plasma calcium when hog, salmon, and codfish calcitonins were injected into the teleost, *Fundulus heteroclitus*, but Louw *et al.* (1967) obtained a decrease in plasma calcium and phosphate after porcine calcitonin was injected into the catfish, *Ictalurus melas*. The lack of effect in *Fundulus* may have been due to the fact that its bone, unlike that of *Ictalurus*, is acellular and hence not capable of being normally resorbed.

The corpuscles of Stannius appear to be involved in the regulation of serum electrolytes in fishes. Fontaine (1964) showed that stanniectomy results in hypercalcemia, hyperkalemia, and hyponatremia in the eel, *Anguilla anguilla*. Pang (1971b) made similar observations with the killifish, *Fundulus heteroclitus*, and found that stanniectomy resulted in increases in serum calcium, chloride, and potassium and decreases in sodium and inorganic phosphate in the killifish. Pang *et al.* (1971) reported a pituitary hypercalcemic mechanism essential for the maintenance of normal calcium levels in a hypocalcemic environment. This pituitary factor is independent of those regulating sodium, potassium, and chloride.

B. PIGMENTARY EFFECTORS

1. *Chromatophores*

Chromatophores are cells or organs found mainly in the integument that effect color changes of their bearer by dispersion and concentration of their pigment. The subject has been reviewed recently by Fingerman (1970a). The functions attributed to the color change system are protective and aggressive coloration, thermoregulation, protection of body tissue from deleterious illumination, and mating color displays. The reflex pathways leading from the eyes which regulate color changes ultimately involve hormones in all crustaceans and, with few possible exceptions, in all fishes. There is also some evidence for endocrine control of color changes in cephalopod mollusks, but the recent research on the latter group has emphasized the nervous rather than the endocrine control of cephalopod chromatophores. Cephalopod chromatophores are organs consisting essentially of a pigment-containing cell surrounded by muscle cells which, by their contraction, enlarge the former thereby making its pigment more conspicuous. In contrast, chromatophores of crustaceans and fishes have a fixed cell outline, the pigment merely flowing in and out of fixed branches.

a. Cephalopod Mollusks. Color changes of cephalopods are primarily due to nervous stimulation of the muscle cells of the chromatophores. Sereni (1930) obtained evidence that showed there is a slower endocrine control superimposed on the nervous control which causes rapid color changes. Blood transfused from a darker species (*Octopus macropus*) into a light species (*Octopus vulgaris*) caused darkening of the latter. Tyramine appeared to be responsible for the darkening. It is more concentrated in the blood of the darker species than in the lighter one. Furthermore, injections of tyramine caused darkening. The posterior salivary gland is the chief source of tyramine in cephalopods. Tyramine appears to act both directly on the chromatophores and by stimulating the motor centers in the central nervous system. However, the entire matter of endocrine control of color changes in cephalopods needs to be investigated further. For example, Sereni reported that removal of the posterior salivary gland resulted in his experimental animals becoming paler, whereas Taki (1964) found that removal of this gland had no effect on cephalopod chromatophores.

b. Crustaceans. Crustaceans exhibit some of the most spectacular color changes in the animal kingdom. The fiddler crab, *Uca pugilator*, whose chromatophores have been the subject of many investigations, has, for example, white, red, yellow, and black chromatophores. The sinus glands and postcommissural organs have been shown to be release sites for the hormones that mediate the color changes.

Crustaceans have been divided into three groups on the basis of the effects of eyestalk removal upon their general body coloration (Brown, 1952). One group, exemplified by the prawn, *Palaemonetes vulgaris*, darkens primarily because its red pigment disperses after eyestalk ablation. Also included in this group are *Penaeus*, *Hippolyte*, *Palaemon*, and the freshwater crayfishes. A second group is exemplified by the genus *Crangon* alone. Just after eyestalk removal there is a transitory darkening of the telson and uropods and a blanching of the rest of the body. Then, 30–60 min later the telson and uropods blanch but the rest of the body assumes an intermediate and mottled coloration. The third group consisting of all the true crabs (Brachyura) is represented by the fiddler crab, *Uca pugilator*. It becomes pale after eyestalk removal, primarily because of the concentration of its black pigment. In about a dozen species of crustaceans there is now evidence for pigment-dispersing and pigment-concentrating hormones (see the review of Fingerman, 1970a, for details). Most recently, by the aid of the technique of gel chromatography, Skorkowski (1971) and Fingerman and Fingerman (1972) showed the presence of these hormones in the eyestalks of the shrimps *Crangon crangon* and *Crangon septemspinosa*, respectively. Fernlund and Josefsson (1972) have shown that the structure of the red pigment-concentrating hormone in the eyestalks of the shrimp, *Pandalus borealis*, is pGlu-Leu-Asn-Phe-Ser-Pro-Gly-Trp-NH_2.

Adenosine 3′, 5′-monophosphate (cyclic AMP) has been proposed as the intracellular mediator of the action of several vertebrate peptide hormones. Fingerman *et al.* (1968) found that this substance evoked pigment dispersion in the red chromatophores of *Palaemonetes vulgaris*. Cyclic AMP may serve as a "second messenger" for the red pigment-dispersing hormone (the "first messenger"). Freeman *et al.* (1968) using *Palaemonetes vulgaris* found that the red pigment-concentrating hormone caused hyperpolarization of the membrane potential of the red chromatophore as the pigment was concentrating. Fingerman (1969) postulated that the primary action of this red pigment-concentrating hormone is stimulation of a pump in the chromatophore cell membrane which exchanges sodium ions from inside the chromatophore for potassium ions from the outside.

c. Fishes. Fishes exhibit a wide diversity of mechanisms of control of their chromatophores. In some species there is endocrine control alone, in very few others there appears to be under normal conditions nervous control alone by direct innervation of the chromatophores, while in still others there is a combination of both nervous and endocrine controls. Melanophores (containing blackish or brownish pigment) are the principal chromatophores involved in the color changes of fishes. Other commonly found pigments are yellow, red, and white in color. Most of the experiments which have been done on fish chromatophores involved the melanophores alone. The specific subject of fish chromatophores has been reviewed recently by Fujii (1969).

In agnathans the chromatophores appear to be controlled by hormones alone. Hypophysectomy with its consequent removal of the intermediate lobe of the pituitary gland and the intermedin (a melanin-dispersing hormone) it contains results in blanching (Young, 1935; Eddy and Strahan, 1968). Pituitary extracts evoke a darkening reaction. In addition, the pineal complex of the ammocoete larva of the lamprey, *Geotria australis*, appears to be the source of a melanin-concentrating hormone which may be melatonin (Eddy and Strahan, 1968). Implantation of the pineal complex produced localized pallor, injections of melatonin induced blanching, and after pinealectomy the animals remained permanently dark. However, the same investigators got negative results when similar experiments were performed with the ammocoete larva of another lamprey, *Mordacia mordax*, and concluded that the pineal complex of the latter species is not involved in its pigmentary effector system. The chromatophores of some elasmobranchs appear not to be innervated while in others there is evidence for a pigment-concentrating nerve fiber. It is generally agreed that darkening of elasmobranchs is due to intermedin. But there is no agreement on the mechanism of blanching. Lundstrom and Bard (1932) suggested that in *Mustelus canis* blanching was simply due to disappearance of intermedin from the blood; Parker and Porter (1934) and Parker (1935) concluded that their experiments showed the presence of melanin-concentrating fibers innervating the melanophores in *Mustelus canis*; and Hogben (1936) concluded that the melanophores of *Scyllium canicula*, *Scyllium catulus*, and *Raja brachiura* are not innervated and that blanching is caused by a hormone from the adenohypophysis. Further support needs to be provided for the existence of both the melanin-concentrating fiber and this blanching hormone in elasmobranchs.

It is agreed that the melanophores of all teleost fishes are innervated by a pigment-concentrating fiber. In addition, in some species there is evidence for a pigment-dispersing fiber as well. So far as hormones are concerned, the melanophores of some intact fishes such as *Fundulus heteroclitus* are not affected by injections of intermedin. However, if melanophores of *Fundulus* are first denervated they then will respond to intermedin (Kleinholz, 1935). Hypophysectomy has no apparent effect on the pigmentary responses of *Fundulus*. It appears that the routine background responses of *Fundulus* which occur even with hypophysectomized specimens are mediated by nerves alone, but that the production of extreme darkness as obtained by keeping specimens in black containers for prolonged periods of time would require intermedin as well. The melanophores of other fishes such as the catfish, *Ameiurus nebulosus*, respond well to intermedin by darkening. In some teleosts (the catfish, *Parasilurus asotus*, for example) there is also evidence for a melanin-concentrating hormone in the pituitary, but its existence has not been universally accepted as yet. Olivereau (1972) has obtained evidence for intermedin release-stimulating and release-inhibiting controls exerted by the hypothalamus over the intermediate lobe of the pituitary gland in *Anguilla anguilla*.

Adrenocorticotropic hormone (ACTH), melatonin, epinephrine, norepinephrine, and prolactin have also been implicated in the control of teleost chromatophores. Köhler (1952) reported that ACTH caused red pigment dispersion in *Phoxinus laevis*. At least part of this response to ACTH may be an indirect one due to stimulation of the interrenal organs. Practically all of the teleost fishes which have been injected with epinephrine become pale through melanin concentration (see review by Fingerman, 1963). Norepinephrine and melatonin cause melanin concentration in *Phoxinus phoxinus* (Healey and Ross, 1966). However, Reed (1968) using the freshwater golden pencil fish, *Nannostomus beckfordi anomalus*, which has separate day and night markings, found that melatonin caused concentration of the pigment in some melanophores and dispersion of the melanin in others causing an animal having day coloration to assume the night coloration. Sage (1970) has found that prolactin dispersed the yellow pigment of the mudsucker, *Gillichthys mirabilis*. Hypophysectomy did not affect the melanophores but caused permanent concentration of the yellow pigment. Consequently, Sage concluded that prolactin is normally involved in the control of the yellow chromatophores in this species. Reed and Finnin (1970) have evidence that the melanophores of *Pterophyllum eimekei* are innervated by the sympathetic nervous system and have α-adrenoreceptors which mediate pigment concentration and β-adrenoreceptors which mediate pigment dispersion. They could find no evidence for a role of the parasympathetic nervous system in the control of these melanophores.

2. *Retinal Pigments*

In addition to chromatophores crustaceans have other pigmentary effectors, the retinal pigments, whose function is to regulate the amount of light striking the photosensory portion of the compound eye, the rhabdom. There are three retinal pigments, the distal, proximal, and reflecting. The distal and proximal pigments screen the rhabdom in bright light and uncover it in darkness or dim light, whereas the reflecting pigment reflects light onto the rhabdom. Much more information is available concerning the control of the distal retinal pigment whose position can be observed in intact eyestalks (Sandeen and Brown, 1952; Fingerman, 1970b; Fielder *et al.*, 1971) than for the others. Crustaceans vary with respect to the number of retinal pigments which migrate. In some eyestalks only one or two of the three pigments undergo these movements. There is evidence for endocrine involvement in the control of all three pigments, but the hormones controlling the distal retinal pigment are the best established ones (see review of Kleinholz, 1966). Kleinholz (1936) found a distal retinal pigment light-adapting hormone in the eyestalks of the prawn, *Palaemonetes vulgaris*. This substance causes the distal retinal pigment to migrate proximally toward the base of the eye thereby shielding the rhabdom. Since then evidence has also been obtained for a distal retinal pigment dark-adapting hormone in the eyestalks of this prawn (Brown *et al.*, 1952; Fingerman *et al.*, 1959). In some crustaceans the proximal pigment may not be under hormonal control, but an independent effector instead.

C. CARBOHYDRATE METABOLISM

1. *Crustaceans*

In their eyestalks crustaceans possess a hormone which when released from the sinus gland causes a rise in the blood glucose concentration. This hormone which was originally called "diabetogenic factor" by its discoverers (Abramowitz *et al.*, 1944) is now known as "hyperglycemic hormone" (Kleinholz *et al.*, 1967). Kleinholz *et al.* (1967) and Keller (1968) found that hyperglycemic hormone from both freshwater and marine species is destroyed by heat and is inactivated by proteolytic enzymes. Keller (1968) estimated the molecular weight of the hormone from the freshwater crayfish, *Orconectes limosus*, to be 9000–11,000. He (Keller, 1968, 1969) also found that there is some species specificity in the response to hyperglycemic hormone from different crustaceans which could be due to differences in the chemical (presumably amino acids) composition of the hormone from different sources. For example, he found that the hormone from this crayfish had little or no effect in the crabs, *Cancer pagurus* and *Carcinus maenas*, and in turn eyestalk extracts from these crabs had no significant effect on the blood glucose of this crayfish. On the other hand, eyestalk extracts of this crayfish and another crayfish, *Astacus leptodactylus*, were effective in each other. One mode of action of this hormone may be to activate the enzyme phosphorylase which converts glycogen to glucose 1-phosphate. Keller (1965) found that eyestalk removal resulted in a decrease in the amount of active phosphorylase in homogenates of the abdominal muscle of a crayfish but there was a rise in glycogen content. Injection of eyestalk extract into eyestalkless crayfish activated nearly all the phosphorylase and reduced the glycogen content of the abdominal muscle.

Another enzyme involved in carbohydrate metabolism, uridinediphosphate glucose–glycogen transglucosylase (UDPG–GT) which catalyzes glycogen synthesis, is also affected by eyestalk extracts. Wang and Scheer (1963) found that this enzyme which occurs in crab muscle (*Cancer magister* and *Hemigrapsus nudus*) is inhibited by a factor present in eyestalk extracts. Ramamurthi *et al.* (1968) using a partially purified preparation of this inhibitor found no significant increase in the blood glucose titer and no effect on the phosphorylase activity in muscles of eyestalkless *Hemigrapsus nudus*. They concluded that the inhibitor of UDPG–GT is not the substance in eyestalk extracts which activates phosphorylase, but suggested as had Keller (1965) that the activator could be the same as the hyperglycemic hormone. This partially purified UDPG–GT inhibitor of Ramamurthi *et al.* (1968) seems to be a smaller molecule than the hyperglycemic hormone, but when Wang and Scheer (1963) first reported its existence the properties they ascribed to the crude preparation (nondialyzability and partial inactivation by heat) were suggestive of a close relationship to the hyperglycemic hormone which is also nondialyzable and heat labile in aqueous solution.

The hepatopancreas of the lobster, *Homarus americanus*, and freshwater crayfish, *Orconectes virilis*, is capable of metabolizing glucose both by the Embden-Meyerhof pathway and by the pentose shunt. McWhinnie and Chua (1964) using this crayfish showed that the relative use of these two means of carbohydrate metabolism in the hepatopancreas varies with the stage of the molting cycle. The level of carbohydrate metabolism through the shunt is 25–30% higher in intermolt specimens than in premolt individuals. Eyestalk extracts from intermolt crayfish caused an increase in carbohydrate utilization through the shunt, whereas premolt eyestalk extracts decreased it. These eyestalk extracts would appear to have an effect on the enzyme glucose 6-phosphate dehydrogenase. Activation or synthesis of it would direct carbohydrate metabolism into the shunt; inhibition of its activation or synthesis would lead to an increase in metabolism via the Embden-Meyerhof pathway. Two hormones were suggested by McWhinnie and Chua, one secreted during the intermolt period which stimulates synthesis or activation of this enzyme thereby directing glucose 6-phosphate into the pentose shunt and another hormone released at premolt which acts to direct glucose 6-phosphate into the Embden-Meyerhof pathway.

2. *Fishes*

The physiological roles of the two pancreatic hormones, insulin and glucagon, are well established in mammals but are not very clear in fishes. Most investigators found that insulin caused a decrease in the blood glucose level of fishes but in one species no response was observed and in another a slight hyperglycemia (Epple, 1969). Insulin may either cause an increase or decrease of liver glycogen, but muscle glycogen is either unaffected or increases. The effects of glucagon administration which in mammals raises the blood glucose level are very variable in fishes, producing decreases, increases, or no effect at all. In fact, the poor effects that hormones which primarily affect carbohydrate metabolism in mammals have on the dogfish, *Squalus acanthias*, and the ratfish, *Hydrolagus colliei*, and the abundance of lipid in contrast to the small glycogen deposits in the livers of these fishes led Patent (1970) to suggest that these hormones may be concerned instead with the control of lipid metabolism.

Interrenal hormones have also been implicated in carbohydrate metabolism in fishes. In general, steroids of the cortisol type promote gluconeogenesis in fishes as in other vertebrates (Chester Jones *et al*., 1969).

Epinephrine appears to induce hyperglycemia in all fishes. Grant and Hendler (1965) found that norepinephrine is twice as potent as epinephrine in causing hyperglycemia in the skate, *Raja erinacea*. In contrast, Patent (1970) reported that norepinephrine had no effect in the dogfish but caused hypoglycemia in the ratfish.

D. HEART RATE

1. *Mollusks*

Cardio-excitor substances have been found in ganglia of mollusks. One, serotonin (5-hydroxytryptamine) which has been found in gastropods, bivalves, and cephalopods appears normally to exert its role by acting as a neurotransmitter substance (Welsh, 1953). A second factor (Substance X) has been found in gastropod (Kerkut and Laverack, 1960) and bivalve ganglia (Cottrell and Maser, 1967). In ganglia of the clam, *Mercenaria mercenaria*, it appears to be associated with granules 1000–3000 Å in diameter (Cottrell and Maser, 1967). Frontali *et al*. (1967) showed that Substance X of *Mercenaria mercenaria* is actually a mixture of cardio-acceleratory substances. The pooled cerebral, pedal, and visceral ganglia of this clam were fractionated and four active substances were obtained. Three of the four were inactivated by proteolytic enzymes. Berry and Cottrell (1970) found that extracts of the vena cava of the cephalopod, *Eledone cirrosa*, have a prolonged, strong cardio-acceleratory effect normally accompanied by an increase in amplitude of the beat. It will be recalled that the vena cava contains terminals of neurosecretory axons. The greatest activity was obtained from the portions of the blood vessel richest in nerves. This vena cava cardio-acceleratory factor is not epinephrine, norepinephrine, or serotonin. Berry and Cottrell suggested that it might, however, be the Substance X previously found in ganglia of gastropod and bivalve mollusks.

2. *Crustaceans*

Extracts of the pericardial organs were found to have a striking effect on the beating of isolated hearts (Alexandrowicz and Carlisle, 1953). The amplitude of the beat increased in all the species tested, but whereas the frequency of the beat also increased in *Cancer pagurus*, *Homarus vulgaris*, and *Squilla mantis*, the frequency decreased in *Maia squinado*. The heart of decapod crustaceans is also innervated by excitatory and inhibitory neurons. The effects of these pericardial organs tend to be long lasting, whereas those of the excitatory and inhibitory neurons are more transitory. The activity of pericardial organ extracts has been ascribed both to peptides (Maynard and Welsh, 1959; Belamarich, 1963) and indolealkylamines (Carlisle, 1956). Maynard and Welsh (1959) found that pericardial organs contain the indolealkylamine serotonin but they concluded there was not enough present to account for the action of the extracts. The effect of serotonin on the heart is very similar to that of pericardial organ extracts (Cooke, 1966). Belamarich isolated two active peptides having molecular weights of approximately 900–1400 daltons. More recently, Berlind and Cooke (1970) analyzed the medium in which isolated pericardial organs from the spider crabs, *Libinia emarginata* and *Libinia dubia*, had been stimulated electrically and found only peptide hormone. They concluded that responses to the pericardial organs are normally due to peptides alone. Berlind *et al.* (1970) provided evidence to show that the release of the peptide neurosecretory material from the pericardial organs is controlled by the electrical activity of the neurosecretory cells and is not influenced by the serotonin present in the pericardial organs.

E. GROWTH AND REGENERATION

1. *Coelenterates and Platyhelminths*

The roles of hormones in control of growth and regeneration in these two phyla have only been investigated with freshwater species. For this reason the experimental findings obtained with them will be only briefly discussed, but they are presented for comparison with the mechanisms to be described in higher organisms.

The normal growth region of *Hydra pseudoligactis* is just proximal to the hypostome which is close to the highest concentration of neurosecretory cells. When the hypostome is excised the number of droplets in the neurons at the tentacle bases at first increases and 4–6 h later they decrease and are gone by the end of the sixth hour. It appears then that a neurohormone is released which stimulates regeneration of the missing portion of the *Hydra* (Burnett *et al.*, 1964). Presumably the neurosecretory material is also released when normal growth is about to be initiated. More direct evidence for a growth hormone in *Hydra* was obtained by Lesh and Burnett (1964) who reported that extracts of *Hydra viridis* and *Hydra pirardi* stimulated excessive head growth during the regeneration of annuli of tissue from the gastric budding region of *Hydra pirardi*. The hormone appears to be a peptide, being inactivated by proteolytic enzymes, and is much more concentrated in the distal half than the proximal half of normal animals. Presumably this growth factor is a product of the neurosecretory cells which are also concentrated in the distal end of the body. With respect to the platyhelminths, evidence from studies of planarians reveals that the brain (presumably a neurohormone from it) is essential for eye regeneration from the neoblasts (Stéphan-Dubois and Lender, 1956; Török, 1958).

2. *Nematodes*

Nematodes periodically shed their cuticle as they grow, a process similar to that which occurs in crustaceans. The new cuticle is deposited under the old one. The actual shedding is known as ecdysis. As stated above, *Phocanema decipiens* has its last larval stage in the muscles of cod. In the intestine of a seal it undergoes ecdysis to the adult. Two groups of neurosecretory cells (one in the dorsal ganglion, the other in the ventral ganglion) undergo a cycle of secretion, being most active at the time the new cuticle is being deposited (Davey, 1966). But neurosecretory material is not needed for the deposition of new cuticle as shown by the use of ligatures which prevent this material from reaching the posterior end of the worm. Such ligatured worms secrete new cuticle equally well at both ends. However, it appears that the production and release of the enzyme, leucine aminopeptidase (which is produced by the excretory gland at about the time that shedding of the old cuticle occurs and is thought to be necessary for the splitting off of the old cuticle) depends upon the presence of the neurosecretory material. The production and release of this enzyme appears to be abolished in ligatured worms.

3. *Annelids*

In nereid polychaetes the brain is probably the source of a hormone which is required for both normal and regenerative growth (Casanova, 1955; Durchon, 1956; Clark and Bonney, 1960; Clark and Scully, 1964). The possibility that this brain hormone might in fact be produced by the attached infracerebral gland and not the brain was referred to on page 170. This brain hormone exerts a continuous action on segment formation during regeneration (Fig. 3). Segment formation slows and stops whenever the brain is removed during regeneration (Golding, 1967b,c). Growth and posterior regeneration are essentially similar processes, each involving the formation

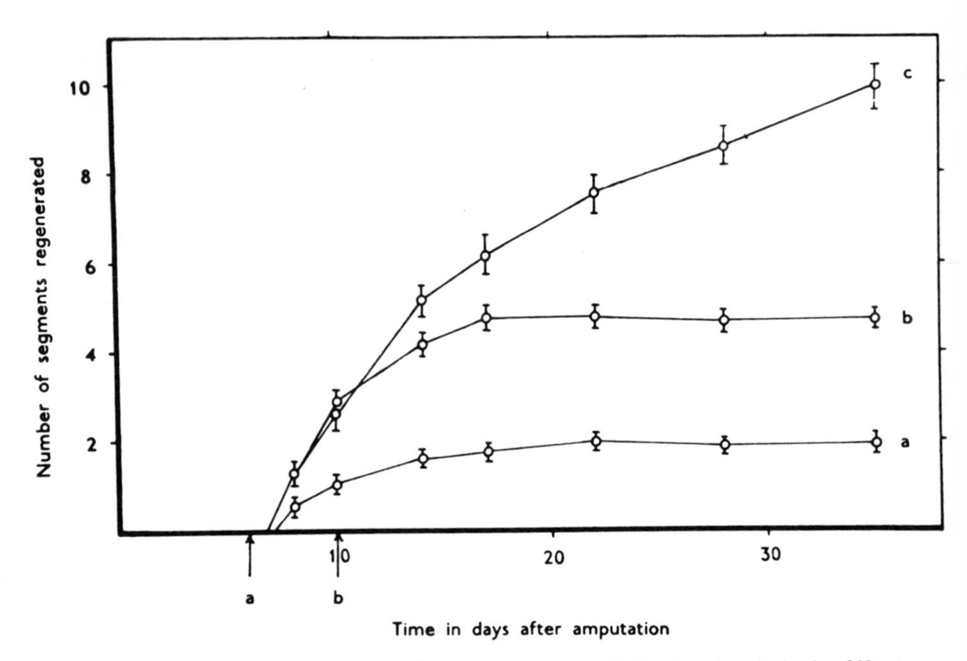

Fig. 3. Effect upon the number of segments regenerated of extirpating the brain of *Nereis diversicolor* at different times after amputating posterior segments; a, brain removed 6 days after amputation of the posterior segments; b, brain removed 10 days after amputation of the posterior segments; c, control with brain. (From Golding, 1967c). The "brain" when referring to nereids herein actually includes the infracerebral gland.

of new posterior segments, but the rate of segment formation is more rapid in regenerating worms (Golding, 1967d). The number of segments regenerated approximates the number removed, whether the worm's own brain is in place or if it is removed and a brain from an intact donor is implanted into the worm's coelom. Furthermore, when an unequal number of segments are removed from a host worm and a brainless graft the two worms do not regenerate the same number of segments; host and graft regenerate their lost segments independently in spite of the fact that both are exposed to the same concentration of hormone. A graft from which posterior segments have been removed will even regenerate attached to an intact host (Golding, 1967e). Regeneration of a large number of segments does not signify a high concentration of hormone in the body fluids. Instead, there appears to be an axial gradient of growth potential in *Nereis diversicolor* declining posteriorly. Regeneration requires a continuous supply of the hormone but the number of segments regenerated depends upon the level of the body (growth potential) at which the segments were removed.

4. *Crustaceans*

The term "molting" is being used more and more frequently in a broader sense than merely as a synonym for ecdysis, the actual shedding of the old exoskeleton, and includes all the events in the preparation for ecdysis, ecydsis itself, and the events such as hardening of the new exoskeleton after ecdysis. Some crustaceans reach a particular size and then cease molting activity, whereas others continue molting and growing throughout their life. Several schemes have been used to define the stages from one ecdysis to the next. However, that of Drach (1939) is now used by virtually all investigators of molting.

Eyestalk removal results in precocious ecdysis. Brown and Cunningham (1939) showed that the sinus gland contains a molt-inhibiting hormone. It appears to be a peptide (Rao, 1965), synthesized primarily in the medulla terminalis X-organ (Passano, 1953). In addition to the molt-inhibiting hormone, crustaceans have the molt-promoting hormone(s) or a prohormone produced by the molting glands. Ablation of both Y-organs, the molting glands of the crab, *Carcinus maenas*, results in permanent cessation of molting when performed during the intermolt period or very early in the premolt period (Echalier, 1954). But if extirpation of the Y-organs occurs later in the molting cycle then the specimens will undergo ecdysis and then permanently cease all molting activity. Ecdysones appear to be normal molting hormones in all arthropods. They shorten the intermolt period in a variety of crustaceans (Carlisle, 1965; Lowe *et al.*, 1968; Krishnakumaran and Schneiderman, 1969, 1970; Maissiat, 1970; Rao *et al.*, 1972. For example, injection of either α-ecdysone or 20-hydroxyecdysone into fourth stage larval lobsters, *Homarus americanus*, early in the premolt period resulted in an accelerated premolt and led to the precocious initiation of ecdysis (Rao *et al.*, 1973).

Adelung (1969) has determined the amount of molting hormone (20-hydroxyecdysone) in the crab, *Carcinus maenas*, from one ecdysis to the next and found the highest quantity was present just before ecdysis. Faux *et al.* (1969) did the same experiment with the blue crab, *Callinectes sapidus*, but found the highest concentration of molting hormone was in crabs which had just undergone ecdysis. Using crabs which were in early premolt Faux *et al.* were only able to extract a substance they called "callinecdysone A" which may be one or more isomers of inokosterone. Later in premolt this compound was accompanied by a lesser quantity of 20-hydroxyecdysone, and after ecdysis the latter represented the major component accompanied by a smaller quantity of still another compound (callinecdysone B) which is either makisterone A or its C_{24} isomer. In view of the unexpected high titer of molting hormone after ecdysis, Faux *et al.* suggested that the events leading to ecdysis may be sequentially triggered partly by different ecdysones and partly by a rising hormone titer with the final hardening of the new exoskeleton, at least in the blue crab, occurring only at the highest titer. In fact, Adelung found some secondary peaks of molting hormone titer. The shrimp, *Crangon nigricauda*, and the fiddler crab, *Uca pugilator*, are able to convert α-ecdysone to 20-hydroxyecdysone (King and Siddall, 1969).

The existence of an "ecdysis factor" in the isopod, *Orchestia cavimana*, and the isopods, *Proasellus cavaticus*, *Idotea balthica*, and *Ligia oceanica*, has been postulated by Graf (1972) and Maissiat and Graf (1973). This factor which induces ecdysis is presumed to be released normally during the premolt period, but 20-hydroxyecdysone appears to inhibit its release because injections of the latter during part of the premolt period can block ecdysis However, 20-hydroxyecdysone does evoke precocious ecdysis in these animals when administered during the intermolt period. Presumably, release of this ecdysis promoting factor normally occurs only when ecdysone (s) is absent from the blood or in low concentration there.

Some crustaceans may produce in their eyestalks an additional hormone, a molt-accelerating hormone, whose function is presumed to be activation of the molting glands. But the evidence for this hormone has not yet been as well accepted as that for the other hormones which regulate the molting cycle. An example of the sort of evidence adduced for this hormone with a marine species is that of Carlisle and Dohrn (1953) who used the shrimp, *Lysmata seticaudata*. They found that eyestalk extracts prepared at pH 3.5–3.9 from female shrimp collected during the summer when molting activity was at its peak induced precocious ecdysis during December and January when the natural rate of molting activity is lowest.

The ability of decapod crustaceans to regenerate lost appendages has been carefully investigated (Bliss, 1956, 1960). Within 6 days after a limb has been removed from the land crab, *Gecarcinus lateralis*, the scab covering the wound is pushed aside by an outgrowing papilla which enlarges for 2–3 weeks (basal limb growth). Further growth may then cease (growth plateau) or continue very slowly (advancing growth plateau). Then at the onset of the premolt period, under the influence of molting hormone, there is a rapid increase in the size of the limb bud (premolt growth) ending just before ecdysis. Removal of the Y-organs inhibits premolt limb regeneration, whereas implantation of Y-organs into crabs lacking these organs induces limb regeneration (Echalier, 1956).

5. *Fishes*

Growth hormone (somatotropin) is a protein produced in the adenohypophysis. Hypophysectomy results in cessation of growth in elasmobranchs (Vivien, 1941) and teleosts (Pickford, 1953). Presumably hypophysectomy would have the same effect in agnathans as well but the experiment does not appear to have been done as yet. Correspondingly, injections of purified growth hormone from fishes will cause growth in the killifish, *Fundulus heteroclitus* (Pickford, 1954). The hypothalamus may control the release of growth hormone in fishes by means of a growth hormone release-stimulating hormone as in higher vertebrates (Ball, 1969b).

The thyroid gland has also been implicated as a growth stimulator in teleosts, but its role is not clear as yet (Gorbman, 1969). Some investigators found some growth stimulation, whereas others found that the growth rate was retarded by thyroid hormone. Ultimately, it may be demonstrated that the thyroid hormone, while not directly stimulating growth, does play a permissive role in the growth regulation of fishes as it does in mammals.

F. REPRODUCTION

1. Coelenterates, Platyhelminths, and Nemerteans

Sexuality in hydras as evidenced by the formation of gametes from interstitial cells occurs only when growth ceases or is reduced. In hydras, growth and sexual reproduction appear to be antagonistic to one another. On the other hand, asexual reproduction in hydras by budding is part of the normal growth pattern and does depend on the growth hormone from the hypostomal region. Growth and budding cease completely in hydras approaching complete sexuality. It appears that sexuality commences when the hormone which stimulates growth is no longer being produced or is produced at a low concentration only (Burnett and Diehl, 1964).

Among the flatworms asexual reproduction by fissioning appears to be under the control of neurosecretory material from the brain. There is evidence that fissioning is initiated by neurosecretory material in the planarian, *Dugesia gonocephala* (Lender and Zghal, 1968; Lender, 1970). Destruction of the brain suppresses fissioning. Furthermore, the brain begins to release its neurosecretory material just before fission occurs. On the other hand, in *Dugesia dorotocephala* the brain (presumably by release of a neurohormone from it) suppresses fissioning (Best *et al.*, 1969).

Sexual differentiation in the nemertean, *Lineus ruber*, has recently also been shown to be under endocrine control (Bierne, 1970). A substance produced by the brain inhibits sexual differentiation in immature individuals. There is also an androgenic factor produced in the male which is necessary for its normal sexual development and is capable of causing sex reversal in females. This androgenic factor does not have a cerebral origin; it is probably of testicular origin.

2. *Annelids*

The endocrine control of reproduction in nereids has been studied in detail. A juvenile hormone, probably produced by the brain, which inhibits maturation of the gametes and transformation to the heteronereis form. The heteronereis is the reproductive individual and differs from the nonsexual form in the size and shape of various body parts such as its eyes and parapodia. Release of this hormone is gradually reduced at the start of the breeding season (Durchon, 1960; Hauenschild, 1966). Baskin (1970), using the viviparous polychaete, *Nereis limnicola*, found that the thickness of the glandular epithelium of the infracerebral gland is correlated with the stage of reproduction, being thickest about the time of fertilization. He therefore concluded that it is unlikely that this gland is the source of the juvenile hormone, but suggested that the infracerebral gland may affect the amounts of metabolites used in the early development of the fertilized eggs. Nereid juvenile and growth hormones could be the same substance.

The endocrine control of reproduction in *Arenicola marina*, a polychaete which is not a nereid, is different from that described for nereids in that it is stimulatory rather than inhibitory. There is no evidence for the juvenile hormone in *Arenicola*. Instead, the brain in females produces a hormone which causes maturation of the eggs (completion of meiosis). Without this hormone the eggs remain in arrested prophase of the first maturation division. In males the brain hormone controls spermatogenesis by stimulating gonadal mitoses and sperm maturation divisions. Release of this hormone is somehow inhibited by the accumulation of sperm in the coelom (Howie, 1963; Howie and McClenaghan, 1965). Decerebration prevents ripening of the gametes.

With respect to annelids other than polychaetes, in earthworms maturation of the gonads, egg laying, and the development of accessory reproductive structures such as the clitellum to a functional state are dependent upon a stimulatory hormone (gonadotropin) from neurosecretory cells of the brain (Herlant-Meewis, 1957). Spermatogenesis in leeches also requires a brain gonadotropin (Hagadorn, 1969).

3. *Mollusks*

Using the terrestrial slugs, *Arion subfuscus*, and *Arion ater*, Pelluet and Lane (1961) found that removal of the optic tentacles resulted in an increase in the number of eggs in the ovotestis, but there was a reduction in the volume of the ovotestis in young animals despite the increase in number of eggs. This reduction was presumably due to inhibition of sperm formation. Furthermore, brain extracts promoted egg formation, but tentacle extracts had no effect on the number of eggs formed. These slugs are protandrous, young specimens forming sperm and no eggs. These investigators concluded that egg production is stimulated by a brain hormone, but a tentacle hormone can suppress the brain hormone while stimulating sperm production so that sperm only are formed. Consequently, egg and sperm production appear to depend upon the balance between these two hormones. More recently Gottfried and Dorfman (1970) using another land slug, *Ariolimax californicus*, found that removal of the optic tentacles resulted in rapid maturation of the male portion of the ovotestis; precocious spermatogenesis was induced. Optic tentacle extracts inhibited this precocious spermatogenesis. Steroids from the ovotestis may have a negative feedback on the optic tentacles modifying release of that hormone. The observations of Gottfried and Dorfman differ from those of Pelluet and Lane and may be due to species differences. Pelluet and Lane suggested that the optic tentacles contain a spermatogenesis stimulator, whereas on the basis of their experiments Gottfried and Dorfman favored the interpretation of a spermatogenesis inhibitor. Choquet (1965, 1971) also has evidence for a tentacular substance which inhibits spermatogenesis in the gastropod, *Patella vulgata*, as well as for a factor from the cerebral ganglia that stimulates this process.

Kupfermann (1970, 1972) has shown that the bag cells in the abdominal ganglion of *Aplysia californica* contain a neurohormone that stimulates egg-laying. Coggeshall (1970, 1972) has reported that the latter hormone appears to exert its action by causing contraction of the muscle cells which surround the egg follicles. These muscle cells do not appear to be innervated. The data of Arch (1972) support the hypothesis that this egg-laying hormone is first synthesized as part of a 25,000 dalton precursor which is subsequently enzymatically cleared to yield two products, one being the 6000 dalton hormone.

Maturation of the ovary or testis in the octopus, *Octopus vulgaris*, is due to stimulation by a hormone from the optic glands (Wells and Wells, 1959). These glands are in turn controlled by an inhibitory nerve supply from the brain. Cutting the inhibitory nerve leading to an optic gland (even on only one side of the body) causes the gland to become swollen with secretion and enlargement of the gonad follows. Enlargement of the ovary or testis is always accompanied by enlargement of at least one of the optic glands. Maturation of the gonad does not occur in the absence of the optic glands. The optic glands also enlarge when the optic nerves leading from the eyes to the brain are cut. The inhibitory center in the brain is apparently activated by light stimuli through the eyes and optic nerves. Presumably, changes in the photoperiod govern the inhibitory center and consequently the entire optic gland system.

4. *Crustaceans*

The eyestalks of crustaceans contain a gonad-inhibiting hormone which appears to originate in the medulla terminalis X-organ. Removal of the eyestalks results in precocious maturation of the ovaries (Panouse, 1943) and testes (Demeusy, 1953). In addition, there is evidence for a gonad-stimulating hormone in the central nervous system of male and female crabs (Ōtsu, 1963; Gomez, 1965).

Although the sex of an individual, male or female, is genetically determined, normal sexual differentiation of male crustaceans requires the presence of the androgenic gland (Charniaux-Cotton, 1954). This gland, found only in the male, secretes a hormone which determines the development of the testes, male accessory reproductive structures, and male secondary sexual characteristics. Androgenic glands do not develop in females; differentiation of the ovaries appears to occur without the intervention of a hormone. Implantation of an adrogenic gland into an immature female results in masculinization; the potential ovaries are transformed into testes which produces viable sperm. On the other hand, implantation of a testis into a female has no effect on the host. The crustacean testis does not appear to produce any hormone. An ovary transplanted into an androgenic glandless individual survives without modification while the host's testes degenerate. The ovaries are, however, the source of a hormone that controls the differentiation of accessory sexual structures in females. In females the actions of the gonad-inhibiting and gonad-stimulating hormones are directly on the ovary, but in males it is presumably on the androgenic glands rather than on the testes (Adiyodi and Adiyodi, 1970). The chemical nature of the gonad-stimulating and gonad-inhibiting hormones are not known but because of their neurosecretory origin they may well be polypeptides. The androgenic gland hormone may be a steroid (Sarojini, 1964), perhaps in combination with a protein (Highnam and Hill, 1969) or a polypeptide or protein alone (Adiyodi and Adiyodi, 1970). It will be recalled that the ultrastructure of this gland suggests a proteinaceous product.

Using cultures of testes from *Orchestia gammarella*, Berreur-Bonnenfant (1970) obtained evidence for a factor produced by the brains of males that appears to be necessary to prevent degeneration of the germinative zone from which the sperm are produced. Androgenic glands alone could not prevent this degeneration. This brain factor was found in male brains only after they had been exposed to the androgenic hormone. Brains from normal females did not contain this factor. But if androgenic glands were first implanted into females, 2 months later their brains were producing this substance. On the other hand, brains of males whose androgenic glands had been ablated 3 months previously no longer were capable of producing this factor.

5. *Echinoderms*

The radial nerves of both sexes of sea stars produce a substance, presumably in the neurosecretory cells, which acts both on their ovaries and testes to induce spawning (Chaet and McConnaughy, 1959). This substance has so far been extracted from about 20 species of sea stars, and with few exceptions this substance is cross-reactive. The active factor was originally called the "gamete-shedding substance" but it is now commonly known as the "gonad-stimulating substance" (Kanatani, 1969) or "radial nerve factor" (Schuetz, 1969). That from the sea star, *Asterias amurensis*, is a polypeptide with a molecular weight about 2100 (Kanatani *et al.*, 1971). In the case of females it has been found that the gonad-stimulating substance acts on the ovary to cause the formation of a second substance which causes maturation of the oocytes (Kanatani and Shirai, 1967; Schuetz and Biggers, 1967; Schuetz, 1969). This substance, known as the "meiosis-inducing substance" or the "ovarian factor," has now been identified as 1-methyladenine by Kanatani *et al.* (1969). It appears to mediate the activity of the gonad-stimulating substance (Kanatani and Shirai, 1970). Synthetic 1-methyladenine causes spawning in both sexes. Its action does not appear to be species specific. Presumably, it mediates the action of the gonad-stimulating substance in males as well as in females. In the female sea star, *Asterina pectinifera*, the follicle cells surrounding the full-grown oocytes seem to be the site of production of this meiosis-inducing substance (Hirai and Kanatani, 1971). In addition, a substance that inhibits the spawning action of the gonad-stimulating substance is also produced by the gonads of both sexes of sea stars. The inhibitor from *Asterina pectinifera* has been identified as L-glutamic acid (Ikegami *et al.*, 1967), but its action shows species specificity, having little effect in *Asterias amurensis* whose own inhibitor substances are asterosaponins A and B (Ikegami *et al.*, 1972). In addition, an antimitotic substance which blocks meiosis has been found in the ovaries of sea stars (Heilbrunn *et al.*, 1954) but does not occur in their testes (Kanatani and Ohguri, 1970). It seems to arrest meiosis until the meiosis-inducing substance is released and is able to overcome its inhibitory action.

6. *Fishes*

The pituitary gland is essential for reproduction in all fishes. This gland regulates gametogenesis and steroidogenesis in jawed fishes (Hoar, 1969). Removal of their pituitary gland results in arrest of ovarian development and cessation of spermatogenesis. But in agnathans hypophysectomy results only in retardation but not suppression of spermatogenesis, spermiogenesis, and ovarian growth (Dodd *et al.*, 1960; Larsen, 1965, 1969). In agnathans these processes appear to be autonomous ones which are only speeded by the pituitary gland. However, in both agnathans and jawed fishes the production of gonadal hormones is fully dependent upon the pituitary gland. Hypophysectomy results in degeneration of the steroid hormone-producing cells in the gonads. The secondary sexual structures whose normal postnatal development depends upon gonadal hormones fail to develop in hypophysectomized individuals (Matty, 1966). Prolactin, in addition to testosterone, has recently been implicated in the control of the seminal vesicles of the goby, *Gillichthys mirabilis* (DeVlaming and Sundararaj, 1972). Castration or hypophysectomy caused regression of the seminal vesicles. A combination of testosterone propionate and prolactin was necessary to maintain the secretory activity of the seminal vesicles of hypophysectomized fish at the high level seen in the sham-operated controls. Although neither hormone alone was able to induce hypophysectomized fish to maintain the high level of seminal vesicle activity seen in the controls, each hormone by itself was at least partially successful in this regard. Berlind (1972b) found that extracts of the urophysis of the teleost, *Gillichthys mirabilis*, induced contraction of its sperm duct and suggested that the active factor may have a normal role in spawning.

All tetrapods have a follicle-stimulating hormone (FSH) and a luteinizing hormone (LH) in their pituitary gland. However, although both FSH-like and LH-like effects have been observed when extracts of fish pituitary glands were assayed on tetrapods, both actions seem to be due to a single protein (Yamazaki and Donaldson, 1968). Estradiol-17β is probably the most physiologically important estrogen in fishes. Testosterone (an androgen) has been found in the blood of some male fishes as well as in their testes. The role of the corpora lutea in fishes is not clear. Although Chieffi (1961) has found progesterone in the corpora lutea of the electric ray, *Torpedo*, a normal endocrinological role for this substance has not been established. Furthermore, the corpora lutea of some fishes do not appear to be sites of hormone production (Lambert and van Oordt, 1965).

IV. Materials and Methods

This section will be devoted to a description of some of the basic techniques used by comparative endocrinologists. It is obvious that only a small percentage of them can be referred to in the allotted space. They will, furthermore, reflect to some extent the interests of the author. The techniques to be described are essentially the classic endocrinological ones of extirpation and replacement therapy. Readers more interested in working with fishes than invertebrates have the advantage of being able to obtain large numbers of purified hormones of mammalian origin from several companies. While these preparations may not in all cases have the same chemical structure as the fish hormones, the commercial preparations can usually mimic their actions quite well.

A. ANNELIDS

The necessity of a hormone from the head for the occurrence of posterior regeneration in nereid polychaetes can be readily demonstrated (see Golding, 1967b,c,d,e). The animals can be anesthetized by placing them in 0.5% tricaine methanesulfonate in seawater or in a 7.3% $MgCl_2$ solution which is roughly isosmotic with seawater. Once the worm has been anesthetized the brain can be removed by decapitating the worm (cutting off the prostomium as far behind the posterior eyes as possible) thereby eliminating the source of the hormone. Posterior segments can then be removed by a transverse cut intersegmentally about halfway along the worm. Regeneration can then be compared among brainless and control worms which were deprived of the same number of segments, the controls still having their brains. The animals quickly recover from the anesthetic when placed in fresh seawater. To minimize mortality the seawater should be pasteurized or Millipore-filtered and 10^6 units of benzylpenicillin should be added to each liter. The possible role of the infracerebral gland was discussed above.

B. CRUSTACEANS

Stalk-eyed crustaceans which have their sinus glands in their eyestalks have been used experimentally to a larger degree than other crustaceans. An especially useful experimental procedure is to remove both eyestalks from a crustacean whose eyestalks contain the sinus glands, observe the effects on the particular system under study, such as the chromatophores, next inject an extract prepared from whole eyestalks or from sinus glands alone, and then observe the effects of the injection. In this laboratory we have found that the highest percentage of survival of crustaceans which have been subjected to bilateral eyestalk ablation (so-called "eyestalkless" individuals) is obtained when the stubs are immediately cauterized and the animals then kept for approximately 10 min in a dry container before being returned to their containers which have seawater in them. This procedure prevents hemorrhaging. The eyestalks are removed by cutting them at their base by use of a pair of fine-pointed scissors or a scalpel with a pointed blade.

1. *Chromatopheres and Color Changes*

The fact that hormones are involved in the control of color changes in crustaceans can be readily demonstrated by injecting an eyestalk or sinus gland extract (one-third of an eyestalk or sinus gland equivalent per dose evokes excellent responses) into an eyestalkless individual and observing its effects on the chromatophores. Crude extracts are prepared by triturating the tissue in the appropriate volume of isosmotic saline or seawater, centrifuging to remove the bits of tissue and exoskeleton, and then injecting the supernatant. For fiddler crabs we routinely inject 0.05 ml per crab, but for small organisms such as the prawn, *Palaemonetes vulgaris*, the dose often is 0.02 ml. The effects of the extracts on the chromatophores can be followed by use of the system of Hogben and Slome (Fig. 4) where the degree of pigment dispersion is divided into five stages ranging from "1" (maximal concentration) to "5" (maximal dispersion). This staging method can also be used with the chromatophores of vertebrates. We ordinarily observe the responses until the effect of the extract has worn off and then use the recorded stage values to calculate the Standard Integrated Response (Fingerman *et al*., 1967) which takes into account both the amplitude of the response and its duration.

An interesting experiment is to compare the chromatophoric responses of fiddler crabs and prawns after their eyestalks have been ablated. The crabs will become pale primarily because the pigment in their melanophores, their predominant chromatophore type, concentrates, whereas the prawns darken primarily because their predominant pigment, a dark red one, becomes dispersed. Freshly prepared extracts of crab and prawn eyestalks or sinus glands alone cause dispersion of the black pigment (melanin) of these eyestalkless crabs, but concentration of the red pigment of these eyestalkless prawns. The eyestalks of the fiddler crab have also been shown to possess pigment-dispersing and pigment-concentrating hormones for its red and white chromatophoric pigments (Brown, 1950; Rao *et al*., 1967). The pigment-concentrating activities can be separated from the pigment-dispersing activities (Fig. 5) by use of the technique of gel filtration with Bio-Gel P-6 (Bio-Rad Laboratories) as described by Fingerman *et al*., 1971a. Through use of eyestalkless individuals and intact ones on black and on white backgrounds, an investigator can obtain one group of fiddler crabs whose red pigment or white pigment is dispersed and another group in which one of these pigments is concentrated thereby allowing assay for the pigment-concentrating and pigment-dispersing substances respectively.

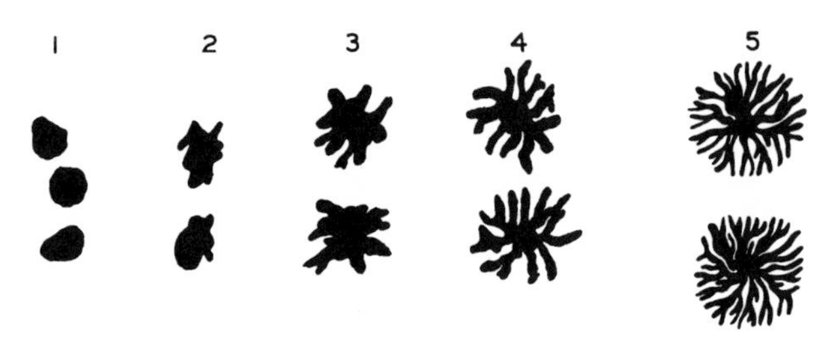

Fig. 4. The Hogben and Slome stages. (From Fingerman and Yoshioka 1968.)

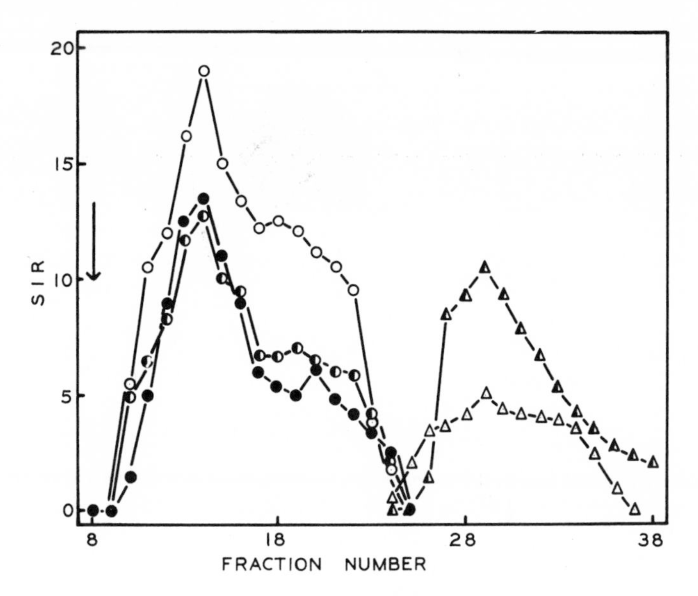

Fig. 5. The Standard Integrated Responses (SIR) of fiddler crabs, *Uca pugilator*, evoked by the fractions obtained by passing an extract of fiddler crab eyestalks through a column of Bio-Gel P-6; filled circles, melanin-dispersing responses; empty circles, white pigment-dispersing responses; half-filled circles, red pigment-dispersing responses; empty triangles, white pigment-concentrating responses; half-filled triangles, red pigment-concentrating responses. The melanin-dispersing, red pigment-dispersing, and white pigment-concentrating responses were determined with eyestalkless crabs; the red pigment-concentrating and white pigment-dispersing responses with intact crabs in black containers. Arrow indicates end of void volume. (From Fingerman *et al.*, 1971a.)

For example, the red pigment of eyestalkless fiddler crabs is concentrated while that of intact specimens in black containers is dispersed.

2. *Retinal Pigments*

Intact prawns and crabs can be readily used to demonstrate the actions of the hormones controlling the distal retinal pigment without resorting to sectioning the eyestalks and preparing slides for microscopic analysis, whereas direct observations of the proximal and reflecting pigments require such sectioning techniques. The method of Sandeen and Brown (1952) allows direct, rapid determination of the position of the distal retinal pigment in prawns. Their technique has been recently modified for use with the fiddler crab (Fingerman, 1970b). The method of Sandeen and Brown (Fig. 6) consists essentially of immersing the prawn, *Palaemonetes*, in a dish of seawater on the stage of a dissecting microscope. Then with the aid of transmitted light and an ocular micrometer (A) the width of the translucent

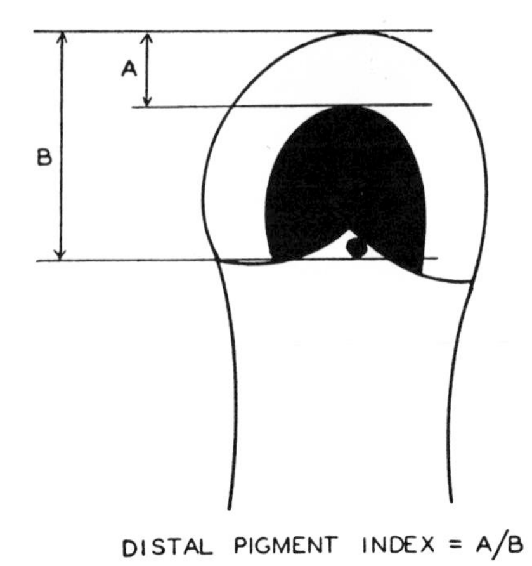

Fig. 6. Diagram of the dorsal aspect of the eyestalk of the prawn, *Palaemonetes vulgaris*, illustrating the method of obtaining the Distal Retinal Pigment Index. (From Sandeen and Brown, 1952.)

area of the distal portion of the compound eye in a direction parallel to the long axis of the eye and (B) the distance from the cornea to the proximal edge of the conspicuous black spot on the dorsal surface of the eyestalk are measured. In a fully dark adapted eye there is virtually no translucent area. But in a fully light adapted eye the translucent area is quite broad because of the proximal migration of the distal retinal pigment. To minimize the effect of size differences among the prawns Sandeen and Brown used the ratio of the two measurements, A/B, which they called the "Distal Retinal Pigment Index." In a fully dark adapted eye of *Palaemonetes* this ratio is less than 0.05 and in a fully light adapted eye about 0.27. By maintaining specimens at a light intensity such that this pigment is approximately midway between the fully light adapted and fully dark adapted positions, injection of an eyestalk extract (two-thirds of an eyestalk equivalent/dose will give a good response) will first evoke a light-adapting response, caused by the distal retinal pigment light-adapting hormone, with a maximum effect about 1 hr after the extract was injected followed by a dark-adapting response caused by the dark-adapting hormone (Fingerman *et al.*, 1959, 1971b; Fingerman and Mobberly, 1960).

3. *Blood Glucose*

The spider crab, *Libinia emarginata*, is an especially useful animal with which to observe the effect of the hyperglycemic hormone because this crab is large and yields enough blood so that each specimen can serve as its own control. That is, the blood glucose level of an eyestalkless spider crab can be determined, eyestalk extract injected, and then blood (0.1 ml) drawn again 90 min afterward and its glucose content determined (Kleinholz *et al.*, 1967). The crabs are so large that the usual dose of an extract is 0.1 ml. An extract containing one eyestalk equivalent per dose will cause a readily detectable response. The glucose oxidase method is now a very popular one for determination of the blood glucose concentration. The necessary reagents for this colorimetric technique are commercially available as "Glucostat" from the Worthington Biochemical Corporation.

4. *Molting and Growth*

Removal of the eyestalks is ordinarily sufficient to initiate premolt activity. One expriment that can be done is to compare the time required for ecdysis to occur in intact and eyestalkless crustaceans. The effects of eyestalk extracts (one eyestalk equivalent per dose) given to eyestalkless specimens every other day could be observed in order to demonstrate the presence of molt-inhibiting hormone (Fingerman and Yamamoto, 1964). In addition, shortening of the intermolt period, evidenced by the onset of the premolt period, by ecdysones can be demonstrated by injecting these steroids into intact crustaceans (Krishnakumaran and Schneiderman, 1970). 20-Hydroxy-ecdysone is commercially available from Schwarz/Mann. A dosage of 20 μg/gm body weight of 20-hydroxyecdysone will induce precocious premolt activity such as deposition of a new exoskeleton in about 14–18 days in the fiddler crab, *Uca pugilator*.

C. ECHINODERMS

Induction of spawning both by radial nerve extracts containing the gonad-stimulating substance and by 1-methyladenine can be readily demonstrated. The radial nerves can be obtained by first removing the aboral surface of a sea star and then after making an incision along each ambulacral groove, the halves of the oral portion of each ray can be separated revealing the radial nerve. Chaet (1966) has described the methods of preparing radial nerve extracts which have proven successful in his laboratory. In an *in vivo* assay, after the extract (one nerve/ml) has been prepared, it is injected into the coelomic cavity of a sea star (0.15 ml/gm) whose gonads are ripe. Release of the gametes from the sea star, *Asterias forbesi*, occurs about 35 min after the extract has been injected. The gonad-stimulating substance can also be demonstrated *in vitro* by using the technique of Moore and Biggers (1964). By weighing a ripe ovary at the start and reweighing it 1hr after shedding begins, one can demonstrate that as the dosage of radial nerve extract increases, more eggs are shed and consequently the weight loss is greater. Good results are obtained by using a concentration range of 0.01–0.1 radial nerve equivalent in 70 ml seawater for each ovary.

A method to induce spawning and meiotic maturation of sea star oocytes by treatment with 1-methyladenine has been published by Stevens (1970). Her technique consists of placing ripe ovarian fragments 2–3 mm long in a 1×10^{-6} *M* solution of 1-methyladenine in seawater. Concentrations as high as 1×10^{-3} *M* can be used and the eggs will still be fertilizable. In the case of the sea star, *Patiria miniata*, spawning begins after 20–30 min of incubation.

D. FISHES

The killifish, *Fundulus heteroclitus*, is one of the euryhaline fishes whose pituitary gland is essential for survival in freshwater. Because of this fish's ready availability along the Atlantic coast from Maine to northeastern Florida it has been used quite often as an experimental animal. Once it has been hypophysectomized it can be used to study a number of different endocrine functions. Abbott and Favreau (1971) have recently described in considerable detail the method of hypophysectomizing this particular fish. The technique consists essentially of first anesthetizing the fish in tricaine methanesulfonate (0.225 gm/liter) and then drilling a hole through the roof of the mouth in order to reach the gland and remove it. A few of the experiments which can be done with fishes will be described.

1. *Prolactin and Survival in Freshwater*

Hypophysectomized *Fundulus heteroclitus* do not survive longer than 6–7 days after transferal from seawater to freshwater (Burden, 1956). Injections of extracts of pituitary glands can be shown to enable hypophysectomized *Fundulus heteroclitus* to survive in freshwater for at least 20 days. The same type experiment can be done using prolactin itself (Pickford *et al.*, 1965).

2. *Color Changes*

Color changes of fishes present a challenge to an investigator because of the wide diversity in the control of their chromatophores that occurs. An interesting series of experiments can be performed by comparing the ability of a number of different fishes to adapt to black and to white backgrounds before and after hypophysectomy. The chromatophorotropic effects of extracts of the pituitary gland or of purified hormone preparations (e.g., intermedin and prolactin; see Sage, 1970) can also be determined.

3. *Carbohydrate and Lipid Metabolism*

The effects of a variety of different hormones which are commercially available can be tested on fishes to determine their effects on blood glucose and muscle and liver glycogen levels. The hormones that Patent (1970) tested in such experiments were insulin, glucagon, epinephrine, norepinephrine, cortisol, and corticosterone. In addition, he determined the effects of adrenocortical hormones on the lipid levels in the liver.

V. Conclusions

Endocrine mechanisms have been investigated in only a small percentage of the marine organisms. The fact that similar functions are under endocrine control in a wide variety of organisms (e.g., growth) is the keystone of comparative endocrinology. In this connection, the demonstration of the widespread importance of neurosecretory mechanisms in the animal kingdom reveals the kind of important generalization to which the comparative approach to experimental biology can lead. This author hopes that many of the experiments described in this chapter will lead to new and exciting publishable findings as a greater variety of marine organisms begins to be utilized in teaching and research laboratories.

Acknowledgment

The author is indebted to the National Science Foundation for support of the work from this laboratory.

References

Abbott, F. S., and Favreau, M. B. (1971). The effect of hypophysectomy on physiological color change in *Fundulus heteroclitus*. *Can. J. Zool.* **49**, 129–131.

Abramowitz, A. A., Hisaw, F. L., and Papandrea, D. N. (1944). The occurrence of a diabetogenic factor in the eyestalks of crustaceans. *Biol. Bull.* **86**, 1–5.

Acher, R., Chauvet, J., and Chauvet, M. T. (1972). Phylogeny of the neurohypophysial hormones. Two new active peptides isolated from a cartilaginous fish, *Squalus acanthias*. *Eur. J. Biochem.* **29**, 12–19.

Adam, H. (1963). The pituitary gland. *In* "Biology of Myxine" (A. Brodal and R. Fänge, eds.), pp. 457–476. Oslo Univ. Press, Oslo.

Adelung, D. (1969). Die Ausschüttung und Funktion von Häutungshormon während eines Zwischenhäutungs-Intervalls bei der Strandkrabbe *Carcinus maenas* L. *Z. Naturforsch.* **24b**, 1447–1455.

Adiyodi, K. G., and Adiyodi, R. G. (1970). Endocrine control of reproduction in decapod Crustacea. Biol. Rev. **45**, 121–165.

Adiyodi, R. G. (1969). Gliosecretion in the crab *Paratelphusa hydrodromous* (Herbst). *Gen. Comp. Endocrinol.* **13**, 306–308.

Alexandrowicz, J. S. (1952). Notes on the nervous system in the Stomatopoda. I. The system of median connectives. *Pubbl. Staz. Zool. Napoli* **23**, 201–214.

Alexandrowicz, J. S. (1964). The neurosecretory system of the vena cava in Cephalopoda. I. *Eledone cirrosa*. *J. Mar. Biol. Ass. U.K.* **44**, 111–132.

Alexandrowicz, J. S. (1965). The neurosecretory system of the vena cava in Cephalopoda. II. *Sepia officinalis* and *Octopus vulgaris*. *J. Mar. Biol. Ass. U.K.* **45**, 209–228.

Alexandrowicz, J. S., and Carlisle, D. B. (1953). Some experiments on the function of the pericardial organs in Crustacea. *J. Mar. Biol. Ass. U.K.* **32**, 175–192.

Andrews, P. M., Copeland, D. E., and Fingerman, M. (1971). Ultrastructural study of the neurosecretory granules in the sinus gland of the blue crab, *Callinectes sapidus*. *Z. Zellforsch.* **113**, 461–471.

Arch, S. (1972). Biosynthesis of the egg-laying hormone (ELH) in the bag cell neurons of *Aplysia californica*. *J. Gen. Physiol.* **60**, 102–119.

Baker-Cohen, K. F. (1959). Renal and other heterotopic thyroid tissue in fishes. *In* "Comparative Endocrinology" (A. Gorbman, ed.), pp. 283–301. Wiley, New York.

Ball, J. N. (1969a). Prolactin and osmoregulation in teleost fishes: a review. *Gen. Comp. Endocrinol. Suppl.* **2**, 10–25.

Ball, J. N. (1969b). Prolactin (fish prolactin or paralactin) and growth hormone. *In* "Fish Physiology" (W. S. Hoar and D. J. Randall, eds.), Vol. II, pp. 207–240. Academic Press, New York.

Ball, J. N., and Baker, B. I. (1969). The pituitary gland: anatomy and histophysiology. *In* "Fish Physiology" (W. S. Hoar and D. J. Randall, eds.), Vol. II, pp. 1–110. Academic Press, New York.

Baskin, D. G. (1970). Studies on the infracerebral gland of the polychaete annelid, *Nereis limnicola*, in relation to reproduction, salinity, and regeneration. *Gen. Comp. Endocrinol.* **15**, 352–360.

Belamarich, F. A. (1963). Biologically active peptides from the pericardial organs of the crab *Cancer borealis*. *Biol. Bull.* **124**, 9–16.

Berlind. A. (1972a). Teleost caudal neurosecretory system: release of urotensin II from isolated urophyses. *Gen. Comp. Endocrinol.* **18**, 557–560.

Berlind, A. (1972b). Teleost caudal neurosecretory system: sperm duct contraction induced by urophysial material. *J. Endocrinol.* **52**, 567–574.

Berlind, A., and Cooke, I. M. (1970). Release of a neurosecretory hormone as peptide by electrical stimulation of crab pericardial organs. *J. Exp. Biol.* **53**, 679–686.

Berlind, A., Cooke, I. M., and Goldstone, M. W. (1970). Do the monoamines in crab pericardial organs play a role in peptide neurosecretion? *J. Exp. Biol.* **53**, 669–677.

Bern, H. A. (1969). Urophysis and caudal neurosecretory system. *In* "Fish Physiology" (W. S. Hoar and D. J. Randall, eds.), Vol. II, pp. 399–418. Academic Press, New York.

Bern, H. A., and Nandi, J. (1964). Endocrinology of poikilothermic vertebrates. *In*: "The Hormones" (G. Pincus, K. B. Thimann, and E. B. Astwood, eds.), Vol. IV, pp. 199–298. Academic Press, New York.

Berreur-Bonnenfant, J. (1970). Organotypic culture *in vitro* and endocrine mechanisms in crustaceans. *In* "Invertebrate Organ Cultures" (H. Lutz, ed.), pp. 211–227. Gordon and Breach, New York.

Berry, C. F., and Cottrell, G. A. (1970). Neurosecretion in the vena cava of the cephalopod *Eledone cirrosa*. *Z. Zellforsch.* **104**, 107–115.

Best, J. B., Goodman, A. B., and Pigon, A. (1969). Fissioning in planarians: control by the brain. *Science* **164**, 565–566.

Bianchi, S. (1969). On the neurosecretory system of *Cerebratulus marginatus* (Heteronemertini). *Gen. Comp. Endocrinol.* **12**, 541–548.

Bierne, J. (1970). Recherches sur la différenciation sexuelle au cours de l'ontogenèse et de la régénération chez le Némertien *Lineus ruber* (Muller). *Ann. Sci. Nat.* (*Zool.*) *Ser. 12* **12**, 181–298.

Björkman, N. (1963). On the ultrastructure of the optic gland in *Octopus*. *J. Ultrastruct. Res.* **8**, 195.

Bliss, D. E. (1951). Metabolic effects of sinus gland or eyestalk removal in the land crab, *Gecarcinus lateralis*. *Anat. Rec.* **111**, 502–503.

Bliss, D. E. (1956). Neurosecretion and the control of growth in a decapod crustacean. *In* "Bertil Hanström. Zoological Papers in Honour of His Sixty-Fifth Birthday, November 20th, 1956" (K. G. Wingstrand, ed.), pp. 56–75. Zoolog. Inst., Lund. Sweden.

Bliss, D. E. (1960). Autotomy and regeneration. *In* "The Physiology of Crustacea" (T. H. Waterman, ed.), Vol. I, pp. 561–589. Academic Press, New York.

Bliss, D. E., and Welsh, J. H. (1952). The neurosecretory system of brachyuran Crustacea. *Biol. Bull.* **103**, 157–169.

Bliss, D. E., Wang, S. M. E., and Martinez, E. A. (1966). Water balance in the land crab, *Gecarcinus lateralis*, during the intermolt cycle. *Amer. Zool.* **6**, 197–212.

Brown, F. A., Jr. (1950). Studies on the physiology of *Uca* red chromatophores. *Biol. Bull.* **98**, 218–226.

Brown, F. A., Jr. (1952). Hormones in crustaceans. *In* "The Action of Hormones in Plants and Invertebrates" (K. V. Thimann, ed.), pp. 171–214. Academic Press, New York.

Brown, F. A., Jr., and Cunningham, O. (1939). Influence of the sinusgland of crustaceans on normal viability and ecdysis. *Biol. Bull.* **77**, 104–114.

Brown, F. A., Jr., Hines, M. N., and Fingerman, M. (1952). Hormonal regulation of the distal retinal pigment of *Palaemonetes*. *Biol. Bull.* **102**, 212–225.

Bunt, A. H., and Ashby, E. A. (1967). Ultrastructure of the sinus gland of the crayfish, *Procambarus clarkii*. *Gen. Comp. Endocrinol.* **9**, 334–342.

Burden, C. E. (1956). The failure of hypophysectomized *Fundulus heteroclitus* to survive in fresh water. *Biol. Bull.* **110**, 8–28.

Burnett, A. L., and Diehl, N. A. (1964). The nervous system of *Hydra*. III. The initiation of sexuality with special reference to the nervous system. *J. Exp. Zool.* **157**, 237–250.

Burnett, A. L., Diehl, N. A., and Diehl, F. (1964). The nervous system of *Hydra*. II. Control of growth and regeneration by neurosecretory cells. *J. Exp. Zool.* **157**, 227–236.

Camp, W. E. (1917). The development of the suprapericardial (postbranchial, ultimobranchial) body in *Squalus acanthias*. *J. Morphol.* **28**, 369–415.

Carlisle, D. B. (1956). An indole-alkylamine regulating heart-beat in Crustacea. *Biochem. J.* **63**, 32–33P.

Carlisle, D. B. (1965). The effects of crustacean and locust ecdysons on moulting and proecdysis in juvenile shore crabs, *Carcinus maenas*. *Gen. Comp. Endocrinol.* **5**, 366–372.

Carlisle, D. B., and Connick, R. O. (1973). Crustecdysone (20-hydroxyecdysone): site of storage in the crayfish *Orconectes propinquus*. *Can. J. Zool.* **51**, 417–420.

Carlisle, D. B., and Dohrn, P. F. R. (1953). Studies on *Lysmata seticaudata* Risso (Crustacea Decapoda). II.—Experimental evidence for a growth and moult-accelerating factor obtainable from eyestalks. *Pubbl. Staz. Zool. Napoli* **24**, 69–83.

Carlisle, D. B., and Knowles, F. (1959). "Endocrine Control in Crustaceans." Cambridge Univ. Press, London and New York.

Carpenter, M. B., and deRoos, R. (1970). Seasonal morphology and histology of the androgenic gland of the crayfish, *Orconectes nais*. *Gen. Comp. Endocrinol.* **15**, 143–157.

Casanova, G. (1955). Influence du prostomium sur la régénération caudale chez *Platynereis massiliensis* (Moquin-Tandon). *C. R. Acad. Sci. Paris* **240**, 1814–1816.

Chaet, A. B. (1966). The gamete-shedding substances of starfishes: a physiological-biochemical study. *Amer. Zool.* **6**, 263–271.

Chaet, A. B., and McConnaughy, R. A. (1959). Physiologic activity of nerve extracts. *Biol. Bull.* **117**, 407–408.

Charniaux-Cotton, H. (1954). Décourverte chez un Crustacé Amphipode (*Orchestia gammarella*) d'une glande endocrine responsable de la différenciation de caractères sexuels primaires et secondaires mâles. *C. R. Acad. Sci. Paris* **239**, 780–782.

Chavin, W. (1966). Adrenal histochemistry of some freshwater and marine teleosts. *Gen. Comp. Endocrinol.* **6**, 183–194.

Chester Jones, I., and Bellamy, D. (1964). Hormonal mechanisms in the regulation of the vertebrate body with special reference to the adrenal cortex. *Symp. Soc. Exp. Biol.* **18**, 195–236.

Chester Jones, I., Phillips, J. G., and Bellamy, D. (1962). Studies on water and electrolytes in cyclostomes and teleosts with special reference to *Myxine glutinosa* L. (the hagfish) and *Anguilla anguilla* L. (the Atlantic eel). *Gen. Comp. Endocrinol. Suppl.* **1**, 36–47.

Chester Jones, I., Chan, D. K. O., Henderson, I. W., and Ball, J. N. (1969). The adrenocortical steroids, adrenocorticotropin and the corpuscles of Stannius. *In* "Fish Physiology" (W. S. Hoar and D. J. Randall, eds.), Vol. II, pp. 321–376. Academic Press, New York.

Chieffi, G. (1961). La luteogenesi nei selaci ovovivipari. Ricerche istologiche e istochimiche in *Torpedo marmorata* e *Torpedo ocellata*. *Pubbl. Staz. Zool. Napoli* **32**, 145–166.

Choquet, M. (1965). Recherches en culture organotypique, sur la spermatogenèse chez *Patella vulgata* L. (Mollusque Gastéropode). Rôle des ganglions cérébroïdes et des tentacules. *C. R. Acad. Sci. Paris* **261**, 4521–4524.

Choquet, M. (1971). Etude du cycle biologique et de l'inversion du sexe chez *Patella vulgata* L. (Mollusque Gastéropode Prosobranche). *Gen. Comp. Endocrinol.* **16**, 59–73.

Clark, R. B., and Bonney, D. G. (1960). Influence of the supra-oesophageal ganglion on posterior regeneration in *Nereis diversicolor*. *J. Embryol. Exp. Morphol.* **8**, 112–118.

Clark, R. B., and Scully, U. (1964). Hormonal control of growth in *Nereis diversicolor*. *Gen. Comp. Endocrinol.* **4**, 82–90.

Coggeshall, R. E. (1970). A cytologic analysis of the bag cell control of egg laying in *Aplysia*. *J. Morphol.* **132**, 461–486.

Coggeshall, R. E. (1972). The muscle cells of the follicle of the ovotestis in *Aplysia* as the probable target organ for bag cell extract. *Amer. Zool.* **12**, 521–523.

Cooke, I. M. (1966). The sites of action of pericardial organ extract and 5-hydroxytryptamine in the decapod crustacean heart. *Amer. Zool.* **6**, 107–121.

Copp, D. H. (1969). The ultimobranchial glands and calcium regulation. *In* "Fish Physiology" (W. S. Hoar and D. J. Randall, eds.), Vol. II, pp. 377–398. Academic Press, New York.

Cottrell, G. A., and Maser, M. (1967). Subcellular localization of 5-hydroxytryptamine and Substance X in molluscan ganglia. *Comp. Biochem. Physiol.* **20**, 901–906.

Couch, E. F. (1971). Mitochondrial transformations in the crayfish Y-organ during molt. *In Proc. Electron Microsc. Soc. Amer.* (C. J. Arceneaux, ed.), pp. 514–515. Claitor's Publ. Div. Baton Rouge, Louisana.

Dahlgren, U. (1914). The electric motor nerve centers in the skates (Rajidae). *Science* **40**, 862–863.

Davey, K. G. (1964). Neurosecretory cells in a nematode, *Ascaris lumbricoides*. *Can. J. Zool.* **42**, 731–734.

Davey, K. G. (1966). Neurosecretion and molting in some parasitic nematodes. *Amer. Zool.* **6**, 243–249.

Davey, K. G., and Breckenridge, W. R. (1967). Neurosecretory cells in a cestode, *Hymenolepsis diminuta*. *Science* **158**, 931–932.

Davis, L. E., Burnett, A. L., and Haynes, J. F. (1968). Histological and ultrastructural study of the muscular and nervous systems in *Hydra*. II. Nervous system. *J. Exp. Zool.* **167**, 295–332.

Demeusy, N. (1953). Effets de l'ablation des pédoncules oculaires sur le développement de l'appareil génital mâle de *Carcinus maenas* Pennant. *C. R. Acad. Sci. Paris* **236**, 974–975.

DeVlaming, V. L. and Sundararaj, B. I. (1972). Endocrine influences on seminal vesicles in the estuarine gobiid fish, *Gillichthys mirabilis*. *Biol. Bull.* **142**, 243–250.

Dhainaut-Courtois, N. (1966a). Le complexe cérébro-vasculaire de *Nereis pelagica* L. (Annélide Polychète). Données histologiques et ultrastructurales. *C. R. Acad. Sci. Paris* **262D**, 2048–2051.

Dhainaut-Courtois, N. (1966b). Le complexe cérébro-vasculaire de *Nereis pelagica* L. Origine des cellules infracérébrales et structure de la paroi du réseau vasculaire. *C. R. Soc. Biol.* **160**, 1232–1234.

Dhainaut-Courtois, N. (1968a). Étude histologique et ultrastructurale des cellules nerveuses du ganglion cérébral de *Nereis pelagica* L. (Annélide Polychète). Comparaison entre les types cellulaires I-VI et ceux décrits anterieurement chez les Nereidae. *Gen. Comp. Endocrinol.* **11**, 414–443.

Dhainaut-Courtois, N. (1968b). Contribution à l'étude du complexe cérébrovasculaire des Néréidiens. Cycle évolutif des cellules infracérébrales de *Nereis pelagica* L. (Annélide Polychète); étude ultrastructurale. *Z. Zellforsch.* **85**, 466–482.

Dhainaut-Courtois, N., and Warembourg, M. (1967). Étude des cellules sécrétrices de la chaîne nerveuse de *Nereis pelagica* L. (Annélide Polychète). *Gen. Comp. Endocrinol.* **9**, 276–286.

Dittus, P. (1940). Histologie und Cytologie des Interrenalorgans der Selachier unter normalen und experimentellen Bedingungen. Ein Beitrag zum Kenntnis der Wirkungsweise des kortikotropen Hormons und des Verhältnisses von Kern zu Plasma. *Z. Wiss. Zool. Abt. A* **154**, 40–124.

Dodd, J. M., Evennett, P. J., and Goddard, C. K. (1960). Reproductive endocrinology in cyclostomes and elasmobranchs. *Symp. Zool. Soc. London* **1**, 77–103.

Drach, P. (1939). Mue et cycle d'intermue chez les Crustacés Décapodes. *Ann. Inst. Océanogr. Monaco* **19**, 103–391.

Durchon, M. (1956). Influence du cerveau sur les processus de régénération caudale chez les Néréidiens (Annélides Polychétes). *Arch. Zool. Exp. Gén.* **94**, 1–9.

Durchon, M. (1960). L'endocrinologie chez les Annélides Polychètes. *Bull. Soc. Zool. Fr.* **85**, 275–301.

Echalier, G. (1954). Recherches expérimentales sur le rôle de "l'organe Y" dans la mue de *Carcinus maenas* (L.) Crustacé Décapode. *C. R. Acad. Sci. Paris* **238**, 523–525.

Echalier, G. (1956). Influence de l'organe Y sur la régénération des pattes, chez *Carcinides maenas* L. (Crustacé Décapode). *C. R. Acad. Sci. Paris* **242**, 2179–2180.

Eddy, J. M. P., and Strahan, R. (1968). The role of the pineal complex in the pigmentary effector system of the lampreys, *Mordacia mordax* (Richardson) and *Geotria australis* Gray. *Gen. Comp. Endocrinol.* **11**, 528–534.

Enami, M. (1951). The sources and activities of two chromatophorotropic hormones in crabs of the genus *Sesarma*. II. Histology of incretory elements. *Biol. Bull.* **101**, 241–258.

Enami, M. (1955). Melanophore-contracting hormone (MCH) of possible hypothalamic origin in the catfish, *Parasilurus*. *Science* **121**, 36–37.

Enami, M. (1959). The morphology and functional significance of the caudal neurosecretory system of fishes. *In* "Comparative Endocrinology" (A. Gorbman, ed.), pp. 697–724. Wiley, New York.

Epple, A. (1969). The endocrine pancreas. *In* "Fish Physiology" (W. S. Hoar and D. J. Randall, eds.), Vol. II, pp. 275–319. Academic Press, New York.

Falkmer, S. (1966). Quelques aspects comparatifs des cellules A pancréatiques et du glucagon. *Ann. Endocrinol.* **27**, 321–330.

Falkmer, S., and Winbladh, L. (1964). An investigation of the pancreatic islet tissue of the hagfish (*Myxine glutinosa*) by light and electron microscopy. *In* "The Structure and Metabolism of the Pancreatic Islets" (S. E. Brolin, B. Hellman, and H. Knutson, eds.), pp. 17–32. Pergamon, Oxford.

Faux, A., Horn, D. H. S., Middleton, E. J., Fales, H. M., and Lowe, M. E. (1969). Moulting hormones of a crab during ecdysis. *Chem. Commun.* 175–176.

Fernlund, P. and Josefsson, L. (1972). Crustacean color-change hormone: amino acid sequence and chemical synthesis. *Science* **177**, 173–175.

Fielder, D. R., Rao, K. R., and Fingerman, M. (1971). Control of distal retinal pigment migration in the fiddler crab *Uca pugilator*. *Mar. Biol.* **9**, 219–223.

Fingerman, M. (1963). "The Control of Chromatophores." Pergamon, Oxford and Macmillan, New York.

Fingerman, M. (1966). Neurosecretory control of pigmentary effectors in crustaceans. *Amer. Zool.* **6**, 169–179.

Fingerman, M (1969). Cellular aspects of the control of physiological color changes in crustaceans. *Amer. Zool.* **9**, 443–452.

Fingerman, M. (1970a). Comparative physiology: chromatophores. *Ann. Rev. Physiol.* **32**, 345–372.

Fingerman, M. (1970b). Circadian rhythm of distal retinal pigment migration in the fiddler crab, *Uca pugilator*, maintained in constant darkness and its endocrine control. *J. Interdisciplinary Cycle Res.* **1**, 115–121.

Fingerman, M., and Aoto, T. (1959). The neurosecretory system of the dwarf crayfish, *Cambarellus shufeldti*, revealed by electron and light microscopy. *Trans. Amer. Microsc. Soc.* **78**, 305–317.

Fingerman, M., and Fingerman, S. W. (1972). Evidence for a substance in the eyestalks of brachyurans that darkens the shrimp *Crangon septemspinosa*. *Comp. Biochem. Physiol.* **43A**, 37–46.

Fingerman, M., and Mobberly, W. C., Jr. (1960). Investigation of the hormones controlling the distal retinal pigment of the prawn *Palaemonetes*. *Biol. Bull.* **118**, 393–406.

Fingerman, M., and Yamamoto, Y. (1964). Endocrine control of tanning in the crayfish exoskeleton. *Science* **144**, 1462.

Fingerman, M., and Yoshioka, P. M. (1968). Dispersion of the dark red chromatophoric pigment in the dwarf crayfish, *Cambarellus shufeldti*: a quantitative analysis of the Hogben and Slome stages. *Tulane Stud. Zool.* **14**, 133–136.

Fingerman, M., Lowe, M. E., and Sundararaj, B. I. (1959). Dark-adapting and light-adapting hormones controlling the distal retinal pigment of the prawn *Palaemonetes vulgaris*. *Biol. Bull.* **116**, 30–36.

Fingerman, M., Rao, K. R., and Bartell, C. K. (1967). A proposed uniform method of reporting response values for crustacean chromatophorotropins: the Standard Integrated Response. *Experientia* **23**, 962.

Fingerman, M., Hammond, R. D., and True, R. S. (1968). The response of the red chromatophores of the prawn *Palaemonetes vulgaris* to cyclic 3′,5′-adenosine monophosphate. *Biol. Bull.* **135**, 418.

Fingerman, M., Bartell, C. K., and Krasnow, R. A. (1971a). Comparison of chromatophorotropins from the horseshoe crab, *Limulus polyphemus*, and the fiddler crab, *Uca pugilator*. *Biol. Bull.* **140**, 376–388.

Fingerman, M., Krasnow, R. A., and Fingerman, S. W. (1971b). Separation, assay, and properties of the distal retinal pigment light-adapting and dark-adapting hormones in the eyestalks of the prawn *Palaemonetes vulgaris*. *Physiol. Zool.* **44**, 119–128.

Fontaine, M. (1964). Corpuscles de Stannius et regulation ionique (Ca, K, Na) du milieu interieur de l'anguille (*Anguilla anguilla* L.). *C. R. Acad. Sci. Paris* **259**, 875–878.

Frazier, W. T., Kandel, E. R., Kupfermann, I., Waziri, R., and Coggeshall, R. E. (1967). Morphological and functional properties of identified neurons in the abdominal ganglion of *Aplysia californica*. *J. Neurophysiol.* **30**, 1288–1351.

Freeman, A. R., Connell, P. M., and Fingerman, M. (1968). An electrophysiological study of the red chromatophore of the prawn, *Palaemonetes*: observations on the action of red pigment-concentrating hormone. *Comp. Biochem. Physiol.* **26**, 1015–1029.

Fridberg, G. (1962). The caudal neurosecretory system in some elasmobranchs. *Gen. Comp. Endocrinol.* **2**, 249–265.

Frontali, N., Williams, L. and Welsh, J. H. (1967). Heart excitatory and inhibitory substances in molluscan ganglia. *Comp. Biochem. Physiol.*, **22**, 833–841.

Fujii, R. (1969). Chromatophores and pigments. *In* "Fish Physiology" (W. S. Hoar and D. J. Randall, eds.), Vol. III, pp. 307–353. Academic Press, New York.

Gabe, M. (1953). Sur l'existence, chez quelques Crustacés Malacostracés, d'un organe comparable à la glande de mue des Insects. *C. R. Acad. Sci. Paris* **237**, 1111–1113.

Gabe, M. (1966). "Neurosecretion." Pergamon, Oxford.

Gaskell, J. F. (1912). The distribution and physiological action of the suprarenal medullary tissue in *Petromyzon fluviatilis*. *J. Physiol.* **44**, 59–67.

Gersch, M. (1957). Wesen und Wirkungsweise von Neurohormonen im Tierreich. *Naturwissenschaften* **44**, 525–532.

Gersch, M., and Scheffel, H. (1958). Sekretorisch tätige Zellen im Nervensystem von *Ascaris*. *Naturwissenschaften* **45**, 345–346.

Gerschenfeld, H. M., Tramezzani, J. H., and DeRobertis, E. (1960). Ultrastructure and function in neurohypophysis of the toad. *Endocrinology* **66**, 741–762.

Golding, D. W. (1967a). The diversity of secretory neurons in the brain of *Nereis*. *Z. Zellforsch.* **82**, 321–344.

Golding, D. W. (1967b). Neurosecretion and regeneration in *Nereis*. I. Regeneration and the role of the supraesophageal ganglion. *Gen. Comp. Endocrinol.* **8**, 348–355.

Golding, D. W. (1967c). Neurosecretion and regeneration in *Nereis*. II. The prolonged secretory activity of the supraesophageal ganglion. *Gen. Comp. Endocrinol.* **8**, 356–367.

Golding, D. W. (1967d). Regeneration and growth control in *Nereis*. I. Growth and regeneration. *J. Embryol. Exp. Morphol.* **18**, 67–77.

Golding, D. W. (1967e). Regeneration and growth control in *Nereis*. II. An axial gradient in growth potentiality. *J. Embryol. Exp. Morphol.* **18**, 79–90.

Golding, D. W., Baskin, D. G., and Bern, H. A. (1968). The infracerebral gland—a possible neuroendocrine complex in *Nereis*. *J. Morphol.* **124**, 187–216.

Gomez, R. (1965). Acceleration of development of gonads by implantation of brain in the crab *Paratelphusa hydrodromous*. *Naturwissenschaften* **52**, 216.

Gorbman, A. (1969). Thyroid function and its control in fishes. *In* "Fish Physiology" (W. S. Hoar and D. J. Randall, eds.), Vol. II, pp. 241–274. Academic Press, New York.

Gottfried, H., and Dorfman, R. I. (1970). Steroids of invertebrates. IV. On the optic tentacle-gonadal axis in the control of the male-phase ovotestis in the slug (*Ariolimax californicus*). *Gen. Comp. Endocrinol.* **15**, 101–119.

Graf, F. (1972). Action de l'ecdystérone sur la mue, la cuticule et le métabolisme du calcium chez *Orchestia cavimana* Heller (Crustacé, Amphipode, Talitridé). *C. R. Acad. Sci.* (Paris) **274**D, 1731–1734.

Grant, W. C., Jr., and Hendler, F. J. (1965). The response of blood glucose and lactate to catecholamines in the skate, *Raja erinacea*. *Bull. Mt. Desert Is. Biol. Lab.* **5**, 15–16.

Grunewald-Lowenstein, M. (1956). Influence of light and darkness on the pineal body in *Astyanax mexicanus* (Filippi). *Zoologica* **41**, 119–128.

Hagadorn, I. R. (1969). Hormonal control of spermatogenesis in *Hirudo medicinalis*. II. Testicular response to brain removal during the phase of testicular maturity. *Gen. Comp. Endocrinol.* **12**, 469–478.

Hanström, B. (1931). Neue Untersuchungen über Sinnesorgane und Nervensystem der Crustaceen. I. *Z. Morphol. Ökol. Tiere* **23**, 80–236.

Hanström, B. (1933). Neue Untersuchungen über Sinnesorgane und Nervensystem der Crustaceen. II. *Zool. Jb. Anat.* **56**, 387–520.

Hauenschild, C. (1966). Der hormonale Einfluss des Gehirns auf die sexuelle Entwicklung bei dem Polychaeten *Platynereis dumerilii. Gen. Comp. Endocrinol.* **6**, 26–73.

Healey, E. G., and Ross, D. M. (1966). The effects of drugs on the background colour response of the minnow *Phoxinus phoxinus* L. *Comp. Biochem. Physiol.* **19**, 545–580.

Heilbrunn, L. V., Chaet, A. B., Dunn, A., and Wilson, W. L. (1954). Antimitotic substances from ovaries. *Biol. Bull.* **106**, 158–168.

Heller, H., and Bentley, P. J. (1965). Phylogenetic distribution of the effects of neurohypophysial hormones on water and sodium metabolism. *Gen. Comp. Endocrinol.* **5**, 96–108.

Herlant-Meewis, H. (1957). Reproduction et neurosécrétion chez *Eisenia foetida* (Sav). *Ann. Soc. Roy. Zool. Belg.* **87**, 151–185.

Highnam, K. C., and Hill, L. (1969). "The Comparative Endocrinology of the Invertebrates." American Elsevier, New York.

Hirai, S., and Kanatani, H. (1971). Site of production of meiosis-inducing substance in ovary of starfish. *Exp. Cell Res.* **67**, 224–227.

Hoar, W. S. (1957a). Endocrine organs. *In* "The Physiology of Fishes" (M. E. Brown, ed.), Vol. I, pp. 254–285. Academic Press, New York.

Hoar, W. S. (1957b). The gonads and reproduction. *In* "The Physiology of Fishes" (M. E. Brown, ed.), Vol. I, pp. 287–321. Academic Press, New York.

Hoar, W. S. (1969). Reproduction: *In* "Fish Physiology" (W. S. Hoar and D. J. Randall, eds.), Vol. III, pp. 1–72. Academic Press, New York.

Hodge, M. H., and Chapman, G. B. (1958). Some observations on the fine structure of the sinus gland of a land crab, *Gecarcinus lateralis. J. Biophys. Biochem. Cytol.* **4**, 571–574.

Hoffman, R. A. (1970). The epiphyseal complex in fish and reptiles. *Amer. Zool.* **10**, 191–199.

Hogben, L. (1936). The pigmentary effector system. VII—The chromatic function in elasmobranch fishes. *Proc. Roy. Soc. London* **B120**, 142–158.

Hogben, L. and Slome, D. (1931). The pigmentary effector system. VI. The dual character of endocrine co-ordination in amphibian colour change. *Proc. Roy. Soc. London* **B108**, 10–53.

Howie, D. I. D. (1963). Experimental evidence for the humoral stimulation of ripening of the gametes and spawning in the polychaete *Arenicola marina* (L.). *Gen. Comp. Endocrinol.* **3**, 660–668.

Howie, D. I. D., and McClenaghan, C. M. (1965). Evidence for a feedback mechanism influencing spermatogonial division in the lugworm (*Arenicola marina* L.). *Gen. Comp. Endocrinol.* **5**, 40–44.

Ikegami, S., Kamiya, Y., and Tamura, S. (1972). Isolation and characterization of spawning inhibitors in ovary of the starfish, *Asterias amurensis. Agr. Biol. Chem.* **36**, 2005–2011.

Ikegami, S., Tamura, S., and Kanatani, H. (1967). Starfish gonad: action and chemical identification of spawning inhibitor. *Science* **158**, 1052–1053.

Joose, J. (1964). Dorsal bodies and dorsal neurosecretory cells of the cerebral ganglia of *Lymnaea stagnalis* L. *Arch. Neerl. Zool.* **16**, 1–103.

Kamemoto, F. I. (1964). The influence of the brain on osmotic and ionic regulation in earthworms. *Gen. Comp. Endocrinol.* **4**, 420–426.

Kamemoto, F. I., Kato, K. N., and Tucker, L. E. (1966). Neurosecretion and salt and water balance in the Annelida and Crustacea. *Amer. Zool.* **6**, 213–219.

Kanatani, H. (1969). Mechanism of starfish spawning: action of neural substance on the isolated ovary. *Gen. Comp. Endocrinol. Suppl.* **2**, 582–589.

Kanatani, H., and Ohguri, M. (1970). Effect of gonad extract on oocyte maturation in the starfish, *Asterias amurensis*. *Zool. Mag.* **79**, 58–59.

Kanatani, H., and Shirai, H. (1967). *In vitro* production of meiosis inducing substance by nerve extract in ovary of starfish. *Nature* (*London*) **216**, 284–286.

Kanatani, H., and Shirai, H. (1970). Mechanism of starfish spawning. III. Properties and action of meiosis-inducing substance produced in gonad under influence of gonad-stimulating substance. *Develop. Growth Differentiation* **12**, 119–140.

Kanatani, H., Shirai, H. Nakanishi, K., and Kurokawa, T. (1969). Isolation and identification of meiosis inducing substance in starfish *Asterias amurensis*. *Nature* (*London*) **221**, 273–274.

Kanatani, H., Ikegami, S., Shirai, H., Oide, H., and Saburo, T. (1971). Purification of gonad-stimulating substance obtained from radial nerves of the starfish, *Asterias amurensis*. *Develop. Growth Differentiation* **13**, 151–164.

Karlson, P. (1963). New concepts on the mode of action of hormones. *Perspect. Biol. Med.* **6**, 203–214.

Kato, K. N., and Kamemoto, F. I. (1969). Neuroendocrine involvement in osmoregulation in the grapsid crab *Metopograpsus messor*. *Comp. Biochem. Physiol.* **28**, 665–674.

Keller, R. (1965). Über eine hormonale Kontrolle des Polysaccharidstoffwechsels beim Flusskrebs *Cambarus affinis* Say. *Z. Vergl. Physiol.* **51**, 49–59.

Keller, R. (1968). Über Versuche zur Charakterisierung des diabetogenen Augenstielhormons des Flusskrebses *Orconectes limosus*. *Verhandl. Deutsch. Zool. Gesell. Innsbruck* 628–635.

Keller, R. (1969). Untersuchungen zur Artspezifität eines Crustaceenhormons. *Z. Vergl. Physiol.* **63**, 137–145.

Kerkut, G. A., and Laverack, M. S. (1960). A cardio-accelerator present in tissue extracts of the snail *Helix aspersa*. *Comp. Biochem. Physiol.* **1**, 62–71.

King, D. S. (1964). Fine structure of the androgenic gland of the crab, *Pachygrapsus crassipes*. *Gen. Comp. Endocrinol.* **4**, 533–544.

King, D. S., and Siddall, J. B. (1969). Conversion of α-ecdysone to β-ecdysone by crustaceans and insects. *Nature* (*London*) **221**, 955–956.

Kleinholz, L. H. (1935). The melanophore-dispersing principle in the hypophysis of *Fundulus heteroclitus*. *Biol. Bull.* **69**, 379–390.

Kleinholz, L. H. (1936). Crustacean eye-stalk hormone and retinal pigment migration. *Biol. Bull.* **70**, 159–184.

Kleinholz, L. H. (1966). Hormonal regulation of retinal pigment migration in crustaceans. *Proc. Int. Symp. Functional Organization Compound Eye* pp. 89–101, Pergamon, Oxford.

Kleinholz, L. H., Kimball, F., and McGarvey, M. (1967). Initial characterization and separation of hyperglycemic (diabetogenic) hormone from the crustacean eyestalk. *Gen. Comp. Endocrinol.* **8**, 75–81.

Knowles, F. G. W. (1953). Endocrine activity in the crustacean nervous system. *Proc. Roy. Soc. London* **B141**, 248–267.

Knowles, F. (1962). The ultrastructure of a crustacean neurohaemal organ. *In* "Neurosecretion" (H. Heller and R. B. Clark, eds.), pp. 71–88. Academic Press, New York.

Knowles, F. (1965). Neuroendocrine correlations at the level of ultrastructure. *Arch. Anat. Micros.* **54**, 343–358.

Knowles, F., and Vollrath, L. (1966a). Neurosecretory innervation of the pituitary of the eels *Anguilla* and *Conger*. I. The structure and ultrastructure of the neuro-intermediate lobe under normal and experimental conditions. *Phil. Trans. Roy. Soc. London* **B250**, 311–327.

Knowles, F., and Vollrath, L. (1966b). Neurosecretory innervation of the pituitary of the eels *Anguilla* and *Conger*. II. The structure and innervation of the pars distalis at different stages of the life-cycle. *Phil. Trans. Roy. Soc. London* **B250**, 329–342.

Köhler, V. (1952). Die Wirkung des Adrenocorticotropins auf die Lipophoren der Pfrille (*Phoxinus laevis*). *Naturwissenchaften* **39**, 554.

Krawarik, F. (1936). Über eine bisher unbekannte Drüse ohne Ausführungsgang bei den heimischen Knochenfischen. *Z. Mikrosc. Anat. Forsch.* **39**, 555–608.

Krishnakumaran, A., and Schneiderman, H. A. (1969). Induction of molting in Crustacea by an insect molting hormone. *Gen. Comp. Endocrinol.* **12**, 515–518.

Krishnakumaran, A., and Schneiderman, H. A. (1970). Control of molting in mandibulate and chelicerate arthropods by ecdysones. *Biol. Bull.* **139**, 520–538.

Kupfermann, I. (1970). Stimulation of egg laying by extracts of neuroendocrine cells (bag cells) of abdominal ganglion of *Aplysia*. *J. Neurophysiol.* **33**, 877–881.

Kupfermann, I. (1972). Studies on the neurosecretory control of egg laying in *Aplysia*. *Amer. Zool.* **12**, 513–519.

Lambert, J. G. D., and van Oordt, P. G. W. J. (1965). Preovulatory corpora lutea or corpora atretica in the guppy, *Poecilia reticulata*. A histological and histochemical study. *Gen. Comp. Endocrinol.* **5**, 693–694.

Lane, N. J. (1964). The fine structure of certain secretory cells in the optic tentacles of the snail, *Helix aspersa*. *Quart. J. Microsc. Sci.* **105**, 35–47.

Larsen, L. O. (1965). Effects of hypophysectomy in the cyclostome, *Lampetra fluviatilis* (L.) Gray. *Gen. Comp. Endocrinol.* **5**, 16–30.

Larsen, L. O. (1969). Effects of hypophysectomy before and during sexual maturation in the cyclostome, *Lampetra fluviatilis* (L.) Gray. *Gen. Comp. Endocrinol.* **12**, 200–208.

Lechenault, H. (1962). Sur l'existence de cellules neurosécrétrices dans les ganglions cérébroïdes des Lineidae (Hétéronémertes). *C. R. Acad. Sci. Paris* **255**, 194–196.

Lederis, K. (1970a). Teleost urophysis. I. Bioassay of an active urophysical principle on the isolated urinary bladder of the rainbow trout, *Salmo gairdnerii*. *Gen. Comp. Endocrinol.* **14**, 417–426.

Lederis, K. (1970b). Teleost urophysis. II. Biological characterization of the bladder-contracting activity. *Gen. Comp. Endocrinol.* **14**, 427–437.

Lender, T. (1970). Le rôle de la neurosécrétion au cours de la régénération et de la reproduction asexuée des Planaires d'eau douce. *Ann. Endocrinol.* **31**, 463–466.

Lender, T., and Klein, N. (1961). Mise en évidence de cellules sécrétrices dans le cerveau de la Planaire *Polycelis nigra*. Variation de leur nombre au cours de la régénération postérieure. *C. R. Acad. Sci. Paris* **253**, 331–333.

Lender, T., and Zghal, F. (1968). Influence du cerveau et de la neurosécrétion sur la scissiparité de la Planaire *Dugesia gonocephala*. *C. R. Acad. Sci. Paris* **267**, 2008–2009.

Lentz, T. L. (1967). Fine structure of nerve cells in a planarian. *J. Morphol.* **121**, 323–338.

Lentz, T. L., and Barrnett, R. J. (1965). Fine structure of the nervous system of *Hydra*. *Amer. Zool.* **5**, 341–356.

Lesh, G. E., and Burnett, A. L. (1964). Some biological and biochemical properties of the polarizing factor in *Hydra*. *Nature* (*London*) **204**, 492–493.

Lever, J., Jansen, J., and DeVlieger, T. A. (1961). Pleural ganglia and water balance in the freshwater pulmonate *Lymnaea stagnalis*. *Koninkl. Nederl. Acad. Wetens.* **C64**, 532–542.

Louw, G. N., Sutton, W. W., and Kenny, A. D. (1967). Action of thyrocalcitonin in the teleost fish *Ictalurus melas*. *Nature* (*London*) **215**, 888–889.

Lowe, M. E., Horn, D. H. S., and Galbraith, M. N. (1968). The role of crustecdysone in the moulting crayfish. *Experientia* **24**, 518–519.

Lundstrom, H. M., and Bard, P. (1932). Hypophysial control of cutaneous pigmentation in an elasmobranch fish. *Biol. Bull.* **62**, 1–9.

McWhinnie, M. A., and Chua, A. S. (1964). Hormonal regulation of crustacean tissue metabolism. *Gen. Comp. Endocrinol.* **4**, 624–633.

Maetz, J. (1969). Observations on the role of the pituitary-interrenal axis in the ion regulation of the eel and other teleosts. *Gen. Comp. Endocrinol. Suppl.* **2**, 299–316.
Maetz, J., Bourguet, J., and Lahlouh, B. (1964). Urophyse et osmorégulation chez *Carassius auratus*. *Gen. Comp. Endocrinol.* **4**, 401–414.
Maissiat, J. (1970). Anecdysis expérimentale provoquée chez l'Oniscoïde *Ligia oceanica* L. et rétablissement de la mue par injection d'ecdysone ou réimplantation de glande maxillaire. *C. R. Soc. Biol.* **164**, 1607–1609.
Maissiat, J., and Graf, F. (1973). Action de l'ecdystérone sur l'apolysis et l'ecdysis de divers Crustacés Isopodes. *J. Insect Physiol.* **19**, 1265–1276.
Mantel, L. H. (1968). The foregut of *Gecarcinus lateralis* as an organ of salt and water balance. *Amer. Zool.* **8**, 433–442.
Matty, A. J. (1966). Endocrine glands in lower vertebrates. *Int. Rev. Gen. Exp. Zool.* **2**, 43–138.
Maynard, D. M. (1961). Thoracic neurosecretory structures in Brachyura. II. Secretory neurons. *Gen. Comp. Endocrinol.* **1**, 237–263.
Maynard, D. M., and Maynard, E. A. (1962). Thoracic neurosecretory structures in Brachyura. III. Microanatomy of peripheral structures. *Gen. Comp. Endocrinol.* **2**, 12–28.
Maynard, D. M., and Welsh, J. H. (1959). Neurohormones of the pericardial organs of brachyuran Crustacea. *J. Physiol.* **149**, 215–227.
Meusy, J. J. (1965). Contribution de la microscopie électronique à l'étude de la physiologie des glands androgènes d'*Orchestia gammarella* P. (Crustacé Amphipode) et de *Carcinus maenas* L. (Crustacé Décapode). *Zool. Jb. Physiol.* **71**, 608–623.
Meusy, J. J. (1968). Précisions nouvelles sur l'ultrastructure de la glande du sinus d'un Crustacé Décapode Brachyoure, *Carcinus maenas* L. *Bull. Soc. Zool. Fr.* **93**, 291–299.
Miyawaki, M., and Taketomi, Y. (1970). Structural changes induced in the cells of Y gland of crayfish by an administration of ecdysterone. *Zool. Mag.* **79**, 150–155.
Moore, B. D., and Biggers, J. D. (1964). In vitro studies on the shedding activity found in *Asterias* nerve extracts. *Biol. Bull.* **127**, 381–382.
Morita, M., and Best, J. B. (1965). Electron microscopic studies on planaria. II. Fine structure of the neurosecretory system in the planarian *Dugesia dorotocephala*. *J. Ultrastruct. Res.* **13**, 396–408.
Motais, R., and Maetz, J. (1964). Action des hormones neurohypophysaires sur les échanges de sodium (mesurés à l'aide du radio-sodium Na^{24}) chez un téléostéen euryhalin: *Platichthys flesus* L. *Gen. Comp. Endocrinol.* **4**, 210–224.
Nandi, J. (1962). The structure of the interrenal gland in teleost fishes. *Univ. Calif. Publ. Zool.* **65**, 129–211.
Nolte, A. (1965). Neurohämal-"Organe" bei Pulmonaten (Gastropoda). *Zool. Jb. Anat.* **82**, 365–380.
Normann, T. C. (1965). The neurosecretory system of the adult *Calliphora erythrocephala*. I. The fine structure of the corpus cardiacum with some observations on adjacent organs. *Z. Zellforsch.* **67**, 461–501.
O'Dor, R. K., Parkes, C. O., and Copp, D. H. (1969). Amino acid composition of salmon calcitonin. *Can. J. Biochem.* **47**, 823–825.
Olivereau, M. (1967). Observations sur l'hypophyse de l'Anguille femelle, en particulier lors de la maturation sexuelle. *Z. Zellforsch.* **80**, 286–306.
Olivereau, M. (1971). Action de la réserpine chez l'Anguille. I. Cellules à prolactine de l'hypophyse du mâle. *Z. Zellforsch.* **121**, 232–243.
Olivereau, M. (1972). Action de la réserpine chez l'Anguille. II. Effet sur la pigmentation et le lobe intermédiaire. Comparaison avec l'effet de l'adaptation sur un fond noir. *Z. Zellforsch.* **137**, 30–46.

Oosaki, T., and Ishii, W. (1965). Observations on the ultrastructure of nerve cells in the brain of the planarian, *Dugesia gonocephala. Z. Zellforsch.* **66**, 782–793.
Ōtsu, T. (1963). Bihormonal control of sexual cycle in the freshwater crab, *Potamon dehaani. Embryologia* **8**, 1–20.
Pang, P. K. T. (1971a). Calcitonin and ultimobranchial glands in fishes. *J. Exp. Zool.* **178**, 89–100.
Pang, P. K. T. (1971b). The relationship between corpuscles of Stannius and serum electrolyte regulation in killifish, *Fundulus heteroclitus. J. Exp. Zool.* **178**, 1–8.
Pang, P. K. T., and Pickford, G. E. (1967). Failure of hog thyrocalcitonin to elicit hypocalcemia in the teleost fish, *Fundulus heteroclitus. Comp. Biochem. Physiol.* **21**, 573–578.
Pang, P. K. T., Griffith, R. W., and Pickford, G. E. (1971). Hypocalcemia and tetanic seizures in hypophysectomized killifish, *Fundulus heteroclitus. Proc. Soc. Exp. Biol. Med.* **136**, 85–87.
Panouse, J. B. (1943). Influence de l'ablation du pédoncule oculaire sur la croissance de l'ovarie chez la crevette *Leander serratus. C. R. Acad. Sci. Paris* **217**, 553–555.
Parker, G. H. (1935). The electric stimulation of the chromatophoral nerve-fibers in the dogfish. *Biol. Bull.* **68**, 1–3.
Parker, G. H., and Porter, H. (1934). The control of the dermal melanophores in elasmobranch fishes. *Biol. Bull.* **66**, 30–37.
Passano, L. M. (1951). The X organ-sinus gland neurosecretory system in crabs. *Anat. Rec.* **111**, 502.
Passano, L. M. (1953). Neurosecretory control of molting in crabs by the X-organ sinus gland complex. *Physiol. Comp. Oecol.* **3**, 155–189.
Patent, G. J. (1970). Comparison of some hormonal effects on carbohydrate metabolism in an elasmobranch (*Squalus acanthias*) and a holocephalan (*Hydrolagus colliei*). *Gen. Comp. Endocrinol.* **14**, 215–242.
Pelluet, D., and Lane, N. J. (1961). The relation between neurosecretion and cell differentiation in the ovotestis of slugs (Gasteropoda: Pulmonata). *Can. J. Zool.* **39**, 789–805.
Perks, A. M. (1969). The neurohypophysis. *In* "Fish Physiology" (W. S. Hoar and D. J. Randall, eds.), Vol. II, pp. 111–205. Academic Press, New York.
Pflugfelder, O. (1956). Wirkungen von Epiphysan und Thyroxin auf die Schilddrüse epiphysektomierter *Lebistes reticulatus* Peters. *Roux' Arch. Entwicklungsmech.* **148**, 463–473.
Pickford, G. E. (1953). A study of the hypophysectomized male killifish, *Fundulus heteroclitus* (*Linn.*). *Bull. Bingham Oceanogr. Coll.* **14**, 5–41.
Pickford, G. E. (1954). The response of hypophysectomized male killifish to purified fish growth hormone, as compared with the response to purified beef growth hormone. *Endocrinology* **55**, 274–287.
Pickford, G. E., and Atz, J. W. (1957). "The Physiology of the Pituitary Gland of Fishes." New York Zool. Soc., New York.
Pickford, G. E., and Phillips, J. G. (1959). Prolactin, a factor promoting survival of hypophysectomized killifish in fresh water. *Science* **130**, 454–455.
Pickford, G. E., Robertson, E. E., and Sawyer, W. H. (1965). Hypophysectomy, replacement therapy, and the tolerance of the euryhaline killifish, *Fundulus heteroclitus*, to hypotonic media. *Gen. Comp. Endocrinol.* **5**, 160–180.
Pickford, G. E., Griffith, R. W., Torretti, J., Hendler, E., and Epstein, F. H. (1970a). Branchial reduction and renal stimulation of (Na^+, K^+)-ATPase by prolactin in hypophysectomized killifish in fresh water. *Nature* (*London*) **228**, 378–379.
Pickford, G. E., Pang, P. K. T., Weinstein, E., Torretti, J., Hendler, E., and Epstein, F. H. (1970b). The response of the hypophysectomized cyprinodont, *Fundulus heteroclitus*, to replacement therapy with cortisol: effects on blood serum and sodium-potassium acti-

vated adenosine triphosphatase in the gills, kidney, and intestinal mucosa. *Gen. Comp. Endocrinol.* **14**, 524–534.
Quay, W. B. (1970). Endocrine effects of the mammalian pineal. *Amer. Zool.* **10**, 237–246.
Ramamurthi, R., and Scheer, B. T. (1967). A factor influencing sodium regulation in crustaceans. *Life Sci.* **6**, 2171–2175.
Ramamurthi, R., Mumbach, M. W., and Scheer, B. T. (1968). Endocrine control of glycogen synthesis in crabs. *Comp. Biochem. Physiol.* **26**, 311–319.
Rao, K. R. (1965). Isolation and partial characterization of the moult-inhibiting hormone of the crustacean eyestalk. *Experientia* **21**, 593–594.
Rao, K. R., Fingerman, M., and Bartell, C. K. (1967). Physiology of the white chromatophores in the fiddler crab, *Uca pugilator. Biol. Bull.* **133**, 606–617.
Rao, K. R., Fingerman, M., and Hays, C. (1972). Comparison of the abilities of α-ecdysone and 20-hydroxyecdysone to induce precocious proecdysis and ecdysis in the fiddler crab, *Uca pugilator. Z. Vergl. Physiol.* **76**, 270–284.
Rao, K. R., Fingerman, S. W., and Fingerman, M. (1973). Effects of exogenous ecdysones on the molt cycles of fourth and fifth stage American lobsters, *Homarus americanus. Comp. Biochem. Physiol.* **44**A, 1105–1120.
Reed, B. L. (1968). The control of circadian pigment changes in the pencil fish: a proposed role for melatonin. *Life Sci.* **7**, Part II, 961–973.
Reed, B. L., and Finnin, B. C. (1970). Adrenergic innervation of melanophores in a teleost fish. *J. Invest. Dermatol.* **54**, 95–96.
Röhlich, P., Aros, B., and Vigh, B. (1962). Elektronenmikroscopische Untersuchung der Neurosekretion im Cerebralganglion des Regenwurmes (*Lumbricus terrestris*). *Z. Zellforsch.* **58**, 524–545.
Rosenbluth, J. (1963). The visceral ganglion of *Aplysia californica. Z. Zellforsch.* **60**, 213–236.
Sage, M. (1970). Control of prolactin release and its role in color change in the teleost *Gillichthys mirabilis. J. Exp. Zool.* **173**, 121–128.
Sandeen, M. I., and Brown, F. A., Jr. (1952). Responses of the distal retinal pigment of *Palaemonetes* to illumination. *Physiol. Zool.* **25**, 222–230.
Sarojini, S. (1964). A note on the chemical nature of the crustacean androgenic hormone. *Current Sci.* **33**, 55–56.
Sawyer, W. H., Baxter, J. W. M., Manning, M., Heinicke, E., and Perks, A. M. (1970). A fraction resembling oxytocin from *Squalus acanthias:* pharmacological comparisons with synthetic peptides. *Gen. Comp. Endocrinol.* **15**, 52–58.
Scharrer, B. (1935). Ueber das Hanströmsche Organ X bei Opisthobranchiern. *Pubbl. Staz. Zool. Napoli* **15**, 132–142.
Scharrer, B. (1936). Über "Drüsen-Nervenzellen" im Gehirn von *Nereis virens* Sars. *Zool. Anz.* **113**, 299–302.
Scharrer, B. (1969). Neurohumors and neurohormones: definitions and terminology. *J. Neuro-Viscer. Relat. Suppl.* **9**, 1–20.
Scharrer, E., and Scharrer, B. (1954). Neurosekretion. *Handb. Mikro. Anat. Menschen* **6**, 953–1066.
Schipp, R., Höhn, P., and Ginkel, G. (1973). Elektronenmikroskopische und histochemische Untersuchungen zur Funktion der Bronchialdrüse (Parabranchialdrüse) der Cephalopoda. *Z. Zellforsch.* **139**, 253–269.
Schuetz, A. W. (1969). Chemical properties and physiological actions of a starfish radial nerve factor and ovarian factor. *Gen. Comp. Endocrinol.* **12**, 209–221.
Schuetz, A. W., and Biggers, J. D. (1967). Regulation of germinal vesicle breakdown in starfish oocytes. *Exp. Cell Res.* **46**, 624–628.
Sereni, E. (1930). The chromatophores of the cephalopods. *Biol. Bull.* **59**, 247–268.

Shivers, R. R. (1967). Fine structure of crayfish optic ganglia. *Univ. Kansas Sci. Bull.* **47**, 677–733.

Shivers, R. R. (1969). Possible sites of release of neurosecretory granules in the sinus gland of the crayfish, *Orconectes nais. Z. Zellforsch.* **97**, 38–44.

Simpson, L., Bern, H. A., and Nishioka, R. S. (1964). Cytologic observations on the nervous system of the pulmonate gastropod *Heliosoma tenue,* with special reference to possible neurosecretion. *Amer. Zool.* **4**, 407–408.

Simpson, L., Bern, H. A., and Nishioka, R. S. (1966). Survey of evidence for neurosecretion in gastropod molluscs. *Amer. Zool.* **6**, 123–138.

Skorkowski, E. F. (1971). Isolation of three chromatophorotropic hormones from the eyestalk of the shrimp *Crangon crangon. Mar. Biol.* **8**, 220–223.

Sochasky, J. B., Aiken, D. E., and Watson, N. H. F. (1972). Y organ, molting gland, and mandibular organ: a problem in decapod Crustacea. *Can. J. Zool.* **50**, 993–997.

Speidel, C. C. (1919). Gland-cells of internal secretion in the spinal cord of the skates. *Carnegie Inst. Wash., Papers Dept. Mar. Biol.* **13**, 1–31.

Stéphan-Dubois, F., and Lender, T. (1956). Corrélations humorales dans la régénération des *Planaires paludicoles. Ann. Sci. Nat.* (*Zool.*) **18**, Ser. 11, 223–230.

Sterba, G. (1955). Das Adrenal- und Interrenalsystem in Lebensablauf von *Petromyzon planeri* Bloch. I. Morphologie und Histologie einschliesslich Histogenese. *Zool. Anz.* **115**, 151–168.

Stevens, M. (1970). Procedures for induction of spawning and meiotic maturation of starfish oocytes by treatment with 1-methyladenine. *Exp. Cell Res.* **59**, 482–484.

Taki, I. (1964). On the morphology and physiology of the branchial gland in Cephalopoda. *J. Fac. Fish. Anim. Husb, Hiroshima Univ.* **5**, 345–417.

Thomas, N. W. (1970). Morphology of the endocrine cells in the islet tissue of the cod *Gadus callarias. Acta Endocrinol.* **63**, 679–695.

Török, L. J. (1958). Experimental contributions to the regenerative capacity of *Dugesia* (= *Euplanaria*) *lugubris* O. Schm. *Acta Biol. Hung.* **9**, 79–98.

Tullis, R. E. and Kamemoto, F. I. (1971). CNS factors affecting water movement in decapod crustaceans. *Amer Zool.* **11**, 646.

Turner, R. S. (1946). Observations on the central nervous system of *Leptoplana acticola. J. Comp. Neurol.* **85**, 53–65.

Ude, J. (1962). Neurosekretorische Zellen im Cerebralganglion von *Dicrocoelium lanceatum* St. u. H. (Trematoda-Digena). *Zool. Anz.* **169**, 455–457.

Unger, H. (1960). Neurohormone bei Seesternen (*Marthasterias glacialis*). *Symp. Biol. Hung.* **1**, 203–207.

Unger, H. (1962). Experimentelle und histologische Untersuchungen über Wirkfaktoren aus dem Nervensystem von *Asterias* (*Marthasterias*) *glacialis* (Asteroidea; Echinodermata). *Zool. Jb. Physiol.* **69**, 481–536.

Van Noorden, S., Greenberg, J., and Pearse, A.G.E. (1972). Cytochemical and immunofluorescence investigations on polypeptide hormone localization in the pancreas and gut of the larval lamprey. *Gen. Comp. Endocrinol.* **19**, 192–199.

Vicente, N. (1969). Contribution a l'étude des Gastéropodes Opisthobranches du Golfe de Marseille. II- Histophysiologie du système nerveux. Étude des phénomènes neurosécrétoires. Rec. Trav. Stat. Mar. Endoume, *Bull. 46, Fasc.* **62**, 13–121.

Vivien, J. H. (1941). Contribution a l'étude de la physiologie hypophysaire dans ses relations avec l'appareil génital, la thyroïde et les corps suprarénaux chez les poissons sélaciens et téléosteens *Scylliorhinus canicula* et *Gobius paganellus. Bull. Biol. Fr. Belg.* **75**, 257–309.

Wang, D. H., and Scheer, B. T. (1963). UDPG-glycogen transglucosylase and a natural inhibitor in crustacean tissues. *Comp. Biochem. Physiol.* **9**, 263–274.

Weitzman, M. (1969). Ultrastructural study on the release of neurosecretory material from the sinus gland of the land crab, *Gecarcinus lateralis. Z. Zellforsch.* **94**, 147–154.

Wells, M. J., and Wells, J. (1959). Hormonal control of sexual maturity in *Octopus. J. Exp. Biol.* **36**, 1–33.

Welsh, J. H. (1953). Excitation of the heart of *Venus mercenaria. Arch. Exp. Pathol. Pharmak.* **219**, 23–29.

Wingstrand, K. G. (1966). Comparative anatomy and evolution of the hypophysis. *In* "The Pituitary Gland" (G. W. Harris and B. T. Donovan, eds.), Vol. I, pp. 58–126. Butterworth, London and Washington, D.C.

Yagi, K., and Bern, H. A. (1965). Electrophysiologic analysis of the response of the caudal neurosecretory system of *Tilapia mossambica* to osmotic manipulations. *Gen. Comp. Endocrinol.* **5**, 509–526.

Yamazaki, F., and Donaldson, E. M. (1968). The spermiation of goldfish (*Carassius auratus*) as a bioassay for salmon (*Oncorhynchus tshawytscha*) gonadotropin. *Gen. Comp. Endocrinol.* **10**, 383–391.

Young, J. Z. (1935). The photoreceptors of lampreys. II. The functions of the pineal complex. *J. Exp. Biol.* **12**, 254–270.

Zarrow, M. X., Yochim, J. M., and McCarthy, J. L. (1964). "Experimental Endocrinology, A Sourcebook of Basic Techniques." Academic Press, New York.

Chapter 6

Comparative Biochemistry: Marine Biochemistry

James S. Kittredge

I. Introduction

Marine biochemistry has undergone an explosive growth in the past decade due as much to a change in philosophy as to the application of new techniques. The established function of comparative biochemistry has been focused on the detection of known metabolic routes in new species or on the distribution of known compounds in a group of species. This approach has now given way to emphasis of the uniqueness of metabolic systems in invertebrates, evident in papers which range in scope from studies of coelenterate carbon monoxide production as a means of floatation and of vertical migration (Pickwell, 1970) to studies of the role of 1-methyl adenine in the induction of the final step of meiosis in starfish (Kanatani, 1969, 1971). Enzymatic studies now reflect interest in control mechanisms as well as in kinetics. New areas of study include active site structures and the primary structure of proteins. Natural product chemists are no longer satisfied only with the elucidation of the structure of a new compound, but they are now frequently interested in its biological activity as well. In itself the discovery of a new compound contributes little to biochemistry, but its discovery often leads to study of its biosynthesis and function. A challenging area which remains a critical unsolved problem in biology is the nature of the interaction of small "messenger" molecules with receptor sites on a cell membrane as well as the sequence of events which lead to a response. The classic work of Lenhoff (1968) on the hydra glutathione receptor exemplifies the advantage of conducting such studies on a "simple" organism in an aqueous environment.

What then delimits the field of marine biochemistry? One might suggest that "any (marine) area in which a biochemical approach can provide an insight" is marine biochemistry, but a more personal criteria would be "any problem that excites a marine biochemist." This chapter will conform to the latter criterion and cover a potpourri of problems.

II. Conversion of the Inorganic Constituents of the Environment

A. MANGANESE NODULES

The widespread occurrence of these ferromanganous nodules throughout the major ocean basins has long posed a geochemical puzzle. They occur in areas of limited sedimentation as nodular concretions formed about a nucleus (shell fragments, sharks teeth). Growth usually is in easily defined concentric layers and Menard and Shipek (1958) have estimated that from 20 to 50% of the deep sea floor in the southwest Pacific Ocean is covered with these nodules.

Goldberg (1963) attributes the occurrence of the thermodynamically unstable manganous ion in seawater "to a lack of reaction sites where equilibrium might be obtained, i.e., the water-mass containing this type of ion does not encounter a surface of the lithosphere, biosphere or atmosphere at which the energetically possible reactions can proceed." As he points out, manganese occurs in the crustal rocks primarily in the divalent state and most probably enters the marine environment in this state. He does not, however, consider the possible contribution of the complex manganese cycle in the soil elucidated by microbiologists concerned with plant nutrition (Mann and Quastel, 1946; Leeper, 1947).

The assumption that a solid interface is needed to catalyze the oxidation of manganous ion to manganese dioxide seems likely, but the conclusion that the divalent form may exist for thousands of years in disequalibrium is unnecessary. As early as 1894 Adeny (1894) observed the rapid reduction of MnO_2 to $MnCO_3$ in the presence of the microorganisms and organic material contained in liquid sewage. Recently, Ehrlich (1964) and Troshonov (1965) have isolated pure cultures of bacteria that reduce MnO_2 to Mn^{++}: Ehrlich's isolates were from manganese nodules and two of the isolates were examined to determine the mechanism of solubilization of MnO_2 (Ehrlich, 1966). He demonstrated that the organism's capacity to reduce MnO_2 was due to a typical dissimilatory transformation of the MnO_2 in which it served as the terminal acceptor in a respiratory chain. The MnO_2-reductase was shown to be incomplete in cells cultured in the absence of an electron carrier, requiring a 1 day lag period for the development of the fully competent respiratory chain. Physiological studies on these organisms by Trimble and Ehrlich (1968) revealed that the induction was substrate and oxygen dependent, but after induction the rate of solubilization of manganese was virtually the same in the presence or absence of oxygen. Trimble (1969) has recently reported that manganous ion can act as an inducer for the formation of a complete substrate-dependent MnO_2-reductase system. On the basis of results obtained with inhibitors of RNA and protein synthesis, he concluded that a classic induction mechanism exists. Active cell free extracts were obtained and the enzymatic nature of the reductase was confirmed by its inhibition by $HgCl_2$, atebrine, or by heating in a boiling water bath.

Goldberg's isotope dating studies suggested that the growth rate of ferromanganese nodules was extremely slow, of the order of hundredths of a millimeter to millimeters per thousand years. I suggest, however, that the rate may in some cases vary widely. I recall recovering large numbers of cone shaped nodules in which the growth cone vertex angle from the nucleus on the bottom was within only a few degrees of planar. The slope of these cones represents the ratio of the rate of nodule formation to that of sedimentation and this slope overall was rather uniform. The growth rings on the bottom suggested alterations in this ratio. These findings suggest that microbial MnO_2 reduction is ubiquitous but its rate is limited by the supply of organic nutrients. Trimble and Ehrlich (1970) have found that the reduction is minimally dependent on the presence of glucose, peptone or oxygen, but that MnO_2 and O_2 do not compete as terminal electron acceptors. Rittenberg (1969) has reviewed the roles of exogenous organic matter in chemolithotropic bacteria. Manganese may go through continuing cycles of deposition and resolution and the resolubilized manganous ions from the sediment diffuse into the immediate bottom water where they move laterally until contacting a solid nucleating interface which catalyzes the reoxidation. This redeposition may also be catalyzed by microorganisms (Ehrlich, 1971). Krumbein (1971) has found manganese oxidizing fungi and bacteria in 70% of the sediment samples that he examined, always in the uppermost millimeter. These organisms can precipitate MnO_2 from unenriched seawater. Brantner (1970) has reported finding a Mn^{2+}-citrate complex in cells. Tyler (1970) has reviewed the field of manganese oxidation in aquatic systems by hyphomicrobia.

As Trimble (1969) suggests, future research should attempt the extraction and purification of the components of the MnO_2-reductase from large batches of the bacillus. He notes that it will be of interest to find out which component of the electron transport system is present in the induced bacteria for which ferricyanide substitutes in the uninduced bacteria. I would like to suggest an experimental procedure for detecting and isolating the inducible component from two relatively small cultures of the bacillus. If the uninduced bacteria are grown in the presence of ^{3}H-labeled amino acids and the induced bacteria are cultured in a medium containing ^{14}C-labeled amino acids, the two cultures may be mixed, sonicated and the soluble proteins fractionated by the usual techniques of Sephadex gel chromatography, DEAE chromatography, etc. The protein peaks in the effluent may be monitored in the UV and the $^{14}C/^{3}H$ ratio measured for each peak. This ratio will be uniform for all of the proteins except the induced component and this ratio will provide the index of purity in subsequent fractionations.

B. IODIDE AND ARSENITE IN SEAWATER

Two other thermodynamically unstable species exist in seawater, the iodide and arsenite ions. It had been suggested (Goldberg, 1963) that the marine biosphere must be intimately involved in the iodine cycle. Recently Tsunogai and Sase (1969) have demonstrated bacterial reduction of iodate to iodide by five species of marine bacteria. All of these species were capable of reducing nitrate. The reduction of nitrate to nitrite (standard redox potential

+0.42V at pH7) is more difficult than the reduction of iodate to iodide (+0.67V at pH7). These workers also demonstrated that the nitrate reductase isolated from *Escherichia coli*, in the presence of 6mM methylene blue, was capable of reducing iodate to iodide. They propose that this reduction of iodate by nitrate-reducing bacteria is the major source of iodide in the surface waters. The rate of iodate reduction in the surface layers of the ocean has been estimated as 3.4×10^{-3}g at/m^2/year (Tsunogai, 1971).

In a study of the distribution of iodine in the ocean, Tsunogai (1971) found a high concentration of iodide in the surface layer and in a bottom layer. Although the intermediate deep water generally was devoid of iodide, or contained only low concentrations, he found occasional deep layers of water with high concentrations of iodide. In an analysis of this distribution, he concludes that the high concentrations in deep water is not due to reduction by nitrate reducing bacteria. The rate of the reduction in the bottom water was estimated to be 7×10^{-4}g at/m^2/year. There are many large uncertainties in this estimate; i.e., a two-point concentration curve, the estimated eddy diffusion constant, the possibility of recycling.

The reduction of arsenate to arsine by fungi is well established (Woolfok and Whitely, 1962). Recently McBride and Wolfe (1971) have demonstrated the reduction of arsenate to dimethylarsine by *Methanobacterium* strain M.o.H. This bacteria produces methane during anaerobic growth but in the presence of mercury compounds it can synthesize dimethylmercury. McBride and Wolfe found that in this reduction the arsenate is reduced to arsenite which is then methylated to methylarsonic acid. Subsequently dimethylarsinic acid, which is formed by the reductive methylation of methyl arsonic acid, is reduced to dimethylarsine. The same workers also demonstrated that *Desulfovibrio vulgaris*, which reduces sulfate to hydrogen sulfide during anaerobic growth, produces a volatile arsenic derivative when incubated with isotopically labeled sodium arsenate. The bacterial reduction of arsenate to arsenite by mixed cultures of bacteria isolated from Sargasso Sea surface water and from a net phytoplankton sample from Narragansett Bay has been demonstrated by Johnson (1972). The conditions of his culture were aerobic and the rate of reduction, calculated at the log phase of growth, was about 10^{-11} μmol cell^{-1} min^{-1}. This was, however, a mixed culture and the true rate by the active species must have been higher.

The work by Tsunogai (1971) leaves unexplained the high concentrations of iodide found in deep water, while Johnson's (1972) work only demonstrates reduction of arsenate by a mixed culture of marine bacteria. These papers suggest the potential for a detailed study of the role of microbial activity in maintaining the observed concentrations of iodide and arsenite found in sea water. It is likely that species of marine bacteria are capable of utilizing iodate and arsenate as the terminal acceptors in typical dissimilatory transformations during their utilization of organic substrates. The observation by McBride and Wolfe (1971) that *Desulfovibrio vulgaris* can produce volatile arsenic derivatives suggests that the ubiquitous marine species of *Desulfovibrio* may be capable of reducing arsenate to arsine or alkyl arsines.

C. REDUCTION OF PHOSPHATE

A further curious possibility is that of the reduction of phosphate to phosphine by *Desulfovibrio desulfuricans* reported by Iverson (1968). His detection of iron phosphide in the corrosion products after heating them to 2250°F in a vacuum is not convincing because of the suspicion that the reduction may have been due to organic carbon in the precipitate. Of the many older papers reporting this reduction, all are suspect because of the analytical techniques used with the possible exception of that of Barrenscheen and Beckh-Widmanstatter (1923). The possibility of this reduction should be reexamined and the ultra sensitive GLC phosphorus specific alkali thermionic detector would provide an ideal method of monitoring the evolved gasses (Brazhnikov *et al.*, 1970).

III. Biochemical Systematics

A. SECONDARY METABOLITES

The occurrence or distribution of a variety of compounds in the tissues of plants has been examined as an aid to taxonomic decisions (Alston, 1967). Comparatively, much less effort has been expended on the biochemical systematics of marine organisms. In large part this disparity is due to the number and diversity of "secondary metabolites" in plant tissues. There are, however, several groups of secondary metabolites in marine organisms, the distribution of which has provided taxonomic information. The naturally occurring naphthaquinones, the spinochromes of Echinodermata, are the outstanding example. The preponderance of the studies of spinochrome distribution has focused on the echinoids, but Singh *et al.* (1967) have extended these studies to asteroids, ophiuroids and crinoids. Certainly, too few species of these classes have been examined but the present data indicate a close relationship between the echinoids and ophiuroids on one hand and between the asteroids and holothurians on the other. These findings agree with the embryological evidence indicating the degree of relationship of these classes. Anderson *et al.* (1968) have reviewed the literature on the distribution of these pigments in the echinoids and have contributed new data. New spinochromes have recently been reported by Moore *et al.* (1968)

and by Thompson and Mathieson (1971). Other newly described groups of compounds that may eventually provide taxonomic information, though none is apparent among the few species examined thus far, include the highly unsaturated hydrocarbons from various algae (Lee *et al.*, 1970; Blumer *et al.*, 1970; Halsall and Hills, 1971). A variety of brominated compounds has been reported from the algal genus *Laurencia* by Irie and his co-workers (1969a,b, 1970). From the algae *Dictyopteris* Moore *et al.* (1971) have reported 5-(3-oxoundecyl) thioacetate, the disulfide, an unsaturated congener and a cyclic heptanone. Of greater interest than their possible significance as taxonomic indices would be an insight into their function. Many of the "secondary metabolites" of higher plants are believed to have evolved as defense mechanisms against herbivores (Whittaker and Feeny, 1971). Could some of the above compounds have a similar biological activity? Do the spinochromes inhibit the settlement of the larvae of sessile invertebrates? Vevers (1966) has speculated that the spinochromes may function as algistats. Do the brominated compounds inhibit herbivores? Aplysin, one of these compounds, is apparently ingested by the tectibranch mollusk, *Aplysia californica*, stored in specialized glands, and used as a toxin to protect it from predators (through the protective action is unproved by controlled tests). The Monarch butterflies, grasshoppers, and several other insect species derive secondary protectin by storing cardiac glycosides derived from the milkweed plants on which they feed (Whittaker and Feeny, 1971). Does the storage of aplysin by *Aplysia* present a parallel to this in the marine environment?

B. STRUCTURE OF INTERMEDIARY METABOLITES

The chemical nature of active intermediary metabolites may possibly suggest taxonomic clues. The polyunsaturated fatty acids of cryptomonads have been studied by Beach *et al.* (1970) and they have discussed the evolutionary relationship of this group in the light of their results. Recent evidence by Gurr (1971) indicates that stepwise desaturation of long chain fatty acids is controlled by pair of enzymes, one of which "measures" the position of the bond from the carboxyl end of the fatty acid and the other which sets the distance to the methyl end of the chain. Thus the tight genetic control of the position and conformation of double bond synthesis implies that the structure of these compounds may be used as legitimate taxonomic criteria. Examination of the sterols of one species of each of the three classes of Echinodermata has indicated that the sterols of asteroids and holothurians are Δ^7-sterols, while those of the echinoids, ophiuroids, and crinoids are Δ^5-sterols, paralling the relationships indicated by the spinochome studies (Gupta and Scheuer, 1968).

C. QUANTIFICATION OF INTERMEDIARY METABOLITES

Another form of biochemical systematics is based, not on the distribution of unique compounds, but on the relative quantitative distribution of common intermediary metabolites. An example of this form, the distribution of free amino acids, has often been abused, apparently through failure of the investigator to evaluate the basis of these studies. The distribution of free amino acids within a tissue is an expression of the dynamic balance of the metabolism within that cell type and thus reflects the "active gene products" of that cell type. Failure to recognize this basis has led some investigators to extract whole organisms or even to hydrolyze tissues and attempt to discern differences. Either approach results in obliteration of the unique amino acid distribution existing in single cell types. Where individual tissues have been examined the pattern of "free amino acids" constitutes a "fingerprint" of that tissue that is distinct from other tissues of that species (Kittredge *et al.*, 1962). An excellent example of the application of "free amino acid patterns" to systematic problems is the study of Demospongiae by Berquist and Hogg (1969) and Berquist and Hartman (1969). These authors point out that many studies in biochemical systematics have fallen short because "incorrect names and category designations have been applied to the species used in biochemical studies . . . potential solutions to this problem lie in closer co-operation between biochemists and systematists and the active involvement of the latter in obtaining and utilizing biochemical information for classificatory purposes." We would further suggest that this would also solve another shortcoming. Chemical observations have often been utilized as a basis for taxonomic speculations (by biochemists) without consideration for the *modus operandi* of the taxonomist. A prime consideration of the systematist in selecting morphological criteria for a taxonomic decision is not only the observed variation of the criterion among the taxa, but the stability within the taxon.

Berquist *et al.* (1969) carefully considered the factors that might lead to fluctuations in the amino acid patterns. A distinct advantage of this form of biochemical systematics is the ease with which large amounts of data can be accumulated from a number of areas or during different seasons. Alston (1967), from his experience with the chromatography of plant extracts, provides some sound advice for biochemical systematic studies in any field, especially the need to correlate data from many samples. The present state of sponge systematics is such that major problems remain in every order. One of the goals of the present author was to test the classification of Demospongiae established by Levi (1956). A study of the free amino acid pattern of 87 species provided evidence supporting Levi's construction of the subclass Ceractinomorpha but suggested that the subclass Tetractinomorpha was heterogenous. The data indicated the existence of two large groups within the Ceractinomorpha and provided evidence for certain discrepancies which exist at the family level in the present classification.

D. ENZYME POLYMORPHISM

The utility of any biochemical criteria in systematic studies is a function of how well the criteria reflect the genetic constitution of the organism, the specificity of the detection technique, and the ease with which large amounts of data can be accumulated. The technique that most closely approximates these requirements is the electrophoretic study of enzyme polymorphism, for this is in essence a measure of the tolerable missense mutations that have accumulated in the genes specifying the enzymes. A disadvantage of the technique is its power. It has proven most useful in elucidating the heterogeneity of the gene pool in subpopulations of a species, rather than in elucidating the degree of divergence of the higher taxonomic groups.

The preponderance of these studies of marine organisms has been applied to fish rather than to invertebrates and the success of these studies probably reflects what Ohno (1970) has termed "nature's great experiment with gene duplication during evolution from tunicate-like creatures to fish" and the resulting extreme diversity of genome size exhibited by fish. Only through gene duplication and the acquisition of redundant genes could evolution have occurred. This redundancy of genes in fish has resulted in a wealth of isozyme systems and extensive polymorphism. The electrophoretic variants of L-α-glycerophosphate dehydrogenase have been studied in the Pacific Ocean perch by Johnson *et al.* (1970). Utter and Hodgins (1970) have found phosphoglucomutase polymorphism in sockeye salmon. The genetic structure of juvenile populations of saury was elucidated by examining the polymorphism of malate dehydrogenase by Numachi (1970). Studies have also been made of other specific proteins as, for example, the work of Utter (1969) who examined the transferrin variants in the Pacific hake and Pantelouris *et al.* (1970) who studied this protein in the eastern North Atlantic eel. The latter workers also studied the hemoglobins of eel. Utilizing a range of partially specific substrates and inhibitors one may examine several groups of enzymes on the electropherogram. Sprague (1970) examined the esterases of skipjack tuna tissues. Comparative electropherograms which visualize all of the soluble proteins with a nonspecific stain are much less informative, but still provide evidence to substantiate taxonomic decisions (Taniguchi and Nakamura, 1970). A recent effort to make these electropherograms more useful has resulted in a concentration on the soluble proteins of a very specific tissue, the eye lens nucleus (Smith, 1968; Cobb *et al.*, 1968; Peterson and Smith, 1969; Peterson, 1970).

The few examples of the application of this technique to marine invertebrates have provided an indication of some of the promises and problems involved. Bowen *et al.* (1969), studying the hemoglobins of *Artemia salina*, concluded that the phenotypic expression was determined by the genotype and the environment. A study of the population genetics of two species of Ectoprocta by Gooch and Schopf (1970), in which 19 bands representing loci responsible for the formation of esterases, malate dehydrogenase and leucine aminopeptidase were examined, revealed a number of monomorphic loci and several polymorphic diallelic loci. The genotypic frequencies indicated that these ectoprocts are dominantly outbreeding and that panmictic local populations covered an area of $2\,m^2$. A study of the polymorphism of esterase isozymes in the heart tissue of the American lobster, *Homarus americanus* by Barlow and Ridgway (1911) provided evidence for two groups of carboxyl esterases that exhibited polymorphism. One group was present at hatching and the other was not evident until after the seventh molt. Phenotypic frequencies were not correlated with sex, size or stage of the molt cycle. Both groups were hypothesized to be due to codominant allelic systems. In an effort to determine whether the arginine kinase isozymes of the horse shoe crab, *Limulus polyphemus*, are the product of two separate genes Blethen (1971) provided a powerful technique. She isolated the two forms and, following partial hydrolysis, prepared peptide maps. The more anionic form of the kinase exhibited six more acidic peptides than did the neutral form and two of three pairs of tryptophan containing tryptic peptides had different amino acid compositions. These studies border on the amino acid sequencing of specific proteins, the ultimate in biochemical systematics (Margoliash *et al.*, 1969). While the sequencing of a protein is a major undertaking, an expansion of studies such as those of Blethen to the amino acid sequencing of specific small peptides will provide specific information on the extent of base pair mutations during the evolution of one form of the enzyme from the other following the initial gene duplication.

IV. Protein Structure

A. PHYLOGENETIC EVIDENCE FROM PROTEIN PRIMARY STRUCTURE

Protein structure is the key to many biochemical questions. Following the rapid development of new techniques in this area and their application to problems of active site structure and enzyme function, the study of protein structure is emerging as the most fertile field in the comparative biochemistry of marine organisms. The improvement and application of techniques for determining the amino acid sequences of proteins has paralleled the development of molecular genetics and we can now compare the sequences of a number of homologous proteins. These comparisons permit calculation of the base pair exchanges that must have occurred in the codons of the genes during evolution of the proteins. The observed substitutions favor Epstein's (1966) conclusion that the genetic code is capable of favoring mutations between amino acids of similar physical properties. Phylogenetic trees have been constructed from these observed substitutions and the existence of hypervariability of some codons has become evident (Fitch and Margoliash, 1969). The latter authors make the necessary distinction between "hypervariability," those mutations that have become fixed, and "hypermutation," the extent of which may not be observable. Although such phylogenetic trees are revealing, the results are sometimes distasteful to biologists. This will be rectified by a more reasonable representation of the various phyla in regard to the sequences compared when such data become available. One is inclined to conclude that, in the course of evolution, the function of a protein has been the long term immutable factor, with the conservative nature of the genetic code acting as the short term continuity factor. Hypervariability, immutable codons, "excess" double mutations and substitutions of amino acids of similar physical properties are reflections of this. Nonetheless, single and sequential codon changes, as revealed by amino acid substitutions are, along with gene duplication, the basic units of evolution and our best cues as to its course.

The determination of the primary structure of the protein by sequencing, when combined with the determination of its tertiary structure by X-ray diffraction, has provided biochemists with a detailed active site topography of a few enzymes and other proteins. This is advancing our understanding of protein-substrate interaction but the door is just opening. Evidence for considerable primary and tertiary similarity among homologous proteins over an extensive span of evolution is appearing and as pointedly emphasized by Neurath *et al.* (1970) in a consideration of the homology and phylogeny of proteolytic enzymes, "the apparent diversity of enzymes in nature has . . . been regrouped to produce a limited number of classes. Members of each class are structurally and functionally related to each other and appear to be products of a common ancestral gene." Almost all of the sequence data presently available is that of mammalian or microbial material. Yet, as discussed by Neurath later in his paper, examples of serine proteases from mammals and bacteria may represent an example wherein convergent evolution has arrived independently at a remarkably similar active site conformation. It is obvious that an extension of such studies to invertebrates is needed, as it will bring to light not only the long trail back to the "ancestral precursors" but it may well also expose other examples of convergence. In such cases the similarities will be most revealing.

B. PROTEASES

Several workers are at present actively examining the proteases of invertebrates. A trypsin-like enzyme has been extracted from the star fish, *Evasterias trochelli* by Winter and Neurath (1970) and Camacho *et al.* (1970) are examining a similar enzyme from another species, *Dermasterias imbricata*. Zwilling and his collaborators (Herbold *et al.*, 1971) in studies of the proteases of crayfish and a crab have found no chymotryptic activity, but a number of isozymes of trypsin-like proteases. Analysis indicated that these were serine proteases with the active site sequence -Asp-Ser-Gly-, which is identical with that of trypsin, chymotrypsin, elastase, and thrombin of mammalian origin. The amino acid composition of these isozymes was markedly different from that of bovine trypsin. The lack of any cross reaction with bovine trypsin as observed by the above group and by Pfleiderer *et al.* (1970) suggests major conformational differences, but in addition the crayfish and crab trypsins were, surprisingly, found to be immunologically unrelated to each other. Zwilling (1970) has also observed low molecular weight crayfish and crab proteases (MW 9000–10,000) which could be separated by electrophoresis into five enzymatically active components. These proteases have the remarkable specificity of being directed toward the second peptide bond toward the amino terminus of the peptide substrate from each proline residue. A report by De Villex and Lau (1969) of a low molecular weight protease from crayfish gastric juice with wide endopeptidase substrate specificity undoubtedly concerns the same uniquely directed protease. A chymotrypsin-like protease has been characterized from the sea anemone, *Metridium dianthus* (Gibson and Dixon, 1969) and a trypsin-like serine protease has been isolated from the sea pansy, *Renilla reniformis*, by Coan and Travis (1970). A thermostable nonserine protease secreted by sea urchin embryos was reported by Barrett (1970).

C. DEHYDROGENASES

The first protein from a marine invertebrate to be sequenced was the glyceraldehyde-3-phosphate dehydrogenase from lobster muscle. The crystalline enzyme has been prepared from a wide range of species, but lobster muscle enzyme proved to be the most suitable for X-ray diffraction analysis. This study was carried out by H. C. Watson and his colleagues while a concurrent study of its amino acid sequence was made by J. Ieuan Harris (Davidson *et al.*, 1967). This enzyme is a tetramer of four identical peptide chains, each of which contains 333 amino acids. The active site proved to be cysteine-148, and Davidson (1970) has identified the reactive lysine as lysine-182. Harris and Perham (1968) later determined the amino acid sequence of this enzyme from pig muscle. A comparison of these two sequences shows that 241 (72%) of the residues occur in an identical sequence. Of the 90 differences, 54 can be ascribed to a single base change and 35 to two base changes. In most cases, polar residues are substituted for polar residues and nonpolar residues for nonpolar residues. From the data available for hemoglobin and cytochrome c they concluded that the structure of glyceraldehyde-3-phosphate dehydrogenase has been conserved to a greater degree during evolution than that of other proteins. The conservation of structure was also suggested by the theoretical analysis of Fondy and Holohan (1971). The comparative structure of this enzyme was examined in seven species of insects (flies, bees, and bumble bees) by Carlson and Brosemer (1971), using amino acid composition and peptide pattern as criteria. Their results suggest a faster rate of evolution in the honey bee enzyme than occurred in that of the other six insects. Their study also provides the first molecular biological evidence of the point in evolutionary time that a particular behavioral pattern evolved. As relatively much more information is to be derived from concerted efforts on the sequencing of related enzymes, we should note the report of the successful sequencing of bovine glutamate dehydrogenase (another diphospho-pyridine nucleotide-linked dehydrogenase) by Smith *et al.* (1970). Compared with glyceraldehyde-3-phosphate dehydrogenase, only 2 of 12 residue sequences were similar, but the active site of lysine-97 was included. Could this be an example of evolution through gene duplication (Ohno, 1970)? Both of these sequences should provide the basic framework for the expansion of amino acid sequential studies to marine invertebrate tissues, with the absolute knowledge that the results will be equally rewarding in the advance of enzymology and invertebrate phylogeny.

D. TOXIC PROTEINS

The study of the amino acid sequences of toxic proteins has another goal. The structure of sea snake neurotoxins should provide an insight into their interaction with neural membranes and thus, hopefully, yield information on the architecture of these membranes. The sequence of erabutoxins a and b has been determined by Sato and Tamiya (1971) and the positions of the disulfide bonds have been established by Endo *et al.* (1971).

E. RESPIRATORY PROTEINS

The primary structure of the respiratory protein, hemerythrin, from the sipunculid worm, *Golfingia gouldi*, has been determined by Subaramanian *et al.* (1968) and by Klippenstein *et al.* (1968). This primitive non-heme iron-containing protein is composed of eight monomers of 13,500 molecular weight each. Each of the subunits contains two atoms of iron and can bind one molecule of oxygen. The final determination of the sequence of the 113 amino acids in the monomers was determined by succinylating the free amino groups of the lysine residues and thus limiting the tryptic attack to the three arginyl groups. The advantage of this technique is that, after isolation, the blocking groups can be removed and further tryptic digestion yields a limited number of peptides. Two sites of amino acid interchange were identified in the polypeptide chain, substitution of threonine for glycine at residue 79 and of alanine for serine at residue 96. The amino terminal 35 amino acids of the hemerythrin from another sipunculid, *Dendrostomum pyroides*, have been sequenced by Ferrell and Kitto (1971a) and they have also determined the amino acid composition and sequence of its tryptic peptides (1971b). Comparison with the sequence determined for *Golfingia* indicates that there are four sequence differences between these two proteins. The amino terminal sequence was determined by means of automatic sequence analysis followed by identification of each residue by gas liquid chromatography. This combined technique will considerably facilitate sequencing. Another technique allied to this, sequence analysis of protein mixtures by isotope dilution and mass spectrometry, shows great promise in its applicability to mixtures of several peptides (Fairwell *et al.*, 1970).

Information concerning the active site of hemerythrin has been gained by examining the reaction of suspected amino acid residues with specific reagents (Fan and York, 1969). The residues involved in binding the two iron atoms would not be expected to react the same as free residues. These studies revealed that four of the seven histidine residues in *Golfingia* hemerythrin are bound to the iron and that the lysine residues were not coordinated to iron. A further study (York and Fan, 1971) implicated two of the five tyrosine residues in iron binding in hemerythrin. Studies of the circular dichroism and optical rotatory dispersion of hemerythrin have indicated a high degree of α-helical structure (Bossa *et al.*, 1970). In view of the implication of tyrosine in bonding the iron, the data suggest that the aromatic side chains reorient upon oxygen binding.

In contrast with the extensive sequencing of hemoglobins from man and other vertebrates, studies of invertebrate hemoglobins and myoglobins are limited. Padlan and Love (1968) provided the tertiary structure of the hemoglobin of the annelid worm, *Glycera dibranchiata*, from X-ray diffraction. In a comparison with that of sperm whale myoglobin and horse hemoglobin they were able to show not only the close similarity, but to estimate the number of residues in six of the helical regions of the chain. All but the F and H regions were found to have, within one residue, the same lengths as the mammalian hemoglobins. Recently Li and Riggs (1971) have determined the amino acid sequence of the first 45 amino terminus residues of Glycera hemoglobin by the automatic Edman degradation and gas chromatography. They observed a substantial homology with the corresponding segment of sperm whale myoglobin; 16 of the residues were identical. The *Glycera* hemoglobin contained a minor variant in which leucine or isoleucine occupied the tenth position, rather than the valine of the major component, and aspartic acid replaced glutamic acid at position 29 in the sequence. Vinogradov *et al.* (1970) have succeeded in separating the minor from the major hemoglobin of this polychaete worm. The gastropods and the holothurians also contain hemoglobins. Tentori *et al.* (1971) have determined the amino acid sequence of a 63 residue segment at the carboxyl end of the myoglobin from *Aplysia limacina* and have determined that the amino terminal is acetylated. Koppenheffer and Reed (1968) have considered the evolution of the radular muscle myoglobin of gastropods. They found that the myoglobin of the archeogastropod, *Nerita peloronta* is a monomer of 17,600 molecular weight, while that of *Buccinum undatum* and of *Acanthopleura granulata* are composed of two subunits with molecular weights of 33,800 and 34,600 respectively (Terwilliger and Reed, 1969a,b). These workers have also examined the hemoglobins of three species of holothurians which were also found to be composed of two subunits with molecular weights of about 18,000–20,000 (Terwilliger and Reed, 1970).

The hemocyanins of mollusks and crustacea are too large to expect a rational approach to their primary structure in the near future; however, the active site of hemocyanins contains cuprous ion. Might not it be possible to react native hemocyanin with "specific" reagents for a given amino acid residue (as applied to hemerythrin by Fan and York, 1969), remove the copper, repeat the reaction with a radioactive reagent, hydrolyze with a protease and then fractionate the resulting peptides by monitoring the activity? One might thus identify the active site and obtain partial sequences in this region.

F. OTHER PROTEINS

Active site labeling and sequencing of the labeled peptide has been applied to the arginine kinase of lobster muscle (Roustan *et al.*, 1970). They utilized [^{14}C]*N*-ethylmaleimide to label the essential thiol group and compared the resulting tryptic peptide sequence with the homologous peptides from rabbit muscle, brain creatine kinases and earthworm lombricine kinase. All were observed to be very similar.

The technique of comparing the tryptic and chymotryptic peptide maps has been utilized to conclude that there is little homology between the muscular actin and the ciliary A-tubulin from the scallop, *Pectin irradians* (Stephens, 1970). One should note, however, the observation of Harris and Perham (1968) that, despite the very close similarity of the lobster and pig glyceraldehyde-3-phosphate dehydrogenase sequences, this identity is partially hidden in the peptide maps of the tryptic digests due to the frequently different dispositions of the lysine and arginine residues. The difficulties of Carlson and Brosemer (1971) with the interpretations of the peptide maps of insect dehydrogenase may be due to similar vagaries in the substitutions of these basic amino acids.

A renewed interest in invertebrate clotting mechanisms has led to the isolation of spiny lobster (*Panulirus interuptus*) fibrinogen and to a study of its transformation into fibrin (Fuller and Doolittle, 1971a,b). These workers confirmed that, in contrast to vertebrate fibrinogen, there is no proteolysis involved in the transformation. They found that gelation was due to the formation of ϵ-(γ-glutamyl)lysine cross-links and that 2–6 of these cross-links were formed per molecule of fibrinogen.

A unique amino acid, 3,3′-methylenebistyrosine, has been isolated from the hinge ligament protein of pelecypod mollusks, in which it forms the interchain cross-links (Andersen, 1967). This cross-link is analogous to the dityrosine cross-link of resilin, the protein of the wing ligament of insects and to the desmosine cross-link of elastin. Andersen also reported the detection of a second uncharacterized phenolic compound from the hinge ligament and Thornhill (1971) reports on a fluorescent compound from this protein which may be identical. A related problem concerns the nature of the covalent linkage of carbohydrates to proteins, a major problem in the structure of the mucins of marine invertebrates. A study of the connective tissue of the sea anemone, *Metridium dianthus* by Katzman and Oronsky (1971) resulted in the isolation of a 47 residue hydroxyproline-containing peptide. The sequence of nine of these residues has been determined and these include the Gly-Asn-Thr- unit which contained the *N*-asparaginyl residue that is apparently involved in an alkaline-stable linkage with *N*-acetylglucosamine. This study is the first to provide evidence for the nature of this linkage in the great general class of tissues which contain collagen firmly bound to heteropolysaccharide.

V. New Compounds

In general, marine organisms often contain many as yet uncharacterized small molecules, sometimes in quite high concentrations. This chapter is intended to be in part a guide to "Where are the fun problems in marine biochemistry?" and we will therefore descriptively emphasize the techniques that we have found to be easy and direct rather than the literature. The discussion will center chiefly on amino acids and peptides though most of the techniques have a broad application to small polar molecules.

At the risk of offending some of my friends who are natural product chemists, I will first parenthetically note some differences in the philosophical approach between natural product chemists and biochemists. Natural product chemists are organic chemists by training, thoroughly familiar with the powerful analytical techniques applicable to the characterization of complex organic solvent-soluble compounds. Biochemists' concerns are chiefly those of elucidating the distribution of known water-soluble metabolic intermediates. They frequently fail to recognize the presence of new compounds. The natural product chemists often belittle the water-soluble small polar compounds as unworthy of interest except to a biochemist and a host of interesting compounds must await "accidental" discovery by a biochemist who was looking for something else. Both disciplines, of course, recognize the thrill of detecting, isolating, and finally characterizing a new naturally occurring compound. The puzzles involved in attempting to work out the biosynthesis and "function" of a new compound would suggest a closer collaboration of the natural product organic chemists and the biochemists in elucidating the unique features of the biochemistry of marine organisms.

A profusion of excellent books is available covering all phases of chromatography, however we would recommend the handbook on thin-layer chromatography edited by Stahl (1969) and that by Kirchner (1967). The older book on paper chromatography by Smith (1960) is valuable for the ease with which the techniques that he describes can be duplicated in any laboratory. It is well to remember that most of the developing systems employed in thin-layer chromatography for the resolution of small polar compounds were adapted directly from paper chromatography and are readily utilized for either technique.

There are two approaches to exploration for new naturally occurring compounds. The first begins with a biological or biochemical observation, for example, the toxicity or "repugnancy" of an organism (or its eggs?) to a natural predator (microbial or metazoan) or perhaps the suspicion of a precursor or catabolite of a known metabolite from observations with radioactive labeled compounds. The second results from a study of all of a group of related compounds in a given tissue, e.g., a fractionation of the sterols or phospholipids.

When the study initiates from a biological observation, a bioassay must be developed, e.g., behavior of a predator or inhibition of growth of a microorganism. This assay should be sensitive, quantitative if possible, and consume a minimum amount of the active principal. When an active extract of the tissue is obtained, the first determinations should establish the stability of the active component. Is it active after boiling? Does it require protection from oxygen? Is it destroyed by acid or base? If it is reasonably stable, the fraction techniques may be straightforward: partition between solvents, chromatography, and electrophoresis. The bioassay may be used to follow and control each step of the purification method adapted for the active component. As purification progresses, two-dimensional thin layer chromatography with a general detection agent such as charring with sulfuric acid will indicate the nature and quantity of the impurities. The range of the techniques available will depend on the properties of the active component and thus no general suggestions beyond those given are appropriate.

As an example of the second approach to the exploration for new compounds, we will describe the search for new amino acids and peptides. The three texts cited above provide descriptions of every phase of extraction and fractionation and we will suggest the advantages and disadvantages of some of these. Extraction of dissected tissues is advantageous, where possible, because the "unknown" may occur in a high concentration in one tissue but its existence may be masked in an extract of the total organism. Extraction with cold perchloric acid followed by neutralization with potassium hydroxide and precipitation of the potassium perchlorate in the cold gives the "cleanest" extract of the small polar components of a tissue in regards to the exclusion of lipids, proteins, and polysaccharides. However, these reagents may alter some labile compounds and we prefer homogenization in 20 volumes of 70% ethanol as a mild procedure. After concentration of this extract to a small volume in a rotary evaporator the lipids may be extracted with an organic solvent. Chloroform is the most efficient solvent for lipid extraction but often forms difficult emulsions. Extraction with diethyl ether is usually satisfactory. Occasionally we have found that fractionation on a column of Sephadex G-10 provides the best low molecular weight fraction. High concentrations of salts in the extracts of marine organisms necessitate a desalting step before chromatography. The most useful procedure for amino acid extracts is ion exchange on Biorad AG50X8 (200–400 mesh) cation exchange resin. A rough guide to the volume of resin required is to consider the tissue extracted as isoionic with seawater and use a fivefold excess of resin. Five column volumes of water will remove all of the neutral and anionic material from the column and the amino acids may then be eluted with 3 *N* ammonium hydroxide. The progress of the elution can be followed visually and collection of the ammonia front will provide all of the neutral and acidic amino acids in a small volume. The basic amino acids in the extract will trail off in the next few column volumes. This technique possesses two disadvantages: the most acidic amino acids (e.g., taurine, cysteic acid, phosphonoalanine) are not retained on the column and the basic amino acids may be incompletely recovered. The rare amino acids dityrosine and methylenebistyrosine (Andersen, 1967) will be retained by adsorption. Brenner (in Stahl, 1969, p. 737) provides two ion exchange columns for desalting (1) the basic amino acids and (2) the neutral and acidic amino acids. We have run surveys of free amino acids on undesalted extracts with some loss in the resolution of the chromatograms. A preferred technique with an undesalted extract is to use two-dimensional electrophoresis and chromatography since the electrophoresis is relatively insensitive to salts. A further reduction in any "salt effect" can be achieved by applying the extract in a line 2–5 cm long at the origin. After the electrophoretic run, for which we prefer an acetic acid–formic acid buffer at pH 2, the thin layer is subjected to development with dilute acetic acid in a direction perpendicular to the electrophoretic run and in the opposite direction to the subsequent chromatography. This will concentrate the amino acids into spots at the starting line for chromatography. This technique also facilitates the rapid application of a dilute extract without repeated spotting at the origin. There are many useful developing solvents for the chromatography described in the texts. We prefer the slightly undersaturated systems described by Smith (1960). If two-dimensional chromatography is employed one of the solvents should be lutidine–water if the desalting technique has retained the taurine. This amino acid is frequently the dominant free amino acid in the tissues of marine invertebrates. With most developing solvents it will swamp the chromatogram in the glycine–serine area. For visualization we use 0.2% ninhydrin in 95% ethanol containing 15 ml of acetic acid and 5 ml collidine per 100 ml, developing the color in an oven set at 125°C. This provides for a bathochromic shift of the resulting colors and aids in distinguishing partially overlapped spots (see also Krause and Reinbothe, 1970).

One should select a prime chromatographic system for the survey and two or more secondary systems. "Reading" the map of the protein amino acids and the common free amino acids such as taurine and comparing this with the chromatogram of the extract becomes quite rapid when one is scanning for unknowns. Co-chromatography of the extract and the standard mixture of amino acids will ascertain whether slightly shifted positions of spots, especially in the longer running compounds are due to "new" compounds or to variations in the conditions of development. Chromatography before and after hydrolysis will reveal peptides and esters such as ethanolamine phosphate. When one detects a suspected new amino acid, a decision as to whether the compound is an α-amino acid can be based on its migration when the origin and line of development in the first dimension are first dusted with cupric carbonate. The development should be with a nonacidic solvent. Amines and non-α-amino acids will migrate to their expected positions in this system while the α-amino acids will remain at the origin.

When one has decided that a new "spot" is worth investigating it is necessary to isolate a small amount of the compound for the comparison of its R_f value in various solvents with those published in the literature. This can be accomplished by two-dimensional paper chromatography on Whatman 3 MM paper. To detect the spot one may use 0.01% ninhydrin spray (without divalent metals). A preferable nondestructive technique is that of dipping the chromatogram in a solution of iodine in benzene followed by immediate marking of the transient yellow spots. To elute the compound in a minimal volume, one may cut the spot out in the shape of a pennant and suspend it in a beaker by means of a thin glass hook through the point. Developing this pennant with dilute acetic acid overnight in the open beaker will concentrate the material at the tip. The dried pennant is then clamped between two large (5×8 cm) microscope slides with 2–3 cm of the tip extending. The assembly is placed in a rectangular petri dish resting on one edge, the tip of the pennant is bent down and the petri dish filled with water. The first two drops from the tip will elute most of the compound.

Comparison of literature R_f values should be corrected for the values of a common amino acid run in the same system, but at best are only a guide. The elution time on an automatic amino acid analyzer will provide additional evidence. Specific detection reagents may reveal functional groups. If it is available, conversion to a volatile derivative for mass spectrometry should be attempted.

As examples of the frequency with which one may expect to detect unique compounds, in a survey of the free amino acids in marine organisms we detected 20 "unknown" compounds in some 400 tissues extracted (Kittredge *et al.*, 1962). In the two papers by Berquist and Hogg (1969) and Berquist and Hartman (1969) they reported 4 "unknown" amino acids in the first study of 20 species of sponges and 11 "unknowns" in the latter study of 67 species of sponges. One will note many similar references to unknowns in the literature. The structure of the second compound isolated from mollusk hinge ligaments by Andersen (1967) has not been reported. Almost no attempt has been made to identify the small peptides that occur in marine invertebrates or to relate their occurrence to the metabolism of the organism (see, however, Konagaya, 1967; Ito, 1969; Terwilliger *et al.*, 1970; and Dall, 1971).

The relative ease with which one can design column chromatographic systems for the isolation of a single "unknown" bears little relation to the complexities of an analytical column, however, the position of elution from an automatic amino acid analyzer will indicate the appropriate pH to use in the isolation column. For the preparative column, volatile buffers should be employed. The use of 1 *N* to 4 *N* HCl as an eluent from Biorad AGX50 often provides suitable resolution. The simple mixture obtained from the initial column may be resolved on an anion exchange column or by simply reapplying it to the cation exchange column and eluting with an ethanol or isopropanol–HCl system. One may also utilize the appropriate thin-layer or paper chromatographic solvent in a column packed with microcrystalline cellulose or Sephadex. These columns have low capacity and are often sensitive to temperature fluctuations. Most column packings bleed to some extent and the final column should be recycled extensively before use. Even so, a final clean-up by boiling briefly with Norit A and filtration on glass fiber filter paper may be necessary. Preparative thin-layer chromatography may provide the method of choice for the final isolation.

At this stage—crystallization! Then to the armamentarium of the organic chemist: UV and infrared spectrophotometry, nuclear magnetic resonance spectrophotometry and conversion to a volatile derivative for mass spectrometry—hopefully with the help of the friendly neighborhood natural product chemist.

A partial bibliography of reviews and original papers describing non-protein amino acids and related compounds has been included for reference (Fowden, 1970; German, 1971; Konosu *et al.*, 1970; Madgwick *et al.*, 1970; Miyazawa *et al.*, 1970; Takagi *et al.*, 1967, 1970; Thompson *et al.*, 1969). The successful detection and characterization of a new compound logically leads to studies of its biosynthesis and metabolism so that discovery of a new natural product is only the initial entrance to a problem. In itself, discovery alone contributes only minimally to an understanding of marine biochemistry.

VI. Natural Occurrence of the Carbon–Phosphorus Bond

A. DISCOVERY OF AMINOPHOSPHONIC ACIDS

A unique bond, the carbon–phosphorus bond, typifies the phosphonic acids. All but one of the naturally occurring phosphonic acids contain the unit 2-aminoethylphosphonic acid (AEP). The exception is the antibiotic phosphonomycin, an epoxypropylphosphonic acid (Hendlin *et al.*, 1969). The discovery of AEP and the development of this new twig of biochemistry have recently been reviewed (Kittredge and Roberts, 1969). The "quirks of research" surrounding this discovery are as revealing of the nature of the game as the formal account of the outcome.

During a survey of the free amino acids of marine invertebrates we were puzzled by a spot on the chromatogram of the sea anemone, *Anthopleura elegantissima*. This acidic ninhydrin-reactive material migrated to a position normally occupied by ethanolamine phosphate, however, it was still present on chromatograms of the hydrolyzed extract. That the spot was not due to some of the phosphate ester that had escaped hydrolysis was evident from the failure to detect any orthophosphate in hydrolysates of the substance eluted from chromatograms. At this juncture we were convinced that the compound did not contain any phosphorus and concluded that it might be a new acidic amino acid. We were more intrigued by the immense amounts of taurine observed in most of the species examined and had initiated a study of this problem. Well into the taurine problem, we were interrupted by undelivered equipment and turned to the *Anthopleura* amino acid. Isolation was accomplished by ion exchange chromatography and the new compound was crystallized from aqueous ethanol. Elemental analysis for carbon, hydrogen, and nitrogen only accounted for 37% of the weight. After ashing further aliquots of the dwindling cluster of crystals without any success in detecting other elements, in desperation I checked the ash for phosphate. This was absurd as all of the then known naturally occurring phosphorus had been shown to exist as hydrolyzable phosphate esters. The analysis revealed that 25% of the weight of the crystals could be accounted for as phosphorus. The structure of the new compound became apparent as the phosphonic acid analog of ethanolamine phosphate containing a C–P bond (also the phosphonic analog of taurine, 2-aminoethylsulfonic acid). After substantiating this structure, we were composing a manuscript when Dr. J. T. Holden of our staff recalled a paper by Horiguchi and Kandatsu (1960) reporting the isolation of this compound from sheep rumen ciliates.

With the observation that anemones were capable of synthesizing the C–P bond, the likelihood of other compounds containing this bond occurred. Since the biosynthesis of many amines proceeds through the decarboxylation of amino acids, we sought the likely precursor of AEP, 2-amino-3-phosphonopropionic acid (phosphonoalanine). At this point we received a letter from A. F. Isbell of Texas A & M University inquiring about the negative results of our study of copper chelation by AEP on which we had based the decision between a 1-AEP or a 2-AEP structure. His studies with synthetic 1-AEP and 2-AEP indicated that they both chelated copper, the latter however was a weak chelator. Fortunately our technique was sufficiently insensitive to permit us to detect this chelation and our guess concerning the structure was not as clouded as it might have been. Professor Isbell informed us that he had in progress a program for synthesizing the phosphonic acid analogs of the natural amino acids and he kindly sent us crystalline synthetic phosphonoalanine. Utilizing this synthetic standard we were able to detect the new aminophosphonic acid in extracts of the zoanthid *Zoanthus sociatus*, and to demonstrate its synthesis by *Tetrahymena* by cocrystallization of the isolated ^{32}P-labeled compound with the synthetic phosphonoalanine.

Saponification of the total lipids of the anemone liberated AEP. Before we were successful in isolating the lipid we were visited by Eric Baer. The evidence for these new lipids presented a challenge and the response was a highly productive program for synthesizing the phosphonic acid analogs of phospholipids; for which he proposed the name phosphonolipids. The lipid was isolated and proved to be ceramid-AEP.

With the characterization of the phosphonic acid analogs of ethanolamine phosphate and serine phosphate, we examined the prospect that a phosphonic acid analog of choline phosphate existed. Fractionation of extracts of *A. elegantissima* and comparison with a synthetic sample of trimethyl-AEP provided by Dr. A. F. Rosenthal indicated its likely presence. We dispatched a technician to collect 6 kg of *A. elegantissima* with careful instructions of where and what to collect. At the site he observed similar appearing but much larger sea anemones in the tide pools, and, thinking that they were also *A. elegantissima*, he returned with 6 kg of *A. xanthogrammica*. Having these on hand, we extracted an aliquot and fractionated it to determine whether to use them or return for the *A. elegantissima*. Chromatography of the fractions revealed not only AEP and the trimethyl-AEP, but also a large amount of a weakly ninhydrin-reactive phosphonic acid. We therefore designed a fractionation procedure to isolate the trimethyl-AEP and the additional compound. During the course of the isolation we noted that the migration of the unknown phosphonic acid on chromatograms fell on a smooth curve when compared to the migration of AEP, dimethyl-AEP, and trimethyl-AEP and we hypothesized a monomethyl-AEP structure. We inquired of Professor Isbell regarding the synthesis of monomethyl-AEP. He not only suggested a route to this compound, but within 2 weeks three vials containing each of the methyl-AEP's arrived in the laboratory. Professor Isbell, with the neighborliness of a true Texan, has actively collaborated with most of the biochemists interested in the aminophosphonic acids. When the fractions were crystallized, the two new compounds were identical with the synthetic mono- and trimethyl-AEP. A trace of dimethyl-AEP was also detected.

This research thus progressed from a presumed phosphate ester that failed to hydrolyze to a sulfonic acid, to an equipment shortage, and to the discovery of the phosphonic acid analog of the sulfonic acid. It was greatly aided by an insensitive test for copper chelation and a friendly Texan. Having a technician who was not a marine biologist also inadvertently contributed.

B. QUADRUPLE LIGHT ON THE BIOSYNTHESIS OF THE C–P BOND

Four laboratories almost simultaneously reported success in demonstrating that phosphoenolpyruvate is the likely precursor of AEP. Trebst and Geike published the results of their investigation of the biosynthesis of AEP from specifically labeled glucose in July, 1967. The results indicated that the 1-carbon of AEP can be derived from the 1-carbon (or 6-carbon) of glucose and that the 2-carbon of AEP can be derived from the 2-carbon of glucose. They proposed a rearrangement of phosphoenolpyruvate to phosphonopyruvate. In August, 1967, unaware of the July publication, we presented our work at the Seventh International Congress of Biochemistry in Tokyo. We had also followed the incorporation of ^{14}C from specifically labeled compounds into either the 1-carbon or 2-carbon of AEP and concluded that the most likely precursors are either phosphoenolpyruvate or oxaloacetate—compounds which are interconvertible by phosphoenolpyruvate carboxylase (EC 4.1.1.32). At the same symposium session Rosenberg presented the results from the laboratory in Canberra which also pointed to a rearrangement of phosphoenolpyruvate. In addition Professor Kidder informally circulated a then unpublished manuscript by Warren (1968) containing results in agreement with those of the above three laboratories. He also demonstrated labeling of phosphonoalanine from [3,4–^{14}C] glucose and proposed the same rearrangement of phosphoenolpyruvate.

Horiguchi (1972) has recently succeeded in overcoming the problems of preparing active cell-free preparations capable of incorporating [^{32}P] orthophosphate and [3–^{14}C]phosphoenolpyruvate into AEP. Inhibition studies with this preparation support the hypothesis that phosphoenolpyruvate is the precursor of AEP and that the intermediate, phosphonopyruvate, is decarboxylated to phosphonoacetaldehyde prior to amination to AEP.

C. PHOSPHONIC ACIDS AS COMPONENTS OF LIPIDS, PROTEINS, CELL MEMBRANES AND HUMAN TISSUES

The literature published since the 1969 review will be briefly collated. Geike (1969) has detected a new, as yet uncharacterized, phosphonic acid in plant extracts. Hori *et al.* (1967) published a review of the distribution of ceramide-AEP and Shelburne (1968) discussed the C–P bond in marine animals. Baldwin and Braven (1968) reported AEP in a unicellular algae and Sarma *et al.* (1970) detected traces of phosphonolipids in mycobacteria (these traces detected only by difference analysis are unconvincing). DeKoning (1970a,b) found AEP in a marine crab but none in a freshwater crab which probably reflects the dietary sources.

Geike (1969) has overcome a major problem in studying the biosynthesis of AEP with the development of a cell-free preparation. Smith and Law (1970) have demonstrated the uptake of AEP and phosphonoalanine by *Tetrahymena* and incorporation of the AEP, but not the phosphonoalanine, into phosphonolipids. These workers demonstrated that there was no methylation of AEP in this ciliate (Smith and Law, 1970). Lacoste and Neuzil (1969) observed transamination of AEP with pyruvate by *Pseudomonas aeruginosa*. LaNauze *et al.* (1970) have purified and characterized a phosphonatase capable of cleaving the C–P bond.

Arakawa *et al.* (1968) detected three new phosphonolipids from a freshwater mussel and characterized two of these as sphingolipids containing *N*, *N*-acylmethylaminoethylphosphonic acid and *N*-acylaminoethylphosphonic acid (Hori and Arakawa, 1969). Hayashi *et al.* (1969) and Hori *et al.* (1969) have both reported the presence of a ceramide containing monomethyl AEP from mollusks. This same laboratory has also reported on the fatty acid components of ceramide-AEP and followed the incorporation of orthophosphate[^{32}P] into this lipid (Sugita *et al.*, 1968; Itasaka *et al.*, 1969).

Thompson (1969) has investigated the metabolism of phosphonolipids in *Tetrahymena* and the enzymatic hydrolysis of ceramide-AEP has been reported by Hori's group (Arakawa *et al.*, 1968; Hori *et al.*, 1968). The phospholipid metabolism in mollusks is being studied by Liang and Strickland (1969) (Liang *et al.*, 1970).

The localization of phosphonolipids in membranes and primarily in the cilia and surface membrane fractions of *Tetrahymena* has been established (Thompson, 1970; Smith *et al.*, 1970; Nozawa and Thompson, 1971a,b; Jonah and Erwin, 1971). Thompson *et al.* (1971) have detected a proteinaceous factor that mediates the transport of lipids from their site of synthesis to the surface membranes. An interesting paper by Ricketts (1971) reports the increase in acid phosphatase during endocytosis by *Tetrahymena*. It has been hypothesized that phosphonolipids may have an important role in imparting phosphatase resistance to membranes. Since active endocytosis results in the synthesis of an area of membrane greater than the total cell surface in a 150 min period, an investigation of the likely stimulation of phosphonolipid biosynthesis and transport during active endocytosis might reveal the importance of this role.

A macromolecular mixture that contains phosphonate phosphorous but no ester phosphorous has been isolated from the sea anemone *Metridium dianthus* (Hildebrand *et al.*, 1971). In a more thorough investigation Kirpatric and Bishop (1971) have isolated a globular protein and a fibrous protein containing both AEP and phosphonoalanine. From the latter they have obtained peptic peptides containing these phosponic acids. The question of the mode of linkage of these amino acids in the protein structure may soon be clarified.

The determination of phosphonate phosphorous has been the subject of several papers. Snyder and Law (1970) have slightly modified the difference technique (ash phosphorous minus hydrolyzable phosphorus). Two sensitive techniques based on gas–liquid chromatography and mass spectrometry have been published (Karlson, 1970; Alhadeff and Daves, 1970). Two laboratories have described the application of nuclear magnetic resonance spectroscopy to this determination (Benezra *et al.*, 1970; Glonek *et al.*, 1970).

Neuzil *et al.* (1969) have developed a technique employing ion-exchange paper for the separation of ethanolamine phosphate and AEP. Two laboratories have reported the successful development of column chromatographic techniques for the separation of phosphonolipids from their phospholipid analogs (Kapoulas, 1969; Berger and Hanahan, 1971).

Erich Baer and his staff have continued their program of the synthesis of phosphonolipids. A synthesis of ceramide-AEP and its enantiomer and comparison with the natural ceramide-AEP isolated by Hori confirmed that the natural compounds are derivatives of D-sphingosine (Baer and Sarma, 1969a,b; Baer and Rao, 1970).

Lacoste and Neuzil (1969) and Alam and Bishop (1969) have further verified the ability of bacteria to cleave the C–P bond.

Alhadeff and Daves (1970, 1971) have unequivocally detected AEP in human brain, liver, heart, and muscle. Dana and Douste-Blazy (1969, 1970) have published an observation that AEP inhibits the biosynthesis of cerebral phospholipids *in vivo* but is not incorporated into phosphonolipids. This work should be repeated with labeled AEP since it conflicts with observations that AEP is probably normally incorporated into mammalian lipids.

VII. Marine Pheromones

A. THE STATUS OF CHEMICAL COMMUNICATION STUDIES IN THE MARINE ENVIRONMENT

The current rapid progress in chemical communication among terrestrial animals is based on half a century of acute observations by naturalists, but the turning point came with the isolation and characterization of some of the "messenger compounds." The bulk of the work has focused on insects; however, recent observations have indicated the importance of chemical communication in fish and mammals. These compounds have a wide range of functions from that of female sex pheromones capable of attracting males over long distances, and male sex pheromones (which are probably aphrodisiacs), to alarm pheromones. There exists a range of recruiting pheromones, trail substances, and territorial markers. Many pheromones function to maintain the social structure of the colony and the "peck order" of social status. Functionally pheromones may be "releasers" which produce an immediate change in behavior, usually stereotypical, or "primers" which cause a physiological change preparing the organism for the later release of a behavioral pattern.

In the field of marine biology we have an equally long record of natural observations indicating the importance of chemoreception in the behavior of members of all major phyla. There is, however, no parallel to the explosive chemical investigation that has occurred in the study of pheromone communication in insects. A review of these observations suggests the diverse functions of chemoreception in the marine environment. There probably exists here a direct parallel of all of the features of chemical communication found in the insect world except that of control of the cast structure in social insects [however, the control of the sex of the individuals in a chain of slipper shells, *Crepidula plana*, by a substance diffusing through the water may be analogous (Gould, 1952)]. A widespread phenomena among sessile marine invertebrates, epidemic spawning, has no parallel in the terrestrial environment.

In most species of marine invertebrates the chemical sense is the dominant sense. While intraspecific communication is the most specialized aspect of chemoreception of knowledge of the environment, interspecific chemical communication may take the form of predator recognition, warning and defense, host selection by symbionts and parasites, and recognition-response in commensalism and mutualism. Beyond this is the general chemical input from the environment: food seeking and selection, location of habitat or niche, and undoubtedly the sensing of an inhospitable environment.

Four fields have been especially well documented:

1. The settlement and metamorphosis of the planktonic larvae of many species of sessile marine organisms have been shown to depend on the presence of a suitable substrate, the detection of which has proved to occur via chemoreception. In some cases the "chemical sign post" can be detected in the water; in others it involves contact chemoreception.

2. The location of suitable hosts by a number of commensals has been shown to depend on chemical cues. On recording the electrical response in the antennule nerves of a shrimp which may be commensal on either sea urchins or abalones, Ache and Case (1969) noted considerable activity when the shrimp was presented with water from a tank containing the host species from which the shrimp had been recovered and none when water containing the alternate host was presented.

3. The "homing" of many mollusks appears to be mediated by chemical cues, and the return to home waters by anadromous fish is the best known example of sensitive "imprinted" chemoresponse.

4. Numerous observations, apparently dating from Louis Agassiz, have suggested the likely existence of chemical communication in Crustacea. In the decapods the male of many species exhibits a display behavior prior to grasping the female and carrying her until she molts. Copulation immediately follows molting. One experimental observation, that of Ryan (1966) describing the induction of display in male crabs by water from a tank containing a premolt female, and the elimination of this response on plugging

the antennule gland pores of the female, strongly implicates the role of a sex pheromone in this behavior.

There are four areas of preliminary or partially complete investigation of the chemical nature of this communication and one successful characterization of an algal gamone:

1. Though many efforts to demonstrate chemotactic response to ova by sperm have yielded negative results, Miller (1966) has conclusively demonstrated a response by sperm of the thecate hydroids, *Campanularia flexuosa* and *C. calceolifera*, to a substance issuing from the aperature of the female gonangium. The response is species-specific. The active principle from *C. calceolifera* is soluble in water and alcohol, heat stable, nonvolatile, dialyzable, polar, and appears to be a singular molecular species of less than 5000 MW. Observations by Dan (1950) suggest the activity of a similar substance from near the germinal vesicle of the eggs of the medusa, *Spirocodon saltatrix*, on the sperm of this species. What may be the first examples of sperm chemotaxis in vertebrates are described in papers on fertilization in the herring, *Culpea*, by Yanagimachi (1957) and in the bitterling, *Acheilognathus*, by Suzuki (1960, 1961). In the latter the induction of sperm and attraction are effected by a substance released by the egg jelly surrounding the micropyle of the egg.

2. A number of species of sessile marine invertebrates which cast their sexual products free into the water are known to exhibit "epidemic spawning" in which the entire local population will shed their products simultaneously. Thorson (1950) reviews these observations and lists the species in which the male has been observed to spawn first and the sperm to stimulate the release by the female. Reiswig (1970) reports observations of epidemic spawning in the sponge, *Neofibularia nolitangere*. At the time, they were conducting long term measurements of the pumping rate of another species of sponge, *Verongia* in the area. At the onset of sperm release by *N. nolitangere*, the pumping rate of the *Verongia* decreased to negligible values and remained depressed for 2 days.

Galtsoff (1930), working with two species of oysters, *Ostrea virginica* and *O. gigas*, demonstrated that the spawning of the females was triggered by the presence of sperm in the water, with a lag of 6–38 min. The active principle in the sperm was heat labile and nondialyzable. Reciprocally, the eggs induced instant release of sperm and the active principle was heat stable and dialyzable.

3. Marine gastropods are known to possess facile chemosensory faculties. The most impressive example is that of the piscivorous species of the genus *Conus* (Kohn, 1956). Chemoreception is likely the primary sense in gregarious spawning behavior, homing, mating, and alarm or escape response (Kohn, 1961). A number of herbiverous gastropods exhibit a violent escape response when touched by a predaceous starfish or carnivorous gastropod

(Clark, 1958). Mackie and Turner (1970) have isolated and partially characterized a biologically active steroid glycoside from the starfish *Marthasterias glacialis*, and a similar substance from the starfish *Asterias rubens* which elicits the escape response in the snail *Buccinum undatum* (Mackie *et al.*, 1968; Mackie, 1970).

4. We have tentatively identified the sex pheromone of the crab, *Pachygrapsus crassipes* as being either identical with or very similar to the molting hormone, crustecdysone (Kittredge *et al.*, 1971). This polar steroid also induces sexual behavior in several other grapsoid crabs. Atema and Engstrom (1971) have completed an analysis of the behavior of the lobster, *Homarus americanus*, when presented with seawater from tanks containing newly molted male or female lobsters. Their results implicate a sex pheromone released by the female moults. The biological activity survived boiling for 5 min under nitrogen and they are proceeding with the chemical characterization of this pheromone.

The male gametes of green and brown algae orient to substances released by the female gametes. In red algae, the female gametes are retained in the gametophyte and little work has been done on the mechanism by which the male gametes orient to this structure. In an early study Cook and Elvidge (1951) attempted to isolate the gamone of *Fucus serratus* and *F. vesculosa* and identify it by mass spectrometry. Their analysis was complicated by a high "background" from inactive organic material in the seawater. They did note that dilute solutions of many simple hydrocarbons, ethers, and esters mimicked cell free "egg water" in attracting sperm. Hlubucek *et al.* (1970) have detected hexane and an unidentified compound from the ripe fruiting female thallus tips of *Fucus vesiculosus*. Studying a dioecious strain of the brown alga, *Ectocarpus siliculosus*, Müller *et al.* (1971) inoculated and harvested 14,900 culture dishes to obtain 1041 gm of gametophyte material. From this, gas flushing into a cold trap and preparative gas chromatography yielded 92 mg of gamone. Mass spectrometry and proton magnetic resonance spectroscopy combined with the infra red spectrum suggested a cyclic C_{11} hydrocarbon with an unbranched side chain. Comparison of the hydrogenated gamone with synthetic C_{11} cyclic hydrocarbons indicated an identity with butylcycloheptane and permitted the conclusion that the gamone is *allo-cis*-1-(cycloheptadien-2′, 5′-yl)-butene-1.

B. AN INVESTIGATION OF THE SEX PHEROMONE OF A CRAB, PACHYGRAPSUS CRASSIPES

Ryan's (1966) observations provided convincing evidence of sex pheromone communication in crabs. We examined the behavior of the lined shore crab, *Pachygrapsus crassipes*, and found that the males displayed a typical precopulatory stance when placed in an aquarium with a premolt female.

They elevate their bodies, walking on the tips of the first three pairs of walking legs, while extending their chelae and the last pair of walking legs. Using this behavior as a bioassay, we examined the chemical nature of the pheromone. "Active seawater" from a tank containing a premolt female, after boiling for 10 min, stimulated a male crab. The active component was not retained on charcoal nor on ion exchange resins. The activity could, however, be removed from the water by isopropanol–diethyl ether extraction and could be recovered from the organic phase. It was not soluble in hexane. These observations suggested a polar lipid. The premolt condition of the active females suggested that the molting hormone, crustecdysone, might be released from the antennule glands of the females and function as a sex pheromone. We soon confirmed that dilute solutions of crustecdysone (β-ecdysone, ecdysterone, isoinokosterone) elicited a typical response from male crabs. The threshold for response was found to be 10^{-13} M.

Liquid–liquid extraction is inefficient for the recovery of traces of polar lipids. Columns of Amberlite XAD-2 have been employed for the recovery of steroids from urine (Bradlow, 1968) and Hori (1969) has employed a column of this resin for the fractionation of phytoecdysones. We have used columns of this resin to recover an active component from "active seawater" and have fractionated the extract on a column eluted with an ethanol gradient. Purification of the UV absorbing active peak on a column of silicic acid yielded a fraction with the UV absorbtion of crustecdysone (Kittredge *et al.*, 1971).

During this study we have also noted that another active component is only poorly retained on the recovery column and elutes from the gradient column near the front. Heinrich and Hoffmeister (1970) have reported that the flesh fly, *Calliphora erythrocephala* rapidly converts the molting hormone into a glycoside as an inactivation mechanism. Crustacea may employ a similar inactivation mechanism and we are examining the as yet unresolved polar active substance for evidence of a steroid glycoside.

The observation that the molting hormone may also function as a sex pheromone may indicate the sequence in the evolution of sex pheromone communication. The improbability of the simultaneous *de novo* origin of both the ability to synthesize a messenger compound and also the receptor site architecture and neuronal pathways necessary to translater the "message" into behavior has long puzzled evolutionary biologists. The Crustacea, having evolved steroid hormones to regulate molting, upon externalization of the receptor site onto chemoreceptor organs and alteration of the timing of the release of this hormone by the female were then capable of signaling the nubial molt (Kittredge and Takahashi, 1973).

We have been concerned that the true biological effects of petroleum pollution were being overlooked in many of the studies currently in vogue.

We have found that minute traces of the seawater soluble component of crude petroleum (Boylan and Trip, 1971) completely inhibit the feeding response of crabs that is normally induced by dilute solutions of amino acids. Even lower dilutions of this petroleum fraction will inhibit the response to the sex pheromone. This, combined with the observations of Cook and Elvidge (1951), Müller (1968), and Atema *et al.* (1971) recording both false responses of algal gametes to lower hydrocarbons and a "confusion" response of lobsters to kerosene fractions, strongly suggests that in the marine environment the most detrimental effect of petroleum pollution is an interference with the organisms' facilities for chemoreception. This effect may well result in an inhibition of their ability to detect food or sexual partners at concentrations 10^{-9} below those nominally determined as lethal.

References

Ache, B., and Case, J. (1969). An analysis of antennular chemoreception in two shrimps of the genus *Betaeus*. *Physiol. Zool.* **42**, 361–371.

Acher, R., Chauvet, J. and Chauvet, M. T. (1971). Phylogeny of the neurohypophysial hormones. Two new active peptides isolated from a cartilaginous fish, *Squalus acanthias*. *Eur. J. Biochem.* **29**, 12–19.

Adeny, W. E. (1894). On the reduction of manganese peroxide in sewage. *Sci. Proc. Roy. Soc. Edinburg, Natur. Sci.* **3**, 247–251 (from Trimble, 1969).

Alam, A. U., and Bishop, S. H. (1969). Growth of *Escherichia coli* on some organophosphonic acids. *Can. J. Microbiol.* **15**, 1043–1046.

Alhadeff, J. A., and Daves, G. D., Jr. (1970). Occurrence of 2-aminoethyl-phosphonic acid in human brain. *Biochemistry* **9**, 4866–4869.

Alhadeff, J. A., and Daves, G. D., Jr. (1971). 2-Aminoethylphosphonic acid: Distribution in human tissues. *Biochem. Biophys. Acta* **244**, 211–213.

Alston, R. E. (1967). Biochemical systematics. *Evol. Biol.* **1**, 197–305.

Anderson, H. A., Mathieson, J. W., and Thomson, R. H. (1968). Distribution of spinochrome pigments in echinoids. *Comp. Biochem. Physiol.* **28**, 333–345.

Andersen, S. O. (1967). Isolation of a new type of cross link from the hinge ligament protein of molluscs. *Nature* (*London*) **216**, 1029–1030.

Arakawa, I., Sugita, M., Itasaka, O., and Hori, T. (1968). Occurrence of three phosphospingolipids other than ceramide 2-aminoethylphosphonate in the fresh water bivalve, *Corbicula sandi*. *Shiga Diagaku Kyoiku Gakabu Kiyo Shizenkagaku* **18**, 41–46.

Arakawa, I., Sugata, M., and Hori, T. (1968). Enzymatic hydrolysis of ceramide 2-aminoethylphosphonate and sphingoethanolamine. *Seikagaku* **40**, 154–157.

Atema, J., and Engstrom, D. G. (1971). Sex pheromone in the lobster, *Homarus americanus*. *Nature* (*London*) **232**, 261–263.

Atema, J., Boylan, D. B., and Todd, J. H. (1971). Importance of chemical signals in stimulation of marine organism behavior. Effects of altered environmental chemistry on animals communication. Abstr. Amer. Chem. Soc. 162nd Nat. Meeting, Div. of Water, Air and Waste Chem., No. 8.

Baer, R., and Rao, K. V. J. (1970). Phosphonolipids. XXI. Synthesis of phosphonic acid analogues of diether L-α-(N,N-dimethyl) cephalins. *Can. J. Biochem.* **48**, 184–186.

Baer, E., and Sarma, G. R. (1969a). Phosphonolipids. XIX. Synthesis of a naturally occurring ceramide aminoethylphosphonate and its enantiomer. *Can. J. Biochem.* **47**, 224–225.

Baer, E., and Sarma, G. R. (1969b). Phosphonolipids. XX. Total synthesis of a naturally occurring ceramide aminoethylphophonate and of its enantiomer. *Can. J. Biochem.* **47**, 603–610.

Baldwin, M. W., and Braven, J. (1968). 2-Aminoethylphosphonic acid in *Monochrysis. J. Mar. Biol. Ass. U.K.* **48**, 603–608.

Barlow, J., and Ridgway, C. J. (1971). Polymorphisms of esterase isozymes in the American lobster (*Homarus americanus*). *J. Fish. Res. Bd. Can.* **28**, 15–21.

Barrenscheen, H. K., and Beckh-Widmanstetter, H. A. (1923). Über backterielle reduktion organisch gebundener Phosphorsaure. *Biochem. Z.* **140**, 279–283.

Barrett, D. (1970). Occurrence and properties of a thermostable protease secreted by sea urchin embryos. *Biochem. J.* **117**, 61–64.

Beach, D. H., Harrington, G. W., and Holtz, G. G. (1970). Polyunsaturated fatty acids of marine and fresh water cryptomonads. *J. Protozool.* **17**, 501–510.

Benezra, L., Pavanaram, S. K., and Baer, E. (1970). Detection of carbonphosphorus bonds by proton nuclear magnetic resonance spectroscopy. *Can. J. Biochem.* **48**, 991–993.

Berger, H., and Hanahan, D. J. (1971). Isolation of phosphonolipids from *Tetrahymena pyriformis. Biochim. Biophys. Acta* **231**, 584–587.

Berquist, P. R., and Hartman, W. D. (1969). Free amino acid patterns and the classification of the Demospongia. *Mar. Biol.* **3**, 347–268.

Berquist, P. R., and Hogg, J. J. (1969). Free amino acid patterns in Demospongiae: A biochemical approach to sponge classification. *Cah. Biol. Mar.* **10**, 205–220.

Blasco, F., Gaudin, C. and Jeanjean, R. (1971). Absorption of arsenate ions by Chlorella. Partial reduction of arsenate to arsenite. *C. R. Acad. Sci. Ser. D.* **273**, 812–815.

Blethen, S. L. (1971). Are the arginine kinase isoenzymes of *Limulus polyphemus* the product of two separate genes? *Biochem. Genet.* **5**, 275–286.

Blumer, M., Mullin, M. M., and Guillard, R. R. L. (1970). Polyunsaturated hydrocarbon (3,6,9,12,15,18-heneicosahexaene) in the marine food web. *Mar. Biol.* **6**, 226–235.

Bossa, F., Brunori, M., Bates, G. W. Antonini, E., and Fassella, P. (1970). Hemerythrin. II. Circular dichroism and optical rotatory dispersion of hemerythrin from *Sipunculus nudus. Biochim. Biophys. Acta* **207**, 41–48.

Bowen, S. T., Lebherz, H. G., Poon, Man-Chiu, Chow, V. H. S., and Grigliatti, T. A. (1969). Hemoglobins of *Artemia salina*. I. Determination of phenotype by genotype and environment. *Comp. Biochem. Physiol.* **31**, 733–747.

Boylan, D. B., and Tripp, B. W. (1971). Determination of hydrocarbons in sea water extracts of crude oil and crude oil fractions. *Nature* (*London*) **230**, 44–47.

Bradlow, H. L. (1968). Extraction of steroid conjugates with a neutral resin. *Steroids* **11**, 265–272.

Brantner, H. (1970). Biological iron and manganese oxidation. *Zentralbl. Bacteriol. Parasitenk. Infektionskr. Hyg., Abt. 2* **124**, 412–426.

Brazhnikov, V. V., Gur'ev, M. V., and Sakodynsky, K. I. (1970). Thermionic detectors in gas chromatography. *Chromatog. Rev.* **12**, 1–41.

Bulmer, M. G. (1971). Protein polymorphism. *Nature* **234**, 410–411.

Bundy, H. F. and Gustafson, J. (1973). Purification and comparative biochemistry of a protease from the starfish *Pisaster giganteus. Comp. Biochem. Physiol. B.* **44**, 241–251.

Camacho, Z., Brown, J. R., and Kitto, G. B. (1970). Purification and properties of trypsin-like proteases from the starfish. *Dermasterias imbricata. J. Biol. Chem.* **245**, 3964–3972.

Carlson, C. W. and Brosemer, R. W. (1971). Comparative structural properties of insect triose phosphate dehydrogenases. *Biochemistry* **10**, 2113–2119.

Cassaigne, A., Lacoste, A. M. and Neuzil, E. (1971). Nonenzymatic transamination of aminophosphonic acids. *Biochim. Biophys. Acta* **252**, 506–515.

Cifonelli, J. A. and Mathews, M. B. (1972). Structural studies on spisulan. Mucopolysaccharide from clams. *Connect. Tissue Res.* **1**, 231–241.

Clark, W. C. (1958). Escape responses of herbivorous gastropods when stimulated by carnivorous gastropods. *Nature* (*London*) **181**, 137–138.

Coan, M. H., and Travis, J. (1970) Comparative biochemistry of proteases from coelenterates. *Comp. Biochem. Physiol.* **32**, 127–139.

Cobb, B. F., III, Carter, L., and Koenig, V. L. (1968). The distribution of soluble protein components in the crystalline lenses of fishes. *Comp. Biochem. Physiol.* **24**, 817–826.

Cook, A. H., and Elvidge, J. A. (1951). Fertilization in the Fucaceae: Investigations on the nature of the chemotactic substance produced by eggs of *Fucus*. *Proc. Roy. Soc. Ser. B.* **138**, 97–114.

Dall, W. (1971). Role of homarine in decapod Crustacea. *Comp. Biochem. Physiol. B* **39**, 31–44.

Dan, J. C. (1950). Fertilization in the medusan, *Spirocodon saltatrix*. *Biol. Bull.* **99**, 412–415.

Dana, R., and Douste-Blazy, L. (1969). Effect of 2-aminoethylphosphonic acid on the incorporation of phosphorus-32 into cerebral phospholipids *in vitro*. *C. R. Acad. Sci. Paris Ser. D.* **268**, 185–187.

Dana, R., and Douste-Blazy, L. (1970). Effect of amino-2-ethylphosphonic acid on cerebral phospholipid biosynthesis *in vitro*. *Bull. Soc. Chem. Biol.* **52**, 405–410.

Davidson, B. E. (1970). The identification of a reactive lysine residue in lobster glyceraldehyde-3-phosphate dehydrogenase. *Eur. J. Biochem.* **14**, 545–548.

Davidson, B. E., Sajgo, M., Noller, H. F., and Harris, J. I. (1967). Amino acid sequence of glyceraldehyde-3-phosphate dehydrogenase from lobster muscle. *Nature* (*London*) **216**, 1181–1185.

DeKoning, A. J. (1970a). Detection of 2-aminoethylphosphonic acid in the phospholipids of the crab (*Cyclograpsus punctatus*). *Biochim. Biophys. Acta* **202**, 187–188.

DeKoning, A. J. (1970b). Phospholipids of marine origin. V. Crab: Comparative study of marine species (*Cyclograpsus punctatus*) and a fresh water species (*Potomon*). *J. Sci. Food Agr.* **21**, 290–293.

DeKoning, A. J. (1971). Separation of ciliatine and phosphorylethanolamine. *J. Chromatog.* **59**, 185–187.

DeKoning, A. J. (1972). Phospholipids of marine origin. VI. Octopus (*Octopus vulgaris*). *J. Sci. Food Agr.* **23**, 1471–1475.

De Villez, E. J., and Lau, L. (1969). Specificity and digestive function of an alkaline protienase of crayfish gastric juice. *Fed. Proc.* **28**, 788.

Doyle, R. W. (1972). Genetic variation in *Ophiomusium lymani* (Echinodermata) populations in the deep sea. *Deep-Sea Res. Oceanogr. Abstr.* **19**, 661–664.

Ehrlich, H. L. (1964). Bacterial release of manganese from manganese nodules. *Bact. Proc.* 42–53 (abstract).

Ehrlich, H. L. (1966). Reactions with bacteria from marine ferromanganese nodules. *Develop. Ind. Microbiol.* **7**, 43–60.

Ehrlich, H. L. (1971). Bacteriology of manganese nodules. V. Effect of hydrostatic pressure on the bacterial oxidation of Mn (II) and reduction of manganese dioxide. *Appl. Microbiol.* **21**, 306–310.

Endo, Y., Sato, S., Ishii, S., and Tamiya, N. (1971). Disulfide bonds of erabutoxin a, a neurotoxic protein of a sea snake (*Laticauda semifasciata*). *Biochem. J.* **122**, 463–467.

Epstein, C. J. (1966). Role of the amino acid "Code" and of selection for conformation in the evolution of proteins. *Nature* (*London*) **210**, 25–28.

Fairweather, R. B., Tanzer, M. L. and Gallop, P. M. (1972). Aldol-histidine, a new tri-functional collagen crosslink. *Biochem. Biophys. Res. Commun.* **48**, 1311–1315.

Fairwell, T., Barnes, W. T., Richards, F. F., and Lovins, R. E. (1970). Sequence analysis of complex protein mixtures by isotope dilution and mass spectrometry. *Biochemistry* **9**, 2260–2267.

Fan, C. C., and York, J. L. (1969). Implications of histidine at the active site of hemerythrin. *Biochem. Biophys. Res. Commun.* **36**, 365–372.

Ferrell, R. E., and Kitto, G. B. (1971a). Amino terminal sequence of *Dendrostomum pyroides* hemerythrin. *Fed. Eur. Biochem. Soc. Lett.* **12**, 322–324.

Ferrell, R. E., and Kitto, G. B. (1971b). Structural studies on *Dendrostomum pyroides* hemerythrin. *Biochemistry* **10**, 2923–2929.

Fitch, W. M. (1971). Evolutionary variability in hemoglobins. *Haematol. Bluttransfus.* **10**, 199–215.

Fitch, W. M., and Margoliash, E. (1969), The construction of phylogenetic trees. II. How well do they reflect past history? *In* "Structure, Function and Evolution in Proteins." *Brookhaven Symp. Biol.* **I**, No. 21, 217–242.

Fondy, T. P., and Holohan, P. D. (1971). Structural similarities within groups of nucleotide linked dehydrogenases. *J. Theor. Biol.* **31**, 229–244.

Fowden, L. (1970). Nonprotein amino acids of plants. *Prog. Phytochem.* **2**, 203–266.

Fuller, G. M., and Doolittle, R. F. (1971a). Invertebrate fibrinogen. I. Purification and characterization of fibrinogen from the spiny lobster. *Biochemistry* **10**, 1305–1311.

Fuller, G. M. and Doolittle, R. F. (1971b) Invertebrate fibrinogen. II. Transformation of lobster fibrinogen to fibrin. *Biochemistry* **10**, 1311–1315.

Galtsoff, P. S. (1930). The role of chemical stimulation in the spawning reactions of *Ostrea virginica* and *Ostrea gigas*. *Proc. Nat. Acad. Sci. U.S.* **16**, 655–659.

Geike, F. (1969). Column and paper chromatographic detection of an acid and phosphatase-stable phosphorus containing compound in plants. *J. Chromatog.* **44**, 181–183.

Geike, F. (1969). Biosynthesis of phosphonoamino acids. II. Incorporation of glucose-U-^{14}C into aminoethylphosphonic acid by cell-free *Tetrahymena* preparations. *Naturwissenschaffen* **56**, 462.

Geike, F. (1971). Biochemistry of phosphonoamino acids. *Naturwiss. Rundsch.* **24**, 335–340.

German, V. F. (1971). N^{α}, N^{α}-dimethylhistamine the hypotensive principle of the sponge *Ianthella* species. *J. Pharm. Sci.* **60**, 495–496.

Gibson, D., and Dixon, G. H. (1969). Chymotrypsin-like protease from the sea anemone, *Metridium senile*. *Nature* (*London*) **222**, 753–756.

Glonek, T., Henderson, T. O., Hilderbrand, R. L., and Myers, T. C. (1970). Biological phosphonates: Determination by phosphorus-31 nuclear magnetic resonance. *Science* **169**, 192–194.

Goldberg, E. D. (1963). The oceans as a chemical system. *In* "The Sea" (M. N. Hill, ed.), Vol. 2, pp. 3–25. Wiley (Interscience), New York.

Gooch, J. L., and Schopf, T. J. M. (1970). Population genetics of marine species of the phylum Ectoprocta. *Biol. Bull.* **138**, 138–156.

Gould, H. M. (1952). Studies on the sex in the hermaphrodite mollusc *Crepidula plana*. IV. Internal and external factors influencing growth and sex development. *J. Exp. Zool.* **119**, 93–163.

Guha, A., Lai, C. Y. and Horecker, B. L. (1971). Lobster muscle aldolase. Isolation, properties, and primary structure at the substrate-binding site. *Arch. Biochem. Biophys.* **147**, 692–706.

Gupta, K. C., and Scheuer, P. J. (1968). Echinoderm sterols. *Tetrahedron* **24**, 5831–5837.

Gurr, M. I. (1971). Biosynthesis of polyunsaturated fatty acids in plants *Lipids* **6**, 266–273.

Halsall, T. G., and Hills, I. R. (1971). Isolation of heneicosa-1,6,9,12,15,18-hexaene and -1,6,9,12,15-pentaene from the algal *Fucus vesiculosus*. *J. Chem. Soc.* D (**9**), 448–449.

Harris, J. I., and Perham, R. N. (1968). Glyceraldehyde-3-phosphate dehydrogenase from pig muscle. *Nature* (*London*) **219**, 1025–1028.

Hayashi, A. and Matsura, F. (1973). 2-Hydroxy fatty acid- and phytosphingosine-containing ceramide (2-N-methylaminoethyl) phosphonate from *Turbo cornutus*. *Chem. Phys. Lipids* **10**, 51–65.

Hayashi, A., Matsubara, T., and Matsura, F. (1969). Biochemical studies in the lipids of *Turbo cornutus*. I. Conjugated lipids of viscera. *Yukagaku* **18**, 118–123.

Heinrich, G., and Hoffmeister, H. (1970). Bildung von hormonglykosiden als inaktivierungs-mechanisus bei *Calliphora erythrocephala*. *Z. Naturforsch.* **25**, 358–362.

Henderson, T. O., Glonek, T., Hilderbrand, R. L. and Myers, T. C. (1972). Phosphorus-31 nuclear magnetic resonance studies of the phosphonate and phosphate composition of the sea anemone, *Bunadosoma* species. *Arch. Biochem. Biophys.* **149**, 484–497.

Hendlin, D., Stapley, E. O., Jackson, M., Wallick, H., Miller, A. K., Wolf, F. J., Miller, T. W., Chaiet, L., Kahan, F. M., Foltz, E. L., Woodruff, H. B., Mata, J. M., Hernandez, S., and Mochaeles, S. (1969). Phosphonomycin, a new antibiotic produced by strains of strepto-myces. *Science* **166**, 123–124.

Herbold, D., Zwilling, R., and Pfleiderer, G. (1971). Evolution of endopeptidases. XIII. Bio-chemical and immunological studies on trypsin and a low molecular weight protease from the decapode *Carcinus maenas*. *Hoppe-Seyler's Z. Physiol. Chem.* **352**, 583–592.

Hilderbrand, R. L., Henderson, T. O., Glonek, T., and Myers, T. C. (1971). Characterization of a phosphonate-rich macromolecular complex from *Metridium dianthus* utilizing ^{31}P NMR. *FASEB Proc.* **30**, 1072.

Hlubucek, J. R., Hora, J., Toube, T. P., and Weedon, B. C. L. (1970). Gamone of *Fucus vesiculosus*. *Tetrahedron Lett.* **59**, 5163–5164.

Hori, M. (1969). Automatic column-chromatographic method for insect-molting steroids. *Steroids* **14**, 33–46.

Hori, T. (1971). Biochemical aspects of compounds with a carbon-phosphorus bond in nature. *Seikagaku* **43**, 1009–10021.

Hori, T., and Arakawa, I. (1969). Isolation and characterization of new sphingolipids con-taining N,N-acylmethylaminoethylphosphonic acid and N-acylaminoethylphosphonic acid from the mussel, *Corbicula sandai*. *Biochim. Biophys. Acta* **176**, 898–900.

Hori, T., Itasaka, O., Sugita, M., and Arakawa, I. (1967). Distribution of ceramide 2-amino-ethylphosphonate in nature and its quantitative correlation to sphingomyelin. *Shiga Daigaku Kyoiku Gakubu Kiyo, Shizenkagaku* No. 17, 23–26.

Hori, T., Arakawa, I., Sugita, M., and Itasaka, O. (1968). Biochemistry of shellfish lipids. IX. Enzymatic hydrolysis of ceramide (2-aminoethyl) phosphonate and sphingoethanolamine. *J. Biochem.* (*Tokyo*) **64**, 533–536.

Hori, T., Sugita, M. and Itasaka, O. (1969). Biochemistry of shellfish lipids. X. Isolation of a sphingolipid containing 2-monomethylaminoethylphosphonic acid from shellfish. *J. Biochem.* (*Tokyo*) **65**, 451–457.

Horiguchi, M. (1972). Natural Carbon-phosphorus compounds. In "Anal. Chem. Phosphorus Compounds" (Halmann, M., ed.) pp. 703–724. Interscience. N.Y.

Horiguchi, M. (1972). Biosynthesis of 2-aminoethylphosphonic acid in *cell*-free preparations from Tetrahymena. *Biochem. Biophys. Acta* **261**, 102–113.

Horiguchi, M., and Kandatsu, M. (1960). Ciliatine: A new aminophosphonic acid contained in rumen ciliate protozoa. *Bull. Agr. Chem. Soc. Japan* **24**, 565–570.

Horowitz, J. J. and Whitt, G. S. (1972). Evolution of a nervous system specific lactate dehy-drogenase isozyme in fish. *J. Exp. Zool.* **180**, 13–31.

Imamura, T., Baldwin, T. O. and Riggs A. (1972). Amino acid sequence of the monomeric hemoglobin component from the bloodworm, *Glycera dibranchiata*. *J. Biol. Chem.* **247**, 2785–2797.

Irie, T., Suzuki, M., and Hayakawa, Y. (1969a). Constituents of marine plants. XII. Isolation of aplysin, debromaplysin and aplysinol from *Laurencia okamurai*. *Bull. Chem. Soc. Japan* **42**, 843–844.

Irie, T., Fukuzawa, A., Izawa, M., and Kurosawa, E. (1969b). XIII. Laurenisol, a sesquiterpenoid containing bromine from *Laurencia nipponica*. *Tetrahedron Lett.* 1343–1346.

Irie, T., Izawa, M., and Kurosawa, E. (1970). XV. Laureatin and Isolaureatin, constituents of *Laurencia nipponica*. *Tetrahedron* **26**, 851–870.

Itasaka, O., Hori, T., and Sugita, M. (1969). Biochemistry of shellfish lipids. XI. Incorporation of orthophosphate-^{32}P into ceramide ciliatine (2-aminoethylphosphonic acid) of the fresh water mussel, *Hyriopsis schlegelii*. *Biochim. Biophys. Acta* **176**, 783–788.

Ito, K. (1969). Free amino acids and peptides in marine algae. *Nippon Suisan Gakkaishi* **35**, 116–129.

Iverson, W. P. (1968). Corrosion of iron and formation of iron phosphide by *Desulfovibrio desulfuricans*. *Nature* (*London*) **217**, 1265–1267.

Iwamori, Masao; Sugita, Mutsumi and Hori, Taro (1971). Biochemistry of shellfish lipids. X. Isolation and characterization of ceramide N-acylaminoethylphosphonate and its partial synthesis. *Shiga Daigaku Kyoiku Gakubu Kiyo Shizenkagaku* **21**, 24–30.

Jamieson, A. and Josson, J. (1971). Greenland component of spawning cod at Iceland. *Rapp. Proces.-Verb. Reunions, Cons. Int. Explor. Mer.* **161**, 65–72.

Jamieson, A., DeLigny, W. and Naevdal, G. (1971). Serum esterases in mackerel, *Scomber scombrus*. *Rapp. Proces-Verb. Reunions, Cons. Int. Explor. Mer.* **161**, 109–117.

Johnson, A. G., Utter, F. M., and Hodgins, H. O. (1970). Electrophoretic variants of L-alpha-glycerophosphate dehydrogenase in Pacific Ocean perch (*Sebastodes alutus*). *J. Fish. Res. Bd. Can.* **27**, 943–945.

Johnson, A. G., and Utter, F. M. (1973). Electrophoretic variants of aspartate aminotransferase of the bay mussel, *Mytilus edulis*. *Comp. Biochem. Physiol.* B **44**, 317–323.

Johnson, A. G., Utter, F. M. and Hodgins, H. O. (1973). Estimate of genetic polymorphism and heterozygosity in three species of rockfish (genus *Sebastes*). *Comp. Biochem. Physiol.* B**44**, 397–406.

Johnson, D. L. (1972). Bacterial reduction of arsenate in seawater. *Nature* **240**, 44–45.

Jolles, P. and Jolles, J. (1971). Primary sequences of proteins and their evolution. *Prog. Biophys. Mol. Biol.* **22**, 97–125.

Jonah, M., and Erwin, J. A. (1971). Lipids of membraneous cell organelles isolated from the ciliate, *Tetrahymena pyriformis*. *Biochim. Biophys. Acta* **231**, 80–90.

Kanatani, H. (1969). Induction of spawning and oocyte maturation by 1-methyladenine in starfish. *Exp. Cell Res.* **57**, 333–337.

Kanatani, H. and Shirai, H. (1971). Chemical structural requirements for induction of oocyte maturation and spawning in starfishes. *Dev. Growth Differ.* **13**, 53–64.

Kapoulas, U. M. (1969). Chromatographic separation of phosphonolipids from their phospholipid analogs. *Biochim. Biophys. Acta* **176**, 324–329.

Karlson, K. A. (1970). Analysis of compounds containing phosphate and phosphonate by gas-liquid chromatography and mass spectrometry. *Biochem. Biophys. Res. Commun.* **39**, 847–851.

Katzman, R. L., and Oronsky, A. L. (1971). Evidence for a covalent linkage between heteropolysaccharide and an hydroxyproline-containing peptide from *Metridium dianthus* connective tissue. *J. Biol. Chem.* **246**, 5107–5112.

Kirchner, J. C. (1967). "Thin Layer Chromatography." Wiley (Interscience), New York.

Kirpatric, D. S., and Bishop, S. H. (1971). Aminophosphonic acids in proteins. *FASEB Proc.* **30**, 1182.

Kittredge, J. S., and Roberts, E. (1969). A carbon-phosphorus bond in nature. *Science* **164**, 37–42.

Kittredge, J. S., and Takahashi, F. T. (1973) The evolution of sex pheromone communication in the Arthropoda. *J. Theor. Biol.* **35**, 467–471.

Kittredge, J. S., Simonsen, D. G., Roberts, E., and Jelinek, B. (1962). Free amino acids of marine invertebrates. *In* "Amino Acid Pools" (J. T. Holden, ed.), pp. 176–186. Elsevier, New York.

Kittredge, J. S., Terry, M., and Takahashi, F. T. (1971). Sex pheromone activity of the molting hormone, crustecdysone, on male crabs (*Pachygrapsus crassipes, Cancer antennarius* and *C. anthonyi. Fishery Bull.* **69**, 337–343.

Klippenstein, G. L., Holleman, J. W., and Klotz, I. M. (1968). Primary structure of *Golfingia gouldi* hemerythrin. Order of peptides in fragments produced by tryptic digestion of succinylated hemerythrin. Complete amino acid sequence. *Biochemistry* **7**, 3868–3878.

Kohn, A. J. (1956). Piscivorous gastropods of the genus *Conus. Proc. Nat. Acad. Sci. U.S.* **42**, 168–171.

Kohn, A. J. (1961). Chemoreception in marine gastropods. *Amer. Zool.* **1**, 291–308.

Konagaya, S. (1967). Detection of L-glutaminyl-L-glutaminyl-L-alanine as DPN-derivative in a brown algae, *Ecklonia cava. Nippon Suisan Gakkaishi* **33**, 417–420.

Konosu, S., Chen, Y. N., and Watanabe, K. (1970). Atrinine, a new betaine isolated from the adductor muscle of fan-muscle. *Nippon Suisan Gakkaishi* **36**, 940–944.

Koppenheffer, T. L., and Read, K. R. H. (1968). Radular muscle myoglobin of the gastropod mollusk *Nerita peloronta. Comp. Biochem. Physiol.* **26**, 753–756.

Korn, E. D., Dearborn, D. G., Fales, H. M., and Sokoloski, E. A. (1973). Phosphonoglycan. Major polysaccharide constituent of the ameba plasma membrane contains 2-aminoethylphosphonic acid and 1-hydroxy-2-aminoethylphosphonic acid. *J. Biol. Chem.* **248**, 2257–2259.

Krause, G. J., and Reinbothe, H. (1970). Polychromic detection of amino acids with different ninhydrin spray reagents after separation on MN 300 HR cellulose thin layers. *Biochem. Physiol. Pflanz.* **161**, 577–592.

Krumbein, W. E. (1971). Manganese-oxidizing fungi and bacteria in recent shelf sediments of the Bay of Biscay and the North Sea. *Naturwissenschaften* **58**, 56–57.

Lacoste, A. M., and Neuzil, E. (1969). Transamination of (2-aminoethyl) phosphonic acid by *Pseudomonas aeruginosa. C.R. Acad. Sci. Paris Ser. D* **269**, 254–257.

LaNauze, J. M., Rosenberg, H., and Shaw, D. C. (1970). Enzymatic cleavage of the carbon-phosphorus bond: Purification and properties of phosphonatase. *Biochim. Biophys. Acta* **212**, 332–350.

Lee, R. F. Nevenzel, J. C., Paffenhoefer, G. A., Benson, A. A., Patton, S., and Kavanagh, T. E. (1970) Unique hexaene hydrocarbon from a diatom (*Skeletonema contatum*). *Biochim. Biophys. Acta* **202**, 386–388.

Leeper, G. W. (1947). The forms and reactions of manganese in the soil. *Soil Sci.* **63**, 79–94.

Lenhoff, H. M. (1968). Behavior, hormones and hydra. *Science* **161**, 434–442.

Lévi, C. (1956). Étude des Halisarca de Roscoff. Embryologie et systématique des Démosponges. *Arch. Zool. exp. gén.* **93**, 1–181.

Li, S. L., and Riggs, A. F. (1971). Partial sequence of the amino-terminal segment of *Glycera* hemoglobin. Homology with sperm whale myoglobin. *Biochim. Biophys Acta* **236**, 208–210.

Li, S. L. and Riggs, A. (1972). Partial sequence of the first 30 residues from the amino-terminus of hemoglobin B in the hagfish, *Eptatretus stoutic*. Homology with lamprey hemoglobin. *J. Mol. Evol.* **1**, 208–210.

Liang, C. R., and Strickland, K. P. (1969). Phospholipid metabolism in the mollusks. I. Distribution of phospholipids in the water snail, *Lymnaea stagnalis. Can. J. Biochem.* **47**, 85–89.

Liang, C. R. (1970). Enzymatic synthesis of [^{14}C]-phosphoenolpyruvate from [^{14}C]-pyruvate for use as a precursor in the biosynthesis of phosphonates. *Bull. Inst. Chem., Acad. Sinica*, **18**, 70–73.

Liang, C. R., Segura, M., and Strickland, K. P. (1970). Phospholipid metabolism in the mollusks. II. Activities of the choline kinase, ethanolamine kinase, and CTP: Phosphorylethanolamine cytidyl-transferase in the mollusk *Helix lactea*. *Can. J. Biochem.* **48**, 580–584.

McBride, B. C. and Wolfe, R. S. (1971). Biosynthesis of dimethylarsine by *Methanobacterium*. *Biochemistry* **10**, 4312–4317.

McLachlan, A. D. (1972). Repeating sequences and gene duplication in proteins. *J. Mol. Biol.* **64**, 417–437.

Mackie, A. M. (1970). Avoidance reactions of marine invertebrates to either steroid glycosides of starfish or synthetic surface-active agents. *J. Exp. Mar. Biol. Ecol.* **5**, 63–69.

Mackie, A. M., and Turner, A. B. (1970). Partial characterization of a biologically active steroid glycoside isolated from the starfish, *Marthasterias glacialis*. *Biochem. J.* **117**, 543–550.

Mackie, A. M., Lasker, R., and Grant, P. T. (1968). Avoidance reactions of a mollusc, *Buccinum undatum*, to saponin-like surface-active substances in extracts of the starfish *Asterias rubens* and *Marthsterias glacialis*. *Comp. Biochem. Physiol.* **26**, 415–428.

Madgwick, J. C., Ralph. B. J., Shannon, J. S., and Simes, J. J. H. (1970). Nonprotein amino acids in Australian sea weeds. *Arch. Biochem. Biophys.* **141**, 766–767.

Mann, P. J. G., and Quastel, J. A. (1946). Manganese metabolism in soils. *Nature (London)* **158**, 154–156.

Margoliash, E., Fitch, W. M., and Dickerson, R. E. (1969). Molecular expression of evolutionary phenomena in the primary and tertiary structures of cytochrome *c*, "Structure, Function and Evolution in Proteins," *Brookhaven Symp. Biol.* **2**, No. 21, 259–305.

Mason, W. T. (1972). Isolation and characterization of the lipids of the sea anemone. *Biochim. Biophys. Acta* **280**, 539–544.

Matsubara, T. and Hayashi, A. (1973). Identification of molecular species of ceramide aminoethylphosphonate from oyster adductor muscle by gas-liquid chromatography-mass spectrometry. *Biochem. Biophys. Acta* **296**, 171–178.

Menard, H. W., and Shipek, C. J. (1958). Surface concentrations of manganese nodules. *Nature (London)* **182**, 1156–1158.

Miller, R. L. (1966). Chemotaxis during fertilization in the hydroid *Campanularia*. *J. Exp. Zool.* **162**, 23–44.

Miyazawa, K., Ito, K., and Matsumoto, F. (1970). Occurrence of (+)-2-hydroxy-3-amino-propanesulfonic acid and 3-amino-propanesulfonic acid in a red algae, *Grateloupia livida*. *Nippon Suisan Gakkaish* **36**, 109–114.

Moore, R. E., Singh, H., and Scheuer, P. J. (1968). A pyranonaphthazarin pigment from the sea urchin, *Echinothrix diadema*. *Tetrahedron Lett.* 4581–4583.

Moore, R. E., Roller, P., and Au, K. (1971). Isolation of S-(3-oxoundecyl) thioacetate, bis (3-oxoundecyl) disulfide. (−)-3-hexyl-4,5-dithiacycloheptanone, and S-(trans-3-oxoundec-4-enyl) thioacetate from *Dictyopteris*. *J. Chem. Soc.* D. (**10**), 503–504.

Müller, D. G. (1968). The nature of a chemotactic substance produced by the female gametes of *Ectocarpus siliculosus*. I. Methods, isolation and detection by gas chromatography. *Planta* **81**, 160–168.

Müller, D. G., Jaenicke, L., Donike, M., and Akintobi, T. (1971). Sex attractant in a brown algae: Chemical structure. *Science* **171**, 815–816.

Negishi, Takashi; Ito, Seisuke and Fujino, Yasuhiko (1971). Enzymatic hydrolysis of sphingophosphorylcholine by phospholipase C. *Nippon Nogei Kagaku Kaishi* **45**, 374–377.

Neurath, H., Bradshaw, R. A., and Arnon, R. (1970). Homology and phylogeny of proteolytic enzymes. *In* "Structure-Function Relationships of Proteolytic Enzymes" (P. Desnuelle, H. Neurath, and M. Ottesen, eds.), *Proc. Int. Symp. Copenhagen, 1969* pp. 113–133. Academic Press, New York.

Neuzil, E., Jensen, H., and LePogam, J. (1969). Chromatographic separation of ciliatine [(2-aminoethyl) phosphonic acid] and phosphoethanolamine [mono (2-aminoethyl) phosphate]. *J. Chromatog.* **39**, 238–240.

Nozawa, Y., and Thompson, G. A., Jr. (1971a). Membrane formation in *Tetrahymena pyriformis.* II. Isolation and lipid analysis of cell fractions. *J. Cell. Biol.* **49**, 712–721.

Nozawa, Y., and Thompson, G. A., Jr. (1971b). Membrane formation in *Tetrahymena pyriformis.* III. Lipid incorporation into various cellular membranes of logarithmic phase cultures. *J. Cell. Biol.* **49**, 722–730.

Numachi, K. (1970). Polymorphism of malate dehydrogenase and genetic structure of juvenile population in saury *Cololabis saira. Nippon Suisan Gakkaishi* **36**, 1235–1241.

Numachi, K. (1971). Genetic polymorphism of α-glycerophosphate dehydrogenase in saury, Cololabis saira. I. Seven variant forms and genetic control. *Nippon Suisan Gakkaishi* **37**, 755–760.

Odense, P. H., Leung, T. C. and MacDougall, Y. M. (1971). Polymorphism of actate dehydrogenase (LDH) in some gadoid species. *Rapp. Proces-Verb. Reunions, Cons. Int. Explor. Mer.* **161**, 75–79.

Ohno, S. (1970). "Evolution by gene duplication." Springer-Verlag, New York.

Padlan, E. A., and Love, W. E. (1968). Structure of the haemoglobin of the marine annelid worm, *Glycera dibranchiata*, at 5.5 A. resolution. *Nature* (*London*) **220**, 376–368.

Pantelouris, E. M., Arnason, A., and Tesch, F. W. (1970). Genetic variation in the eel. II. Transferrins, hemoglobins and esterases in the eastern North Atlantic. Possible interpretations of phenotypic frequency differences. *Genet. Res.* **16**, 277–284.

Peterson, G. L. (1970). Effects of urea on solubility and electrophoretic characteristics of protein from eye lens nucleus of four shark species and a teleost. *Comp. Biochem. Physiol.* **35**, 299–302.

Peterson, G. L. and Shehadeh, Z. H. (1971). Subpopulations of the Hawaiian stripped mullet *Mugil cephalus*. Analysis of variations of nuclear eye-lens protein electropherograms and nuclear eye-lens weights. *Mar. Biol.* **11**, 52–60.

Peterson, G. L., and Smith, A. C. (1969). Intraspecific variation in the soluble nuclear eye lens protein of the sandbar shark, *Carcharhinus milberi. Comp. Biochem. Physiol.* **31**, 679–684.

Pfleiderer, G., Lenhe, R. and Reinhardt, G. (1970). Evolution of endopeptidases. VIII. Cross-reactions of trypsins and chymotrypsins of different species. *Comp. Biochem. Physiol.* **33**, 955–967.

Pickwell, G. V. (1970). Physiology of carbon monoxide production by deep-sea coelenterates: causes and consequences. *Ann. N. Y. Acad. Sci.* **174**, 102–115.

Raven, D. J., Earland, C. and Littler, M. (1971). Occurrence of dityrosine in Tussah silk fibroin and keratin. *Biochim. Biophys. Acta* **251**, 96–99.

Reiswig, H. M. (1970). Porifera: Sudden sperm release by tropical Demospongia. *Science* **170**, 538–539.

Ricketts, T. R. (1971). Endocytosis in *Tetrahymena pyriformis*. Selectivity of the uptake of particles and the adaptive increase in cellular acid phospatase activity. *Exp. Cell Res.* **66**, 49–58.

Rittenberg, S. C. (1969). Roles of exogenous organic matter in the physiology of chemolithotrophic bacteria. *Advan. Microbiol. Physiol.* **3**, 159–196.

Roberts, G. P., Bhattacharyya, A. and Jeanloz, R. W. (1972). Invertebrate connective tissue. 11. Carbohydrate moiety of sea cucumber (*Thyone briareus*) gelatin. *Carbohyd. Res.* **25**, 475–487.

Roustan, C., Der Terrossian, E., and Pradel, L. A. (1970). Essential thiol group of arginine kinase from *Homarus vulgaris* muscle studied by difference spectrophotometry and N-ethyl-[1 ^{14}C] maleimide labeling. *Eur. J. Biochem.* **17**, 467–471.

Ryan, E. P. (1966). Pheromone: Evidence in a decapod crustacean. *Science* **151**, 340.

Sampugna, J., Johnson, L., Bachman, K. and Keeny, M. (1972). Lipids of *Crassostrea virginica*. I. Aldehyde- and phosphorus-containing lipids in oyster tissue. *Lipids* **7**, 339–343.

Sarma, G. R., Chandramouli, V., and Venkitasubramanian, T. A. (1970). Occurrence of phospholipids in mycobacteria. *Biochim. Biophys. Acta* **218**, 561–563.

Sato, S., and Tamiya, N. (1971). Amino acid sequences of erabutoxins, neurotoxic proteins of sea-snake (*Laticauda semifaciata*). *Biochem. J.* **122**, 453–461.

Serene, P. (1971). Esterase of the northeast Atlantic albacore stock. *Rapp. Proces-Verb. Reunions, Cons. Int. Explor. Mer* **161**, 118–121.

Shelburne, F. A. (1968). Univ. Microfilms No. 68–14, 327, Duke Univ., Durham, North Carolina. Diss. Abstr. B 1968 (*29*) (4) 1310. Compounds containing the carbon-phosphorus bond in marine animals. Involvement of aminophosphonic acids in macromolecular structures.

Shirai, H., and Kanatani, H. (1971). Induction of oocyte maturation in calcium-free sea water by methyladenine in starfish. *Dobutsugaku Zasshi* **79**, 155–159.

Singh, H., Moore, R. E., and Scheuer, P. J. (1967). Distribution of quinone pigments in echinoderms. *Experientia* **23**, 624–626.

Smith, A. C. (1968). Protein variation in the eye lens nucleus of the mackerel scad (*Decapterus pinnulatus*). *Comp. Biochem. Physiol.* **28**, 1161–1168.

Smith, A. C. (1972). Lens isoprecipitin in yellowfin tuna (*Thunnus albacores*). *Comp. Biochem. Physiol.* B. **42**, 497–499.

Smith, E. L. (1970). Evolution of enzymes. *In* "The Enzymes" (3rd ed.) (Boyer, P. D., ed.) Academic, N.Y. **1**, pp. 267–339.

Smith, E. L., Landon, M., Piszkiewicz, D., Brattin, W. J., Langley, T. J. and Melamed, M. D. (1970). Bovine liver glutamate dehydrogenase: tentative amino acid sequence; identification of reactive lysine; nitration of specific tyrosine and loss of allosteric inhibition by guanosine triphosphate. *Proc. Nat. Acad. Sci. U.S.* **67**, 724–730.

Smith, I. (1960). "Chromatographic and Electrophoretic Techniques," Vol. I, Chromatography. Wiley (Interscience), New York.

Smith, J. D., and Law, J. H. (1970). Phosphatidyl-choline biosynthesis in *Tetrahymena pyriformis*. *Biochim. Biophys. Acta* **202**, 141–152.

Smith, J. D., and Law, J. H. (1970). Phosphonic acid metabolism in *Tetrahymena*. *Biochemistry* **9**, 2152–2157.

Smith, J. D., Snyder, W. R., and Law, J. H. (1970). Phosphonolipids in *Tetrahymena* cilia. *Biochem. Biophys. Res. Commun.* **39**, 1163–1169.

Snyder, W. R., and Law, J. H. (1970). Quantitative determination of phosphonate phosphorus in naturally occurring aminophosphonates. *Lipids* **5**, 800–802.

Sprague, L. M. (1970). Electrophoretic patterns of skipjack tuna tissue esterases. *Hereditas* **65**, 187–190.

Stahl, E. (1969) "Thin-layer Chromatography. A Laboratory Handbook." Springer-Verlag, Heidelberg.

Stephens, R. E. (1970). On the apparent homology of actin and tubulin. *Science* **168**, 845–847.

Subaramanian, A. R., Holleman, J. W., and Klotz, I. M. (1968). Primary structure of *Golfingia gouldi* hemerythrin. Interpeptide overlaps and sequences from chymotryptic peptides. *Biochemistry* **7**, 3859–3867.

Sugita, M., Arakawa, I., Hori, T., and Sawada, Y. (1968). Shellfish lipids. X. Fatty acid components of ceramide 2-amino-ethylphosphonate. *Seikagaku* **40**, 158–162.

Sugita, M. and Hori, T. (1971). Isolation of diacylglycerol 2-aminoethylphosphonate from *Tetrahymena pyriformis. J. Biochem.* (Tokyo) **69**, 1149–1150.

Suzuki, R. (1960). Sperm activation and aggregation in some fishes. IV. Effects of pH, heat and other agents upon sperm-stimulating factor. *Japanese J. Zool.* **12**, 465–476.

Suzuki, R. (1961). Sperm activation and aggregation during fertilization in some fishes. VI. The origin of the sperm stimulating factor. *Annot. Zool. Japan* **34**, 24–29.

Takagi, M. Oishi, K., and Okumura, A. (1967). Free amino acid composition of some species of marine algae. *Nippon Suisan Gakkaishi* **33**, 669–673.

Takagi, N., Hsu, H. Y., and Takemoto, T. (1970). Hypotensive constituents of marine algae. 5. Amino acid composition of *Petalonia fescia. Yakugaku Zasshi* **90**, 899–902.

Tamari, M. (1971). Intracellular distribution of ciliatine (2-aminoethylphosphonic acid) in *Tetrahymena. Agr. Biol. Chem.* **35**, 1799–1802.

Tamari, M. and Kametaka, M. (1972). Isolation and identification of ciliatine (2-aminoethylphosphonic acid) from phospholipids of the oyster, *Crassostrea gigas. Agr. Biol. Chem.* **36**, 1147–1152.

Tamiya, N. and Abe, H. (1972). Isolation, properties, and amino acid sequence of erabutoxin C, a minor neurotoxic component of the venom of a sea snake, *Laticauda semifasciata. Biochem. J.* **130**, 547–555.

Taniguchi, N., and Nakamura, I. (1970). Comparative electropherograms of two species of frigate mackerel. *Nippon Suisan Gakkaishi* **36**, 173–176.

Tentori, L., Vivaldi, G., Carta, S., Marinucci, M., Massa, A., Antonini, E., and Brunori, M. (1971). Primary structure of *Aplysia* myoglobin: Sequence of 63-residue fragment. *Fed. Eur. Biochem. Soc. Lett.* **12**, 181–185.

Terwilliger, R. C., and Read, K. R. H. (1969a). The radular muscle myoglobins of the amphineuran mollusc, *Acanthopleura granulata* Gmelin. *Comp. Biochem. Pysiol.* **29**, 551–560.

Terwilliger, R. C., and Read, K. R. H. (1969b). Quartenary structure of the radular muscle myoglobin of the gastropod mollusc, *Buccinum undatum* L. *Comp. Biochem. Physiol.* **31**, 55–64.

Terwilliger, R. C., and Read, K. R. H. (1970). Hemoglobins of the holothurian echinoderms, *Cucumaria miniata, Cucumaria piperata* and *Molpadia intermedia. Comp. Biochem. Physiol.* **36**, 339–351.

Terwilliger, R. C., Terwilliger, N. B., Clay, G. A., and Belamarich, F. A. (1970). Subcellular localization of a cardioexcitatory peptide in the pericardial organs of the crab, *Cancer borealis. Gen. Comp. Endocrinol.* **15**, 70–79.

Thompson, G. A., Jr. (1969), Metabolism of 2-aminoethylphosphonate lipids in *Tetrahymena pyriformis. Biochim Biophys. Acta* **176**, 330–338.

Thompson, G. A., Jr. (1970). Phosphonolipids: Localization in surface membranes of Tetrahymena. *Science* **168**, 989–991.

Thompson, G. A., Jr., Bambery, R. J. and Nozawa, Y. (1971). Membrane formation in *Tetrahymena pyriformis.* 4 Lipid composition and biochemical properties of *Tetrahymena pyriformis* membrane systems. *Biochemistry* **10**, 4441–4447.

Thompson, G. A., Jr., Nozawa, Y., and Bamery, R. (1971). How the surface membranes of *Tetrahymena pyriformis* become enriched in certain lipids. *FASEB Proc.* **30**, 1116.

Thompson, J. F., Morris, C. J., and Smith, I. K. (1969). New naturally occurring amino acids. *Ann. Rev. Biochem.* **38**, 137–158.

Thompson, R. H., and Mathieson, J. W. (1971). Naturally occurring quinones. XVIII. New spinochromes from *Diadema antillarum*, *Spatangus purpureus*, and *Temnopleurus toreumaticus. J. Chem. Soc. C* (**1**), 153–160.

Thornhill, D. P. (1971). Abductin. Locus and spectral characteristics of a brown fluorescent chromophore. *Biochemistry* **10**, 2644–2649.

Thornhill, D. P. (1972). Elastin. Locus and characteristics of chromophore and fluorophore. *Connect. Tissue Res.* **1**, 21–30.

Thorson, G. (1950). Reproduction and larval ecology of marine bottom invertebrates. *Biol. Rev.* **25**, 1–45.

Tills, D., Mourant, A. E., and Jamieson, A. (1971). Red-cell enzyme variants of Icelandic and North Sea cod (*Gadus morhua*). *Rapp. Proces-Verb. Reunions, Cons. Int. Explor. Mer* **161**, 73–74.

Trebst, A., and Geike, F. (1967). The biosynthesis of phosphonoamino acids. The distribution of the radioactivity in aminoethylophosphonic acid following synthesis from specifically labeled glucose by *Tetrahymena. Z. Naturforsch.* **22b**, 989–991.

Trimble, R. B. (1969). Induction of manganese dioxide-reducing activity in a marine bacillus. Ph. D. thesis, Rensselaer Polytechnic Inst., 1969, Univ. Microfilms No. 70–2652.

Trimble, R. B., and Ehrlich, H. L. (1968). Bacteriology of manganese nodules. III. Reduction of manganese dioxide by two strains of nodule bacteria. *Appl. Microbiol.* **16**, 695–702.

Trimble, R. B., and Ehrlich, H. L. (1970). Bacteriology of manganese nodules. *Appl. Microbiol.* **19**, 966–972.

Troshanov, E. P. (1965). Bacteria which reduce manganese and iron in bottom deposits. *In* "Applied Capillary Microscopy: The Role of Microorganisms in the Formation of Iron-Manganese Deposits" (B. V. Perfil'ev, D. R. Gabe, A. M. Galperina, V. A. Rabinovich, A. A. Sapotnitskii, E. E. Sherman, and E. P. Troshanov, eds.) (from Trimble, 1969).

Tsunogai, S. and Sase, T. (1969). Formation of iodide-iodine in the ocean. *Deep-Sea Res.* **16**, 489–496.

Tsunogai, S. (1971). Iodine in the deep water of the ocean. *Deep-Sea Res.* **18**, 913–919.

Turner, A. B., Smith, D. S. H. and Mackie, A. M. (1971). Characterization of the principal steroidal saponins of the starfish *Marthasterias glacialis*. Structures of the aglycones. *Nature* **233**, 209–210.

Tyler, P. A. (1970). Hyphomicrobia and the oxidation of manganese in aquatic ecosystems. *Antonie van Leewenhoek*; *J. Microbiol. Serol.* **36**, 567–578.

Utter, F. M., Stormont, C. J. and Hodgins, H. O. (1971). Esterase Polymorphism in vitreous fluid of Pacific hake, *Merluccius productus. Amin. Blood Groups Biochem. Gen.* **1**, 69–82.

Utter, F. M. (1969). Transferrin variants in Pacific hake (*Merluccius productus*). *J. Fish. Res. Bd. Can.* **26**, 3268–3271.

Utter, F. M., and Hodgins, H. O. (1970). Phosphoglucomutase polymorphism in sockeye salmon. *Comp. Biochem. Physiol.* **36**, 195–199.

Vevers, H. G. (1966) Pigmentation. *In* "Physiology of the Echinodermata" (R. A. Boolootian, ed.), p. 267. Wiley (Interscience), New York.

Vinogradov, S. N., Machlik, C. A., and Chao, L. L. (1970). The intracellular hemoglobins of a polychaete. *J. Biol. Chem.* **245**, 6533–6538.

Viswanathan, C. V. (1973). Chromatographic fractionation of the lipids of *Tetrahymena. J. Chromatogr.* **75**, 141–145.

Viswanathan, C. V. and Hagabhushanam, A. (1973). Preparative isolation of phosphonolipids by ascending dry-column chromatography. *J. Chromatogr.* **75**, 227–233.

Vuilleumier, F. and Matteo, M. B. (1972). Esterase polymorphism in European and American populations of the periwinkle, *Littorina littorea. Experientia* **28**,

Warren, W. A. (1968). Biosynthesis of phosphonic acids in *Tetrahymena. Biochim. Biophys. Acta* **156**, 304–346.

Whittaker, R. H., and Feeny, P. P. (1971). Allelochemics: Chemical interactions between specific species. *Science* **171**, 757–770.

Wilkins, N. P. (1971). Biochemical and serological studies on Atlantic salmon (*Salmo salar*). *Rapp. Proces-Verb. Reunions, Cons. Int. Explor. Mer* **161**, 91–95.

Wilkins, N. P. (1972). Biochemical genetics of the Atlantic Salmon *Salmo salar*. I. Review of recent studies. *J. Fish. Biol.* **4**, 487–504.

Winter, W. P., and Neurath, H. (1970). Purification and properties of a trypsinlike enzyme from the starfish *Evasterias trochelli*. *Biochemistry* **9**, 4673–4679.

Woolfok, C. A., and Whitely, H. R. (1962). Reduction of inorganic compounds with molecular hydrogen by *Micrococcus lactilyticus*. *J. Bact.* **84**, 647–658.

Yanagimachi, R. (1957). Some properties of the sperm activating factor in the micropyle area of the herring egg. *Annot. Zool. Japan* **30**, 114–119.

York, J. L., and Fan, C. C. (1971). Implications of tyrosine in iron binding in hemerythrin. *Biochemistry* **10**, 1659–1665.

Zwilling, R. (1970). Comments following "Homology and Phylogeny of Proteolytic Enzymes" by Neurath, H., Bradshaw, R. A. and Amon, R. *In* "Structure—Function Relationships of Proteolytic Enzymes" (P. Desnuelle, H. Neurath, and M. Ottesen, eds.), pp. 136–137. Academic Press, New York.

Chapter 7

Toxicology: Venomous and Poisonous Marine Animals

Findlay E. Russell
and
Arnold F. Brodie

I. Introduction

In previous works, Russell has attempted to describe the venomous and poisonous marine animals of the world, their distribution and general biology, the nature of their venom apparatus or method of poisoning, the chemistry and pharmacology of their toxins, topics related to immunological phenomena, and the clinical problem of poisoning by these animals (Russell, 1965, 1968, 1969, 1972). A more detailed consideration of these topics will be found in the three-volume compendium by Halstead (1965–1970). In this chapter we will present a summary of the methods currently employed in the study of the chemical, pharmacological, and immunological properties of marine toxins. Some attention will also be given to the historical development of these methods, to their current uses and their limitations, and to the experimental designs needed for future research on marine toxins. Obviously, some data on the animals themselves will need to be cited, and some consideration will be given to the clinical problem of envenomation, since man is by far the most versatile of the experimental animals.

In such a work as this the authors will need to be somewhat selective in their presentation. We trust that our colleagues will consider that we have endeavored to provide a comprehensive review on a considerable body of data, from a number of different disciplines, on many kinds of marine toxins, and that they will pardon us for any important errors of omission or commission. We are consoled by the wise words of Goethe:

> Everyone who writes a text book on any branch of experimental science must set down as many wrong statements as right; he cannot carry out most experiments himself, he must rely on the testimony of others and often take probability for truth. Thus a compendium is a monument of the time when the facts were collected and it must be renewed and rewritten again and again. But while first discoveries are accepted and a few chapters approved, others perpetrate misleading experiments and erroneous deductions.

Approximately 1000 species of marine organisms are known to be venomous or poisonous. For the most part, these animals are widely distributed throughout the marine fauna from the unicellular protistan, *Gonyaulax*, to certain of the chordates. They are found in almost all seas and oceans of the world. While in most areas they do not usually constitute a problem, in a few scattered regions, such as the South Pacific where ciguatera poisoning sometimes causes a serious public health and economic problem, they present an occasional threat to man's health and economy.

It seems advisable, in a review such as this, to distinguish between "venomous" and "poisonous" marine animals. In general, the word venomous is used to describe a creature which has a gland or group of highly specialized secretory cells, a venom duct (although this may not be a constant finding) and a structure for delivering the venom. While in the past there has been a tendency to employ the term "venom apparatus" to denote only the sting or spine used by the animal to deliver its venom, most biologists now use the term in its broader context, that is, to denote the gland and duct in addition to the spine or sting. Poisonous marine animals, as distinguished from venomous ones, have no such apparatus. Poisoning by these forms usually takes place through ingestion of the tissues, which either in part or in their entirety, are toxic. It can be seen that all venomous marine animals are, in reality, poisonous but not all poisonous animals are necessarily venomous.

Considerable interest in these animals and their toxins has been demonstrated during the past decade. This has been at least partially due to recent improvements in technology, particularly in the fields of chemistry and pharmacology, which have made it possible to characterize a few of these poisons chemically and to determine, to some extent, their specific modes of action. This was not possible even 5 years ago and although our data are still relatively meager, studies during the last few years have suggested experimental designs and techniques which should give us a considerable amount of knowledge about these toxins during the next decade.

It would appear that most marine poisons are complex mixtures of substances which vary considerably in their chemical and physiopharmacological properties. Some are proteins, while others are amines, steroids, lipids, saponins, quaternary ammonium compounds, mucopolysaccharides, etc.; however, the chemical nature of many of the marine poisons is not known. It must be admitted that at this writing we have determined the structures of only a few of these marine poisons, and that we are relatively uninformed on the exact mechanisms by which they produce their biological effects. On the other hand, during the past 2 decades we have accumulated considerable data on which biological systems are involved during poisoning, as well as on the general nature of the deleterious biological changes. We are beyond the experimental period where we were limited to measuring the changes in the dependent variables of a biological system. We are now in the period of determining the changes in the parameters, the sequence of these changes, the relationships between the parameter changes (e.g., pulmonary artery pressure vs pulmonary artery flow in the presence of normal cardiac output, etc.), and in the next few years, we trust, we will find ourselves seeking to determine the specific mechanisms involved in the poisoning, both at the cellular and organ system levels.

Future work should also be directed toward improving our isolation techniques to the point where some of the marine toxins can be characterized as pure products and then synthesized for use as tools in molecular biology or as drugs for mankind. No area of biology has a greater potential for the development of useful experimental and medical products as does that of marine toxins.

II. Chemistry and Pharmacology

As already noted, the marine toxins, for the most part, are complex and relatively dissimilar mixtures. The more toxic components of the fish venoms, for example, appear to be proteins, although nonproteinaceous materials are certainly found in the crude venoms. Tetrodotoxin is an amino perhydroquinazoline, while paralytic shellfish poison appears to contain ring structures with several of the nitrogen atoms being involved in a heterocyclic structure. The shellfish poison contains two basic groups; one may be a guanidine and the other an amine. Most of the toxins from the other marine organisms are equally as unrelated in their chemcial structure. It would seem that the marine toxins are as diversified in their nature as are the terrestrial arthropod toxins, and certainly far more dissimilar than the reptilian venoms.

Another characteristic of some of the marine toxins, particularly the fish venoms, is their instability. Some are very labile, even at temperatures approaching O°C. However, it is quite possible that when a specific lethal or deleterious fraction has been isolated in its pure form it will be found to be relatively stable. It may well be that since many toxic components of fish venoms are often found within isolated groups of cells, rather than within a true venom gland, other components of these cells, perhaps involved in the protection of the cell from the toxin, tend to destroy or modify the toxic component(s) when the venom-containing cells or tissues are extracted. Unfortunately, it is not yet possible to extract a "pure" venom from the venom-producing cells of a fish's sting or spine. Thus, in an extract one obtains, in addition to the venom, which is complex in structure, a number of equally complex normal cellular components, some of which also have important diversified pharmacological properties. It would certainly seem reasonable, in the case of cells producing a fish venom, that they possess some built-in protection against the poison they produce. Although our laboratory has not yet been able to demonstrate a circulating antibody or antibodies against stingray or certain other fish venoms in these animals, it may be that our techniques are not yet adequate. On the other hand, circulating antibodies may not exist. The mechanism of protection may simply be cellular.

Some insight into the chemistry and biological modes of actions of marine toxins can be obtained if one considers the use to which the animal puts its toxin(s). The various components of venoms appear to have evolved and adapted in a remarkable way, in most instances as parts of the animal's offensive or defensive armament. Although there are some fractions of marine poisons for which we have yet to find a specific pharmacological activity that seems related to the design of the venom, there are some synergistic actions of whole venoms that are not present if one separates the various fractions and studies their individual toxicological properties. It is quite possible that the biological properties of some venom fractions played an important role in eons past, but are no longer essential to the function of the venom in the animal's present ecological niche.

In the case of some marine toxins, such as tetrodotoxin, the poison may, in general, be related to a normal metabolic cycle (such as the ovarian cycle) and play no significant role in the offensive or defensive stature of the animal. Further, some poisons, such as ciguatoxin, are a product of the animal's feeding cycle. The fish or other marine animal feeds upon other fishes or marine animals or plants which contain a material that is toxic. The initial or basic material may (or may not) be toxic in the originating species, but it is usually present in such small quantities that it does not cause deleterious biological effects, at least in man. When concentrated through the food-chain cycle it becomes available in large enough amounts to cause "poisoning." It is obvious that new marine poisons may arise from time to time as the feeding patterns of fishes and other marine animals change. Such changes may be expected, since man is continually affecting the contents of our oceans and is, in one way or another, influencing established marine habitats.

The various types of marine poisoning have been elaborately divided by various workers. The reader is referred to the compendium by Halstead (1965–1970), and the lesser works by Russell (1965, 1969, 1972) for a review of these divisions and diversions. There is no comprehensive chemical or physiopharmacological classification for the toxins of marine animals at the present time. Attempts to present such a classification have been made, but they are either too limited in their considerations or too naive in their organization to provide a working model. Our knowledge of the properties of these complex substances it not broad enough or consistent enough, at the present time, to permit the adoption of a single working classification.

It might seem wisest to use a system based on taxonomy. Such a system would be somewhat bulky but it would serve our purposes during the interim in which we attempt to organize our data on the chemical and biological properties in a more thorough manner. It is regrettable that some of the major efforts involving the chemistry and pharmacology of marine toxins have been and are being based on improperly identified animals, or on commercially prepared toxins in which scrupulous control of the involved species has not been exercised. For these and other reasons, it seems best that until the fractions of a marine poison have been isolated and characterized, their pharmacological properties studied individually and in combination, we should exercise extreme care in systematizing data which are based partly on biological assay methods, partly on biochemical studies, partly on clinical observations, and partly on intuitive hunches.

In previous works, and as we have noted elsewhere in the present contribution, some insight into the chemistry and pharmacology of marine venoms can be gleaned if one considers the use to which the animal puts its toxin.

Toxins originating or delivered from the oral pole are usually associated with the animal's offensive posture, as in the gaining of food; while those associated with or delivered from the aboral pole or fin spines are usually used by the animal in its defensive stance. There are some exceptions, and some marine animals may use their venom apparatus for both purposes, but in general this consideration holds true throughout the animal kingdom. It should seem obvious, with this in mind, that one can direct his search for the toxic components on the basis of the venom's design and, indeed, such workers as Endean, Southcott, Shapiro, Tamiya, Lane, Banner, Saunders, Carlson, Schaeffer, Schantz, and others have made use of this approach.

As one might expect, enzymes have been found in a number of marine toxins associated with the animal's offense or oral delivery, and they appear to be either absent or present only in extremely small amounts in those venoms from the fin spines or the venom-containing cells derived from dermal tissues.

Another important consideration, which is more evident with the terrestrial venoms than the marine toxins, is that venom delivered aborally (i.e., away from the mouth), or in defense, tends to produce greater pain than that delivered orally. Again, this should seem evident, since pain is a far more effective deterrent than almost all other biological activities. Even death would not seem a good defensive property in a toxin, if the design of the poison was to deter, for deterrence requires learning, and perhaps memory, and death is free of both of these.

The pharmacological properties of the marine toxins appear to vary as remarkably as do their chemical properties. Some marine toxins provoke rather simple effects, such as transient vasoconstriction or vasodilatation, while others provoke more complex responses, such as parasympathetic dysfunction or multiple concomitant changes in several organ systems. The effects of the separate and combined activities of the various fractions, and of the metabolites formed by their interactions, may be complicated by the response of the envenomated organism. The organism may produce and release several autopharmacological substances which can not only complicate the poisoning, particularly in man, but may also, in themselves, produce more serious consequences than the venom.

The study of the marine toxins is further complicated by the fact that qualitative as well as quantitative differences in toxins may exist not only from species to species within the same genus but also from individual to individual within the same species. A toxin may even vary in its chemical and zootoxicologial properties within the individual animal at different times of the year or under different environmental conditions.

With these things in mind, and rather than review in detail the chemical and biological properties of the marine toxins, which has recently been done by Halstead (1965–70) and Russell (1972), the present monograph has been prepared in an attempt to summarize some of the current methods employed in the study of a marine toxin. Since it would not be possible to detail all the current methods now being used for the hundred or more marine toxins under study, we have selected a representative experimental design with the hope that this will give the reader a more ready knowledge of the present status of study on this poison and act as a guide for developing further experiments on this and other marine toxins. We have chosen the study of a fish venom for this purpose.

III. Venom of the California Scorpionfish, *Scorpaena guttata*

A. INTRODUCTION

The venom apparatus of *Scorpaena guttata* consists of 12 dorsal, three anal, and two pelvic spines and their enveloping integumentary sheaths. When the integumentary sheath is removed from a dorsal spine, a slender, elongated, fusiform strand of grayish tissue is found lying within the distal $\frac{1}{2}$–$\frac{2}{3}$ of the glandular grooves on either side of the spine. The venom is contained within this tissue. Envenomation occurs through mechanical pressure on the spine, which generally ruptures the integumentary sheath. The venom is thus emptied mechanically into the wound produced by the fin spine. In the stonefish, *Synanceja*, there are true venom glands within the anterolateral-glandular grooves that possess ductlike structures. These ducts extend from the glands to the tip of the spine. Thus, on stimulation the venom is forced from the gland through the duct and into the wound produced by the sharp fin spine. A more thorough description of these structures will be found elsewhere (Russell, 1965; 1969; Halstead, 1965–1970).

The first significant study on the venom of a scorpionfish was that by Bottard (1889), who observed that the venom of the stonefish (*Synanceja*) was clear, bluish in color, coagulated by nitric acid, alcohol, ammonium, or by heat. However, it was not until the late 1950's that more definitive studies were initiated on the venom of this and other marine fishes. Saunders and his colleagues, Taylor, Austin, Wiener and Russell, among others, attempted to stabilize, fractionate, and characterize the venom and describe its more deleterious biological properties. All of these workers found that the extracts of the venom-containing tissues, in the case of *Scorpaena* and *Pterois*, and the venom itself, in the case of *Synanceja*, were unstable; and that certain of the biological properties, particularly the lethal effect, were quickly lost on standing, heating, repeated freezing and thawing, and certain other physical changes. These same observations had been made for other fish venoms

(Russell and van Harreveld, 1954; Russell *et al.*, 1958; Russell and Emery, 1960).

Many techniques have been employed in attempts to extract the poison from the venom-producing cells of *S. guttata*. As previously noted, it is not possible at present to obtain "pure" venom from most of the venomous fishes, since the venom-containing cells are only a part, and sometimes a very small part, of the tissue content within the animal's spine grooves. The most commonly employed method of preparing a fish venom is in saline, buffered saline, or a distilled water extract of the macerated tissues containing the venom-producing cells. Thus, in addition to a water-soluble venom, one obtains normal water-soluble components of all the macerated tissues. In determining the biological and chemical properties of this extract the presence of these normal tissue components must be taken into consideration. It may be that some of these components protect the animal against its own venom.

The extract obtained from these tissues is then used in the various chemical or pharmacological studies. It is always more stable (lethal) when freshly prepared, and it loses its lethal effect, and most of its other deleterious activities, with the notable exception of the pain-producing property, with the passing of time. When placed in a saline solution at room temperature, the lethal property may be markedly lost in a matter of hours. At lower temperatures toxicity is more stable, the lethal property being evident, but decreased, up to 12 hr.

Once put into solution, freezing and thawing inactivate the venom. Repeated freezing and thawing of the frozen intact spines also destroys the toxin. On lyophilization, as much as half of the lethal property may sometimes be lost. Lyophilization of the spine and surrounding tissues is impractical. Most lyophilized products have been made from saline extracts of the macerated tissues. All preparations, on heating, lose their toxicity.

Extracts prepared from the spines of live or freshly killed fishes are the most stable. However, spines may be removed and quickly placed at $-35°C$ or lower and stored for more than a year and then extracted. Extracts from stored spines lose their toxic property more quickly than those prepared from fresh spines. Until the past few years, then, the usual procedure for preparing a fish venom was to remove the venom-containing tissues from certain spines of the fish on a cold plate, macerate these in saline or a buffered saline solution, centrifuge the material at 5000 rpm for 10 min, and draw off the supernatant fraction for the chemical or pharmacological studies, or for lyophilization.

Certain biological properties of this fish venom, as well as others, appear to be less susceptible to physical changes. The pain producing property of most fish venoms does not seem to be lost as quickly as does the lethal property. It is well known that the sting of some venomous fishes is as painful after the fish has been dead for 8 or 10 hr as it is when inflicted by a living fish.

Taylor (1963) found that the dialyzable portion of crude venom extracts produced the same degree of pain as did the crude venom and that heated venom extracts produced pain of lower intensity, yet still having the characteristics of the crude material. He felt that the lethal and pain producing properties might be due to a common property. In the weeverfish, Carlisle (1962) found that the dialyzable part of the venom produced the stabbing pain characteristic of the whole venom. The nondialyzable fraction failed to provoke the pain but did produce a rise in pulse pressure and some respiratory distress. He concluded that the systemic effects of the toxin were due to the nondialyzable fraction, while the pain was a consequence of some constituent of the dialyzable part of the venom. He found that the dialyzable fraction contained a large amount of 5-hydroxytryptamine, which he suggested as the "major if not the only pain producing substance" in the venom.

Serotonin has also been found in the venom of the round stingray, *Urolophus halleri* (Russell, 1965), and it is a common constituent of many terrestrial and marine animal toxins. It is not known, however, if serotonin is the principal pain producing substance in these venoms. There have been no definitive studies on the pain producing components of marine toxins. The few published comments on this property are limited to the finding by Carlisle (1962) and the observation that stings by most venomous fishes are painful, even when inflicted by the spines of a dead fish, and after the lethal property has been lost.

Other studies on the stability of *S. guttata* venom (Saunders, 1960; Taylor, 1963; Saunders and Russell, unpublished data), indicate that the lethal and more deleterious fractions were nondialyzable, that extracts prepared from spines which had been stored for 24 hr at 5°C retained only 50% of their lethal activity, that glycerol provided some stability to the extracts (extracts in 40% glycerol at 5°C were lethal after 3 months), and that 60% of the lethal activity of the extracts was lost after 5 min at 40°C. Extracts were also unstable in 1×10^{-3} *M* ethylenediaminotetraacetic acid (EDTA) and in 1×10^{-3} *M* reduced glutathione (GSH). Lethality was completely inactivated by 1×10^{-3} *M* parachlormercuribenzoate (PCMB), following incubation for one hour at 5°C. A proteinase inhibitor, diisopropylfluorophosphate, had a slight stabilizing effect at 1×10^{-3} *M*.

During the course of the various stabilization studies on the venom, some further chemical studies were carried out. Fractionation on Sephadex G-75, equilibrated with 0.05 *M* sodium phosphate, pH 7.4, gave a single peak which contained 53% of the toxicity and 35% of the protein. On a DEAE-cellulose column with 0.05 *M* sodium phosphate, pH 7.4, using a gradient salt solution from 0.05 to 0.4 *M* NaCl, the crude material was separated into two fractions. The first fraction contained the more lethal material but was not pure nor was it nearly as toxic as the original crude product. Both the crude and the DEAE column fractions were studied on acrylamide gel electrophoresis. In the case of the crude material, toxicity was limited to the area near the origin, an area containing at least two distinct bands or one rather diffuse band. The gel showed 7–8 anionic components. The DEAE material gave a lethal band near the origin. We have been unable to confirm these observations. Attempts to separate the active component on DEAE-cellulose columns on six occasions have been unsuccessful.

The pharmacological properties of the venoms of the lionfish, *Pterois*, the stonefish, *Synanceja*, and the California sculpin, *Scorpaena guttata*, are summarized in Table I:

The amount of venom contained within the fin structures of the venomous scorpionfishes varies considerably. In the stonefish *Synanceja trachynis* (= *S. horrida*) we have found the liquid content of the two dissected venom glands to be approximately 0.045 ml, with a range of 0.021–0.056 ml. The average dried weight of venom per fish was 7.35 mg. This compares favorably with Weiner's (1959) finding of a volume of 0.03 ml and a dried weight of 5.1–9.8 mg. In *S. guttata*, using the batch extraction method (Schaeffer *et al.*, 1971), the average yield per spine, and this is the average for all the venomous spines of one fish, is 6.2 mg of dried crude material, or 2.0 mg protein, or approximately 0.3 mg of dried lethal protein.

In humans, envenomation by *S. guttata* produces immediate, intense, sometimes pulsating pain in the area of the injury. Within 3–10 min. the pain may radiate to involve the entire extremity. Although the involved digit, hand or foot usually becomes red and swollen, the area around the wound may appear ischemic, and the edges of the wound discolored. Bleeding does not appear to be affected. The pain may extend into the axilla or inguinal area within 15 min; palpable nodes are reported in 30% of the untreated cases. Nausea, vomiting, weakness, primary shock, an urgency to urinate, increased perspiration, conjunctivitis, headache and diarrhea have all been reported. Parasthesia about the wound and even up the forearm are not uncommon. The severe pain usually subsides in 3–8 hr but the swelling and tenderness may persist for days. In most cases there are no serious sequelae.

In severe stingings the pain may be excruciating, causing the victim to thrash about in agony. Primary shock and even secondary shock may occur. Pulmonary edema and abnormal electrocardiograms have been reported, and in one patient a pulmonary embolism occurred, requiring specific therapy for 24 days (Russell, 1965).

TABLE I
Some Properties of Scorpaenid Venoms[a]

	Pterois	*Synanceja*	*Scorpaena*
Small dose	Decreased arterial pressure Minimal ECG changes Increased respiratory rate Muscular weakness in mice	Decreased arterial pressure Minimal ECG changes Increased respiratory rate Tremor	Slight decrease in arterial pressure Increased then decreased venous pressure Minimal ECG changes Increased respiratory rate with decreased respiratory excursions
Medium dose	Marked fall in arterial pressure Myocardial ischemia, injury or conduction defects Increased respiratory rate Partial paralysis of legs in mice	Marked fall in arterial pressure Myocardial ischemia, injury or conduction defects Increased respiratory rate Muscular weakness in mice Tremor	Fall in arterial pressure Myocardial ischemia, injury or conduction defects Changes in venous and cerebrospinal fluid pressures Increased respiratory rate with decreased respiratory excursions Muscular weakness in mice
Lethal dose	Precipitous, irreversible fall in systemic arterial pressure Extensive ECG changes Markedly decreased respiratory rate → cessation Complete paralysis of legs in mice Intravenous LD_{50} mice, 1.1 mg protein/kg body weight	Precipitous, irreversible fall in systemic arterial pressure Extensive ECG changes Markedly decreased respiratory rate → cessation Some paralysis of legs in mice Possible neuromuscular junction changes Produces tremors, convulsions, marked muscular weakness, coma Intravenous LD_{50} mice, 200 μg protein/kg body weight	Precipitous, irreversible fall in arterial pressure Extensive ECG changes Markedly decreased respiratory rate → cessation Some paralysis of legs in mice Intravenous LD_{50} mice, in excess 2.0 mg protein/kg body weight

[a] After Russell (1969).

This, then, represented the status of our knowledge up until the late 1960's. It appeared that the crude venom was a mixture of ten or more substances, most of which were proteins, that the lethal fraction was a protein having a molecular weight between 150,000 and 1,300,000 and that it was extremely unstable. Activity was rapidly inactivated by PCMB, and EDTA or GSH did not increase its stability. On electrophoresis the lethal fraction was associated with a slow moving anionic band found close to the origin. The crude venom had a marked effect on the cardiovascular system, provoking arterial hypotension, which appeared to be due to a direct effect of the venom on the heart, or an adrenergic block on the vasomotor center. The intravenous LD_{50}, in mice, was in excess of 2.0 mg protein/kg body weight.

The stinging of Commander Scott Carpenter by this fish during the SeaLab II exercise stimulated renewed interest in the fish's habits and venom. In 1968, Dr. Richard Carlson and Mr. Richard Schaeffer, working in the senior author's laboratory, and with the assistance of Dr. Jacob Dubnoff, Dr. Paul R. Saunders, Dr. Max Weil, and Mr. Howard Whigham renewed efforts to stabilize scorpionfish venoms, and to learn more about their chemistry and physiopharmacological properties, particularly the latter. This report, then, is a summary of the work that has been done in our laboratory since 1968, portions of which have not heretofore been reported. Much of the data have been drawn from the theses of Mr. Schaeffer and Dr. Carlson.

It seemed apparent from the previous studies on this venom that there was a distressing lack of continuity in experimental designs, with the result that although interesting data were obtained, they could not be put together in a manner which demonstrated orderly progression in our thinking on the properties of this poison. This is not to say that there has to be orderly progression in our experiments, but rather, that there should be orderly progression in our thinking, or at least thought given to what the experiment tells us in relation to what we already know, and how the new data might direct us to those experiments that need yet to be done. It is common to all of us to employ those techniques with which we are most familiar, and this is fine, but we should always give careful thought to how the data obtained from such experiments contribute to the development of our overall knowledge on the subject. This is not to infer that one can always predict which experiments need to be done, but rather to caution against a certain amount of intellectual smugness in us all, which is too often reflected by our satisfaction with the use of previously mastered techniques when what is needed are new approaches employing new methods.

In 1969 we set down a protocol that we felt encompassed the basic experiments which needed to be done on the venom of *S. guttata*, or which might be used as a guide for any study on a fish venom. We took into account what was already known about this subject, what we were equipped to do, what time we had to do it in, and what basic information we could contribute for subsequent investigations to build upon. This study is still in progress but we feel that strict adherence to this experimental design has been fruitful, and that we have laid down some simple yet basic concepts which will permit other researchers to carry out the definitive chemical and toxicological studies that need yet to be done. The experimental design is shown in Table II.

It was obvious that the first problem to solve before any definitive chemical or physiopharmacological studies could be done was to obtain a stable fraction of the venom or a venom extract. A revewed effort, therefore, was made to evaluate the various procedures that had previously been used to stabilize this venom, and to try a number of new stabilization techniques. We will note some of these trials, since they represent approaches that would normally come to mind when dealing with an unknown, unstable protein of this size.

B. COLLECTION AND PREPARATION OF VENOMOUS FISHES

Experiences with all venomous fishes so far studied indicate that the best results are obtained from freshly prepared venom or venom extracts; that is, extracts of venom prepared immediately following sacrifice of the fish or removal of the venom containing tissues. The longer the delay following

TABLE II
Protocol for Analysis of *Scorpaena guttata* Venom

I. Processing of fish and/or spines
II. Extraction procedure
III. Stabilization studies
IV. Chemistry and pharmacology
   A. Chemistry
      1. Gel filtration
      2. Ion exchange
      3. Electrophoresis
      4. Immunochemistry
      5. Ultracentrifugation
      6. Amino acid analysis
      7. Amino acid sequence
      8. Further characterization
   B. Pharmacology
      1. LD_{50}
      2. Cardiovascular, survey
      3. Cardiovascular, isolated
      4. CNS
      5. Peripheral nervous system
      6. Clotting
      7. Hemolysis
      8. Antivenin
V. Synthesis
VI. Potential

death of the animal or extraction of the venom, the greater the loss in the lethal activity, as well as the more deleterous activities. The pain producing property appears to be the least affected. If extracts must be kept over a 1–12 hr period, they should always be placed on ice.

As it is not always possible to work with extracts prepared from living fishes, we have found that fishes sacrificed, or spines removed, and immediately placed on dry ice and then maintained between −35°C and −90°C until the extraction procedure is instituted is the next best method of assuring a relatively stable product. Repeated freezing and thawing are to be avoided. When only a few venomous fins or spines are needed, these should be separated from the rest while the tissues are still in the frozen stage, and the unused fins or spines immediately returned to the cold box. It has been our practice to remove the venomous spines from the fish, wash them off with seawater and then immediately wrap them individually, as in the case of the stingray, or in groups of 2–12, as with the weevers or some scorpionfishes, or when larger numbers of fishes are available, as with *S. guttata*, remove the spines from 10 to 15 fish and immediately place them in plastic bags on dry ice for transfer to the laboratory. When it is not possible to put the spines on dry ice, the fish or spines can be kept on ice until they can be placed at a lower temperature. However, and depending on the fish involved, the loss of the lethal and other deleterious biological activities will often occur rather rapidly when the fish is on ice. Some spines appear to lose most of their lethal property within 3 hr while on ice, while others may retain part of this property for as long as 12 hr.

It is always best to wrap the fins in aluminum foil, plastic, or paraffin paper when they are to be placed directly on the dry ice. With some venomous fishes, such as the stingray, whose caudal spine is often covered with sand (particularly when taken in a beach seine), the spine should be carefully rinsed with salt water before it is wrapped and placed on the dry ice. Under no circumstances should marine venomous spines be rinsed with fresh water.

C. EXTRACTION OF VENOM

A review of the methods for extracting fish venoms has been presented elsewhere (Skeie, 1962; Russell, 1965; Halstead, 1970). In the case of *S. guttata*, it is best to extract as much fresh material as you have on hand, lyophilize it and then store it in the dark at 5°C. Each lot can then be characterized as (a) LD_{50} mg/kg, (b) LD_{50} mg protein/mg crude material, and (c) mg lethal protein/mg crude material. We have found the following two methods to be the most satisfactory (Schaeffer *et al.*, 1971).

1. *Aspiration Method*

The integumentary sheath covering these spines is stripped to the base of the spine and the venom-containing secretory cells in the slender, fusiform strand of grayish tissue found lying within the distal $\frac{1}{2}$–$\frac{2}{3}$ of the glandular grooves on either side of the sting are aspirated with a micropipette. The pipette is connected to a vacuum source. The collections are carried out at 5°C and the material centrifuged at 5000 rpm for 10 min. The supernate is then removed for the experimental studies or is lyophilized. If lyophilized, some of the lethal property will be lost.

2. *Batch Method*

The integumentary sheath covering each spine is stripped to the base of the spine and the spine immerged in a small beaker of distilled water at 5°C. The beaker is gently agitated for 5 min and the solution decanted. This procedure is carried out three times and the washings combined and lyophilized. As with the aspiration method, some toxicity will be lost on lyophilization.

3. *Comparison of Methods*

LD_{50} studies indicate that the aspiration method yields a product that is more lethal than that obtained by the batch method, but which, unfortunately, is more unstable. It is felt that a purer extract can be obtained by the aspiration method but that, perhaps due to the trauma to the venom-containing cells during the aspiration process, oxidation of the active components is more likely to occur and thus lead to changes affecting the chemical and biological properties. In any event, although the explanation for the differences in the two preparations is in question, the aspiration method yields a more lethal but a more unstable product. Table III shows the differences in the LD_{50} of the two preparations.

It was apparent, even from visual observations, that the batch prepared product contained far more cells and broken tissue than the aspiration prepared material. In the former, much of the material must come from the torn tissues of the stripped integumentary sheath, as well as from the broken cells of the tissues in the glandular grooves. The amount of protein per spine for the batch method is 8.3 mg, while that obtained by the aspiration method is 0.6 mg.

Because of the stability differences in the two products and the ease of the method, most studies have been done using the batch product. At 5°C the reconstituted lyophilized extract, prepared by either method, is stable only between pH 7.4–8.0. At pH 8.0 the extract is relatively stable at 96 hr. The solution tends to be more stable when the dry weight concentration of the

TABLE III

LD_{50} Determinations[a] of Lyophilized *Scorpaena guttata* Venom

Dose (mg/kg)	Aspiration method[b]	Batch method[c]
0.5	0/10	
0.75	2/10	
1.00	4/10	
1.25	7/10	
6.0		0/10
7.0		3/10
8.0		5/10
9.0		10/10

[a] Data are presented as the number of deaths of mice/total number of mice tested.

[b] Aspiration method LD_{50}: 1.1 mg/kg (1.0 mg protein/kg) body weight.

[c] Batch method LD_{50}: 7.7 mg/kg (2.6 mg protein/kg) body weight.

crude material is approximately 10 mg/ml. At 50 mg/ml and at 5 mg/ml, the solution is far less stable. The extract is least stable in distilled water and, on standing, a precipitate forms. Heat rapidly destroys the lethal property of the toxin, even in the presence of sulfhydryl reagents.

Venom samples prepared by the aspiration method were subjected to 10^{-2} *M* GSH, cystine and PCMB in the presence of 0.05 *M* phosphate buffer, pH 8.0, at 5°C. A second group of samples was subjected to Cleland's reagent, EDTA, and a mixture of the two, all at 10^{-2} *M* in the presence of 0.05 *M* phosphate buffer, pH 7.4, at 5°C. The various samples were tested periodically for lethality up to 21 days.

D. STABILIZATION OF VENOM

It was found that GSH, cystine and PCMB provided no stabilizing effect of the lethal property of *Scorpaena guttata* venom over long periods of time. However, GSH and cystine were thought to provide some stabilizing effect during the first 4 days after mixing. Thereafter, lethality was quickly lost. Cleland's reagent gave considerable stability to the extract up to 21 days, which was the end of the test period. The combination of Cleland's reagent and EDTA also provided some stability, as did EDTA alone, although both to a lesser extent than the Cleland's reagent. Most extracts in 0.05 *M* sodium phosphate buffer, pH 7.4, tended to precipitate, unless 0.5% NaCl was added. In all tests over prolonged periods, a physiological saline solution should be added.

These various tests indicate that sulfhydryl groups are necessary for the stability of the lethal activity of the venom. The most effective sulfhydryl reagent appears to be Cleland's reagent. Subsequent studies indicate that a number of chemical and physiopharmacological studies can now be carried out with a stabilized product prepared in the following manner:

1. Extract in distilled water at 5°C from fresh spines using the batch method.
2. Immediately lyophilize.
3. Reconstitute at 5°C in 0.05 *M* sodium phosphate buffer, pH 7.4, with 0.9% NaCl in Cleland's reagent, and centrifuge.

Once the extract has been reconstituted, and is to be kept for a length of time, it may be necessary to add 0.9% NaCl to prevent precipitation. The solution should be kept at as low a temperature as possible without freezing between experiments.

With the assurance of a reasonably stable fish venom for the first time, it became possible to attempt some definitive chemical and physiopharmacological studies. On the basis of previous experiences and with a limited supply of crude material, it appeared that the most productive chemical procedures to follow would be gel filtration, ion exchange chromatography and disc-gel electrophoresis. Concomitant studies were also initiated to determine some of the more important physiopharmacological properties of the crude venom. These studies are still in progress but some significant results have already been obtained, some of which have been reported (Schaeffer *et al.*, 1971; Carlson *et al.*, 1971).

E. CHEMISTRY OF VENOM

When a crude extract of *Scorpaena guttata* venom without Cleland's reagent is added to a Sephadex G-50 column, three peaks are obtained. The first peak is lethal to mice at the 1.5 mg protein per kg body weight level. The remaining two peaks are not lethal. When the first peak is rechromatographed on a G-200 column, three peaks are again obtained. The first peak is a high molecular weight material and not lethal to mice. The second and third peaks, representing proteins with a molecular weight below 800,000, do not cause death when injected into mice, but the material does produce acute distress characterized by tachypnea, hyperpnea and transient prostration. It was felt that these last two peaks contained the lethal material, and although capable of producing significant symptoms, they were not toxic enough to cause death.

When a batch extraction sample is placed on a Sephadex G-200 column, three fractions are observed. The second fraction (tubes 28–37) contains the lethal material. The intravenous LD_{50} for tube 33 is 0.9 mg protein per kg body weight. Analysis of the optical density of the elution profile indicates that the second peak contains approximately 20% of the total protein. This, then, is the first demonstration of a fraction of a fish venom that is more lethal than the crude material. However, even this fraction is not a pure or single substance. On cellulose acetate strips, we found this fraction to contain several proteins.

When an aspiration extraction sample is chromatographed on DEAE cellulose, four peaks are eluted, although none of these peaks showed lethal activity in the series of studies we carried out. However, the amount of crude material applied to the column might have been insufficient and the biological activities may have been lost during the separation, since Cleland's reagent was not used. Studies with extracts containing Cleland's reagent are in progress.

Cellulose acetate strip electrophoresis of venom extracts shows three to four anionic and two to three cationic bands at pH 8.6. At pH 4.0 the crude venom shows four cationic bands. The lethal fraction from the Sephadex G-200 displays one slow-moving anionic band and two slow-moving cationic bands. These are to be studied in greater detail. The venom has no proteolytic or phosphodiesterase activity.

It is apparent that the lethal property of the venom is associated with protein(s) having a molecular weight of more than 50,000 and less than 800,000, that the biological properties are stable in a strong sulfhydryl reagent, and that a semipurified lethal fraction has an intravenous LD_{50} in mice of 0.9 mg protein per kg body weight, while the crude venom has an LD_{50} of 2.6 mg protein per kg. Further studies on the isolation and characterization of the lethal fraction are in progress at this time.

F. PHARMACOLOGY OF VENOM

Since it was known that the venom of *S. guttata* had a marked effect on the cardiovascular system, as reflected by the changes in the dependent variables, studies were initiated to determine the alterations in the vascular parameters and to gain some insight into the specific physiological modes of action responsible for the poisoning. Along with these studies, certain other toxicological investigations were done. These will be summarized to give some indication of the biological properties of the venom and an approach which might be followed in the study of other fish or marine animal venoms.

Extracts of the venom are as lethal to rats as to mice, the intravenous LD_{50} for the same extract being 5.7 mg/kg for the former and 6.6 mg/kg for the latter. The shore crab *Pachygrapsus crassipes* is relatively immune to the venom, showing no deleterious effects at the 40 mg/kg body weight level. The rabbit appears to be the most sensitive mammal, the dog being less so and the cat being the least sensitive. It would not appear that there is sufficient venom in the dorsal spines of one fish to inflict a lethal dose in an adult human, although a sting by multiple spines of the fish might be lethal to a small child.

At one time the rapid local effects following stings by this fish were attributed to bacterial action. However, Dr. Carlson found that cultures of venom extracts showed only light growths of *Staphylococcus epidermis, Bacillus subtillis*. and *Klebsiella aerobacter*. Gram stain of extracts revealed a few gram-negative rods. He also demonstrated that stonefish antivenin afforded no remarkable protection against the venom of *S. guttata*.

In determining the cardiovascular effects of a fish venom, or other marine toxin, a number of different procedures can be used. Unfortunately, the choice of procedures is sometimes limited by the availability of equipment and/or experience of the researcher. One sometimes sees data which are interesting, and obviously important, but which in themselves do not contribute to the direct development of our knowledge on how the venom produces its deleterious biological effects on the parameters of the cardiovascular system. It might be argued that monitoring the major changes in blood pressure, or some other dependent variable, in a number of different kinds of animals is important but unless one also attempts to determine how these changes in the cardiovascular parameters are produced, one wonders if such an argument is valid, at least at this stage of our knowledge, and also, whether this is a justifiable use of animals. A recent study which demonstrated that rattlesnake and certain other reptilian venoms produce the same blood pressure changes in primates as they do in cats and rabbits certainly seems to us an unjustifiable use of animals, particularly since the effects of these venoms in man are already known, if this was the aim of the experimenter.

Perhaps the most useful preparation for gaining insight into the cardiovascular effects of a toxin is a survey preparation; that is, an intact animal preparation in which a number of vascular dependent-variables can be measured simultaneously. The hemodynamic model proposed by van Harreveld and Shadle (1951) is an excellent preparation which requires a minimum of equipment, although considerable surgical skill. This has been used successfully to study the cardiovascular effects of the venom of the stingray (Russell and van Harreveld, 1954).

Another useful technique is the survey preparation that has been employed in the study of snake venoms (Russell *et al.*, 1962). In this preparation, systemic blood pressure, pulmonary artery pressure, central and peripheral venous pressures, the electrocardiogram (ECG), electroencephalogram

(EEG) and respiratory rate and depth can be measured simultaneously in a single intact animal. Pulmonary artery flow can also be monitored, as well as pressures in the portal system. The technique permits the evaluation of the changes in a number of the dependent variables of the cardiovascular system with time. ECG changes can be related with systemic arterial blood pressure changes; pulmonary artery flow can be related to pulmonary artery pressure and systemic arterial pressure; arterial and venous pressures can be compared in various parts of the body or in different organ systems; respiratory changes can be related to vascular changes, and the EEG can be related to changes in blood flow dynamics. In this way the target area(s) can often be pinpointed and more definitive investigations initiated to determine specific mechanisms and sites of action.

Using this survey preparation, some primary data were obtaind on the cardiovascular effects of *S. guttata* venom. In the cat it was found that lethal and sublethal doses produced a precipitous fall in systemic arterial pressure, which was irreversible when fatal doses were given. With sublethal doses there was a period of hypertension following the hypotensive crisis. Systemic venous pressure rose but only subsequent to the initiation of the arterial pressure decline. Cardiac rate remained normal or increased slightly during the fall in arterial pressure, then it fell to 80% of the pre-injection rate. ECG changes did not appear until approximately 30 sec after the injection, and consisted of inverted T waves, ST segment changes, premature atrial and ventricular contractions, and bundle branch block. Respiration usually ceased approximately 4 min following the injection when lethal doses were given or was variable, and irregular, with sublethal doses. Artificial respiration did not improve the animals' cardiovascular deficit. The EEG was normal, except for a decrease in activity in all areas preceding death. Pulmonary artery pressure fell with the systemic arterial pressure, the two falling concomitantly. Pulmonary blood flow was reduced in the single cat studied. Peripheral blood pressures reflected central blood pressures.

From these experiments it seemed evident that the venom had a marked effect on the cardiovascular system, and that this effect was probably due to a sudden change in circulating blood volume. Since peripheral resistance changes occurred secondarily to central resistance changes, it appeared that a reasonable explanation for the hypotensive crisis might be the pooling of blood in the pulmonary or portal systems. The heart changes, for the most part, were secondary to the decreased systemic arterial blood volume. However, there was evidence of some cardiac change that might contribute to the hypotensive crisis and, certainly, in the case of lethal injections the venom had a direct effect on the heart. The EEG alterations could be attributed to cerebral anemia secondary to a decreased blood supply to the brain. The respiratory changes could be due, again, to changes in heart–lung blood volume equilibrium, reflex phenomena, and to a lesser extent, to depression of the respiratory center. Animals which were artificially ventilated following cessation of normal respiratory movements died at about the same dose levels as those who were not so ventilated, although they may have lived a little longer.

In view of these data we felt a more definitive study should be done on the nature of the changes in the circulating blood volume dynamics, and on the efficiency of the poisoned heart. Because various animals, including man, respond differently to hypotensive crises or shock, we employed three different mammals for the vascular dynamics studies: cats, which perhaps more closely resemble humans, since they are both "pulmonary animals"; dogs, in which the portal circulation plays a very important part in the mechanisms of shock; and rabbits. These studies were carried out by Dr. Richard Carlson (1971) and Mr. Richard Schaeffer (1970) of our staff, under the direction of Dr. Max Weil, Mr. Howard Whigham, and Dr. Russell.

Doses for the cardiovascular studies were calculated in mg/kg body weight and mg venom protein/kg body weight. The intravenous sublethal dose for the dog corresponded to approximately 10% of an equivalent mouse LD_{50} (0.5–0.8 mg/kg or 0.1–0.25 mg protein/kg). The larger dose used for the dog corresponds to approximately 20% of an equivalent mouse LD_{50} (1.0–1.9 mg/kg or 0.4–0.6 mg protein/kg). This dose caused death in 9 of 22 dogs used in the cardiovascular experiments, indicating that it approached the LD_{50} for this particular preparation.

Fifteen cats (2.2–4.0 kg) of either sex were anesthetized by intraperitoneal injection of 30 mg/kg pentobarbital sodium. The right internal carotid artery, right cephalic vein and right external jugular vein were surgically isolated and cannulated with polyethylene catheters filled with heparinized (5.0 U/ml) saline. The right jugular catheter was advanced into the superior vena cava. Sanborn pressure transducers, model 26713, series 1212, were attached to the carotid and jugular catheters and to a Sanborn model 964 recorder with a Sanborn model 350 carrier preamplifier. The pressure recording system was calibrated with a mercury monometer and midchest was used as the zero reference. Mean arterial pressure was obtained electrically or by the formula:

$$\text{Mean arterial pressure} = \frac{\text{Systolic pressure} + 2\,(\text{diastolic pressure})}{3}$$

Needle electrodes were inserted into the extremities and bipolar ECG, leads, I, II, and III were recorded on a modified Medcraft EEG recorder. A cantilever pneumograph was applied to the sub-xiphoid region and connected through a bridge into the Medcraft recorder to record ventilatory movements. A 4.0–6.0 mm tracheostomy tube was inserted and secured. Ventilation was not assisted unless respiratory arrest occurred, when a Harvard volume respiratory pump delivering a tidal volume of 75–125 ml of room air at a rate of 15/min was connected to the tracheostomy tube. Venom was injected into the cephalic vein, followed by 2.0 ml of heparinized saline. At least 30 min elapsed between successive injections of venom. Autopsies were performed on seven animals.

Typical tracings in the cat for a lethal and sublethal dose of venom are shown in Fig. 1. The intravenous LD_{50} for the venom in the cat is approximately 0.8 mg protein/kg body weight. It can be seen that the first significant change is the precipitous fall in systemic arterial pressure. Animals given doses in excess of 2.0 mg protein/kg body weight showed immediate profound shock, with cardiac and respiratory standstill occurring within 30 sec. Most of the cats exposed to sublethal amounts of the venom exhibited the

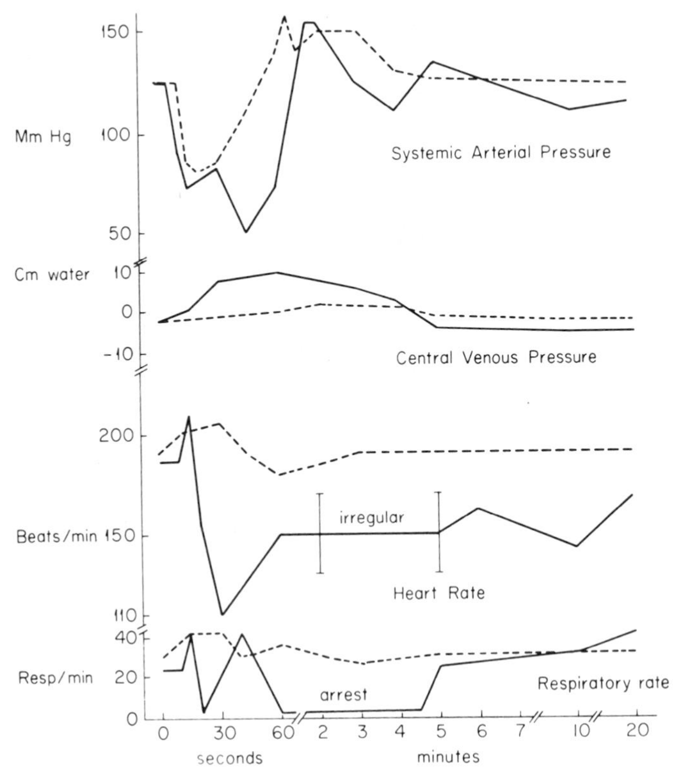

Fig. 1. Blood pressure changes in cat receiving lethal and sublethal doses of *Scorpaena guttata* venom; (——) 1.0 mg/kg, (– – –) 0.1 mg/kg.

initial hypotensive crisis, a period of relative hypertension and a gradual return to the preinjection pressure within 30 min.

Pulse pressure increased moderately during the initial hypotensive crisis and was also elevated during the period of secondary hypertension. Central venous pressure rose from -1.0 to greater than $+5.0$ mm Hg, within the first 2–3 min. Markedly elevated venous pressures were recorded just prior to death. If an animal survived, its arterial and venous pressures returned to normal within 30 min. Sinus tachycardia always developed 30 sec following the injection; some animals displayed sinus tachycardia 10–15 sec after injection. There was no evidence of a decrease in heart rate during the initial period of hypotension.

A number of arrhythmias were produced by these doses, most occurring after the first 30–40 sec of the injection. Alterations included premature atrial and ventricular contractions, inverted T waves, ST segment changes and, not infrequently, ventricular tachycardia and bundle branch block. The larger doses produced periods of apnea in all animal within 3 min. If respiratory failure persisted for longer than 30 sec, the animals were artificially ventilated. Frothy, colorless or slightly pink fluid was often found in the tracheostomy tube within several minutes of the injection. Venom doses in excess of 2.0 mg protein/kg caused cardiac standstill and respiratory arrest within 30 sec.

At necropsy, the lungs and pulmonary vessels were congested, the superior and inferior vena cavae and both ventricles were distended with blood. The larger airways contained frothy fluid. The pleural surfaces showed petechial hemorrhages, and the lung parenchyma exuded fluid, often blood tinged. The abdominal viscera appeared relatively normal.

A similar series of studies was done with rabbits, since this was the animal most often used by previous investigators working with scorpionfish venoms, and since their findings and ours in the cat were not in complete agreement. The techniques for measuring cardiovascular changes were the same as those outlined for the cat. The intravenous LD_{50} for the rabbit is approximately 0.3 mg protein/kg body weight. Respiratory rate was estimated from an analysis of the venous pressure. Doses studied ranged from 0.05 mg protein to 0.3 mg protein/kg body weight.

In general, the findings were much the same as for the cat. Within 10 sec, arterial pressure fell but unlike the findings reported by other workers the initial hypotension persisted for only about 10–20 sec, then pressure returned to normal limits. Respirations increased slightly but appeared more shallow than normal. Changes in the heart rate and central venous pressure were less pronounced than in the cat.

When lethal doses were injected, there was the usual precipitous fall in systemic arterial pressure during the first 10–20 sec and then a more gradual decrease until the death of the animal. Venous pressure rose during the first 20–30 sec and then continued rising more slowly until the death of the rabbit. ECG changes were often marked; they followed the initial phase of the hypotensive crisis. Most ECG showed bundle branch block, premature atrial and ventricular contractions, alterations in the T wave (inversion or increased amplitude), displacement of the ST segment, and ventricular fibrillation. Respiratory rate usually increased early but within 40 sec of the injection there was respiratory arrest in all animals. In several animals, cardiac standstill occurred within 20 sec. Postmortem findings were similar to those seen in the cat, although the pulmonary congestion and distention of the chest vessels was not as severe. This may, of course, be due to the rapidly fatal course in the rabbit.

Although these studies gave us considerable insight into the sequence of the changes occurring in the cardiovascular system following poisoning by this venom, certain data were still lacking for a complete picture of the overall vascular effects. Also, the problem of anesthesia, which is known to affect the action of some venoms, had to be considered. For these reasons, and because the techniques we shall describe are most difficult to do in cats and rabbits, some studies were carried out in the unanesthetized dog.

In the first series of experiments, four dogs were anesthetized with intravenous pentobarbital sodium, 25 mg/kg, 3 days prior to the venom study. Under aseptic conditions the right common carotid artery, left femoral vein and superior vena cava were implanted with catheters. These were filled with heparinized saline and tunneled dorsally to the skin in a sterile dressing. On the day of the study the catheters were flushed with heparinized saline and attached to Statham P23AA pressure transducers and to a Sanborn model 350 recorder with a Sanborn carrier preamplifier, model 350-100B. ECG electrodes were placed for leads I, II, and III and recorded with a Sanborn ECG preamplifier, model 350–3200 and a Sanborn model 1964 recorder. Venom, 0.5–1.0 mg protein/kg body weight was injected into the femoral vein, followed by 5.0 ml of heparinized saline.

The three dogs receiving 0.8–1.0 mg protein/kg died within 4 min of the injection. Death was preceded by coughing, decreased respiratory rate, labored intercostal breathing, restlessness, and nasal flaring and wheezing. The animals were prostrate by 90 sec, and respiratory arrest followed by voiding and/or defecation occurred just prior to death. In the animal receiving 0.5 mg protein/kg, coughing, licking movements, yawning, restlessness, and respiratory difficulties were observed during the first few minutes following the injection. Respiratory rate then decreased and the dog became apathetic, but by 4 hr it appeared perfectly normal. The systemic arterial pressure fell rapidly during the first 20 sec following the injection then rose to normal limits within 20 sec and then became hypertensive for 60 sec, before return to normal levels. Superior vena cava pressure was elevated slightly at 17 sec and had tripled by 45 sec. It then returned toward normal. Heart rate first decreased then increased, and then a significant sinus tachycardia developed after the initial hypotensive crisis. An increase of 35–40 mm Hg occurred within the first 15 sec following the injection but returned to control values by 30 sec. The ECG revealed frequent periods of irregular rhythm with premature ventricular tachycardia and ST-T segment changes. Thus, other than the questionable significance of an initial transient decrease in heart rate following the injection, the findings in the anesthetized cat and unanesthetized dog followed the same general pattern.

In order to establish relationships between pulmonary and portal dynamics and cardiac output, a further series of experiments was carried out in dogs. They were anesthetized with pentobarbital sodium and an endotracheal tube inserted and cuffed, but the animals were allowed to ventilate spontaneously. The right common carotid artery, right external jugular vein and right femoral artery and vein were cannulated. The jugular catheter was advanced to the superior vena cava or the right ventricle or the main pulmonary artery. The femoral artery catheter was positioned in the abdominal aorta. ECG recordings were made. During control periods and at 1, 2, 5, and 30 min after each injection of venom, arterial blood samples were taken. The samples were collected in sealed, heparinized syringes and maintained in an ice water bath awaiting complete blood cell counts, blood gases, and chemistry determinations. A Klett-Summerson Photoelectric colorimeter was utilized for hemoglobin determinations. Microhematocrits, and white and red blood cell counts were performed in the usual manner. Total serum protein was estimated with an American Optical Refractometer and percent oxygen saturation determined with an American Optical Oximeter. A Radiometer Copenhagen Microblood System BMS 1 was used to measure the pH, pO_2, and pCO_2, using a pO_2 electrode type E5046, a pCO_2 electrode type E5036, and an acid–base analyzer BHM 71. Hemolysis was estimated by inspection of centrifuged serum.

Venom was injected via the femoral vein, followed by 5.0 ml of heparinized saline. Two doses were selected: a nonfatal dose of 0.1–0.35 mg protein/kg and a larger dose of 0.4–0.64 mg protein/kg. At least 30 min were allowed to elapse between successive venom doses, and additional doses were administered only if the hemodynamic, ECG, and blood chemistry determinations indicated a return to control or near control values. Post-mortem examinations were performed.

Four dogs were anesthetized as above, an endotracheal tube inserted and cuffed, and the animals ventilated with a Harvard volume respirator delivering a tidal volume of 225–275 ml of room air. The right common carotid artery, external jugular vein and femoral artery and vein were cannulated.

In two dogs the jugular vein was cannulated with a cardiac catheter, which was advanced into the main pulmonary artery under fluoroscopy. In two dogs a Swan-Ganz flow-directed catheter was used. When the Swan-Ganz catheter had entered the right ventricle, the balloon tip was inflated with 0.4 ml of air and the catheter advanced until the pressure tracing indicated position in the pulmonary artery. The balloon was then deflated. A second catheter was then introduced into the jugular vein alongside the pulmonary artery catheter and advanced to the superior vena cava. The carotid artery catheter or a cardiac catheter from the left femoral artery was advanced into the left ventricle under fluoroscopic observation. The opposite femoral artery was also cannulated and the catheter positioned in the abdominal aorta.

In four animals the spleen was exposed and the portal vein cannulated. To obtain ECG recordings similar to standard limb lead ECG's in humans, needle electrodes were inserted according to the method of Weil *et al.* (1956) as follows: The hind limb leads were reversed, the left arm lead was inserted in the left fore-paw and the right arm lead positioned dorsally in the neck.

Cardiac output determinations by indicator dilution technique were performed during control periods and at 15 sec, 3 min and 30 min following injection of the venom, using 2.5 mg indocyanine green, injected manually into the pulmonary artery catheter. This was followed by 10 ml of heparinized saline. The abdominal aorta catheter was connected to a Harvard withdrawal pump (15.3 ml/min) flowing through a Gilford densitometer and indicating the optical density on the Sanborn recorder. A calibration curve was prepared for each animal using the animal's blood and 2.0, 4.0, and 6.0 μg per ml of dye obtained with a calibrated syringe in 10 ml aliquots of blood. The blood was returned to the animal after each calibration. Calculation of the cardiac output was accomplished using the triangle and cosine methods described by Guyton (1956) and Rand *et al.* (1972).

Two additional dogs were anesthetized, ventilated, and a midsternal thoracotomy performed. After adequate hemostasis, 5 U/kg of heparin was administered intravenously. The pericardial sac was incised and the superior and inferior vena cavae isolated and loosely held with umbilical tape. The azygous vein was ligated. The right auricular appendage was grasped with a Satinsky clamp, incised and cannulated with a plastic cannula filled with heparinized saline. The venous reservoir was primed with a 1:1 mixture of heparinized donor dog blood and saline. The superior vena cava was cannulated and the superior vena cava flow diverted through the venous reservoir to a Kay occlusive cardiac pump. The volume of the venous reservoir was monitored by a Statham pressure transducer. The reservoir and transducer had previously been calibrated with 100 ml increments of saline. The pump was started at a slow speed and the superior vena cava flow returned to the right auricle.

After cannulating the inferior vena cava in a similar manner, the pump speed was increased. Thus, all systemic venous return was diverted through the reservoir and then pumped into the right auricle at a constant rate. The right common carotid artery was cannulated. ECG leads were attached. The femoral artery was cannulated and the catheter positioned in the abdominal aorta. A Micron 12 mm electromagnetic flow probe was placed around the ascending aorta and connected to a Micron RC 1000 Electromagnetic flowmeter and to the Sanborn recorder. Zero flow was calibrated by briefly cross-clamping the aorta. At the end of the procedure, the segment of aorta with the flow probe in place was removed and the flowmeter calibrated against the pump. Cardiac output was measured as ascending aorta flow (cardiac output minus coronary flow). Venom, 0.3, 0.5, and 0.6 mg protein/kg, was administered via a small catheter in the return tube.

When the venom was injected into the open chest dog preparation, there was an initial arterial hypotension and secondary hypertension similar to that seen in the cat and rabbit. A slight increase in pulse pressure occurred during the hypotensive period. Pressure in the pulmonary artery increased within 10–15 sec but portal venous pressure did not become elevated until later. At 1 min the portal venous pressure was more than twice the control value. Heart rate remained relatively constant in one dog and initially increased in another.

When a fatal dose of venom was injected into one animal, cardiac arrest occurred at 20 sec. However, by 15 sec pressure in the pulmonary artery had risen from 10 to 16 mm Hg. Arterial pressure declined within the first 20–25 sec in two other studies. This was followed by a brief period of hypertension and then a fatal hypotensive episode. Pressure in the pulmonary artery increased within 10–15 sec and portal venous pressure rose to 18 mm Hg by 90 sec. The initial changes in arterial pressure preceded any decrease in heart rate. ECG alterations included ST segment displacement, increased amplitude or inversion of the T wave, bundle branch block, ventricular tachycardia, premature ventricular contractions, and notched P waves. Within a few seconds after the initial fall in arterial pressure, several premature ventricular contractions were often recorded. Blood from the femoral artery 1 min prior to the fatal dose of venom in one dog had a hematocrit of 50 volume %. Two minutes after the injection of venom, the hematocrit had increased to 61 volume %.

At necropsy, one dog had pulmonary congestion. The right and left ventricles were distended and the right atrium and vena cavae were dilated. The abdominal viscera appeared unremarkable. In another dog the right atrium and ventricle as well as the superior and inferior vena cavae were distended with blood, and the liver and viscera were congested.

In the anesthetized, nonventilated dogs a sublethal dose of venom caused a mean arterial pressure decrease within 20 sec, accompanied by a modest increase in the pulse pressure. Pulse pressure was increased again during the secondary hypertensive period. Pressure in the superior vena cava increased within 30 sec and remained elevated for at least 3 min. In two dogs, pressures were also recorded from the right ventricle, and in one animal from the main pulmonary artery. In these studies it was observed that the right ventricular and pulmonary artery pressures increased shortly after injection of venom. Heart rate was stable until after the initial hypotension, when a moderate sinus bradycardia developed. Several premature ventricular contractions were recorded on the ECG after the first 20–50 sec, and ST and T wave changes were frequently observed during the first 4 min, as were periods of transient bundle branch block. Aside from two instances in which premature ventricular contractions were recorded within 10 sec following injection of venom, ECG alterations were restricted to periods after the first 30 sec. In contrast to the rabbit studies, respiratory rate usually decreased somewhat with sublethal doses of venom, accompanied by an apparent reduction in tidal volume. The respiratory depression was reflected in the arterial blood gas determinations, although several animals were mildly acidotic during control periods, due, no doubt, to the effects of anesthesia. Arterial pH and pO_2 often decreased after venom injection, accompanied by an elevation in the pCO_2. A consistent finding was an increase in the central hematocrit up to 60 volume % within 2 min.

The hemodynamic changes following injection of fatal doses of venom were similar to those produced by the nonfatal dose, although the hypotension and EGG changes were more pronounced. Arterial pressure declined rapidly—accompanied by a modest increase in the pulse pressure. The initial hypotension usually progressed to a fatal outcome. Pressure in the superior vena cava was elevated within 30 sec and pulmonary artery or right ventricular pressure increased within the first 20 sec. Heart rate remained stable for 20 sec then decreased markedly. The bradycardia was more severe than seen with the sublethal dose. Within 40 sec a number of ECG alterations had developed, including inverted T waves, bundle branch block, ventricular tachycardia, ST segment displacement, and notched P waves. Variations in the amplitude of the P wave were also seen. The severe bradycardia often terminated in ventricular tachycardia and/or fibrillation. All animals experienced periods of respiratory arrest, and the depressed ventilation was reflected in the blood gas determinations. In addition, hemoconcentration was a consistent finding. Hemolysis was detected in two serum samples. Although respiratory depression contributed to the death of these animals, evidence of cardiovascular dysfunction (hypotension, severe cardiac arrhythmias) was present before death and could be ascribed to respiratory arrest. However, in these nonventilated animals the mild respiratory acidosis produced by anesthesia may have sensitized the myocardium to arrhythmias.

At post-mortem examination the right ventricle, right atrium and superior vena cava were distended with blood. Several 2–3 mm subendocardial hemorrhages were observed in the ventricles in two of the animals. The lungs were congested. The spleen was small in two animals.

In the anesthetized ventilated dog a sublethal dose of venom produced a decrease in systemic arterial pressure 10 sec after injection, although the most severe hypotension occurred at 15 sec; with a rapid recovery to control pressure by 40 sec. A hypertensive phase then ensued, persisting for up to 12 min. The hypertension was more pronounced than in the nonventilated animal. Mean pressure in the pulmonary artery increased as the systemic arterial (or left ventricular systolic) pressure declined. In fact, the first elevation of pulmonary artery pressure preceded the initial fall in systemic arterial pressure by 1 or 2 sec. The magnitude of the initial change in the pulmonary artery pressure at 10–12 sec was less profound than the fall in arterial pressure during the same period. The maximum pulmonary artery pressure exceeded 21 mm Hg at 2 min. As in previous studies the initial systemic arterial hypotension and secondary hypertension were accompanied by a modest increase in pulse pressure.

After 12 sec the left ventricular end-diastolic pressure increased slightly, and by 30 sec it had risen to nearly twice the control value. It remained elevated for several minutes. Superior vena cava pressure also was elevated, paralleling the pulmonary artery hypertension. But the pressure changes in the superior vena cava were less marked than seen in the nonventilated animal. Portal venous pressure was affected. Mean pressure in the portal vein increased from 9 to more than 20 mm Hg, but this portal hypertension followed the rise in vena cava and right heart pressures by several seconds. Maximum portal venous pressure was recorded during the period from 90 sec to 3 min, and pressure in this vessel returned to control levels by 15 min.

Heart rate was stable during the initial hypotension but decreased after 30 sec. In one study a loss of P waves was recorded on the ECG at 10–25 sec. Most ECG alterations were confined to a period from 40 sec to 3 min following the injection of venom, and consisted of premature ventricular contractions, ST segment displacement, and T wave changes (inversion or increased amplitude). One animal exhibited notched P waves on the ECG at 2 min; another had brief episodes of bundle branch block.

Increases in the central hematocrit were observed in all preparations but the arterial blood gases, blood chemistries and white cell counts were less affected. Some decline in the pO_2 and pH occurred in three preparations. Hemolysis was an inconstant finding in sera following sublethal doses, but some blood samples appeared to clot more readily despite the presence of heparin in the syringe.

IV. Discussion and Conclusions: *Scorpaena guttata* Venom

From these various studies certain conclusions can be drawn. The most dramatic initial cardiovascular change occurring following the administration of *S. guttata* venom is the precipitous fall in systemic arterial pressure. There might be several reasons for this hypotensive crisis if the arterial pressure, the venous pressure and the heart rate were the only variables we had to consider. However, a consistent finding in these experiments is pulmonary artery hypertension, and this occurs a second or so before the initiation of the fall in the systemic arterial pressure. In the earlier experiments the finding of an increase in pulmonary artery pressure appeared related to the finding of a decrease in pulmonary artery blood flow. The time relationship, however, between these two variables was not clear. They appeared to occur at about the same time. The differences in the responses of the measuring gadgets one must employ in determining blood flow and blood pressure contribute to difficulties in interpreting such data. In any event, a comparison between arterial pressures in the systemic, pulmonary and portal circulations indicates that the pulmonary artery pressure is the first to be affected.

If one next considers the early and consistent increase in left ventricular end-diastolic pressure it would appear, again, that the primary site of action is the pulmonary system, since this increase is observed after the onset of the pulmonary arterial hypertension. It seems unlikely that the initial vascular change is directly related to left ventricular function. Also the onset of ECG changes is not seen until after the hypertensive crisis, and when small doses of the venom are injected, no ECG changes are seen at all.

Another consistent finding in the cats, and less so in the dogs, is pulmonary congestion. In view of the hemodynamics already noted, and with no evidence of sudden shifts in plasma proteins, several explanations for the pulmonary system deficit can be advanced: a disturbance of the hydrostatic–osmotic pressure relationships in the pulmonary capillaries, an increase in capillary permeability, or a combination of the two. There appears to be no doubt that the venom produces some increase in capillary permeability. This is also obvious from clinical cases, where local edema is a constant finding, and pulmonary edema has been reported (Russell, 1965). A component which increases capillary permeability and hyaluronidase have been detected in stonefish venom (Austin *et al.*, 1965). However, it does not seem likely, in view of the suddenness of the hypotensive crisis, that capillary permeability is the most important early parameter to be affected. It seems more probable that this is a lesser and perhaps more important later change, and that the initial sudden hemodynamic shift is due to changes in the resistances of the pulmonary vessels, including the venules and arterioles. The most plausible explanation for the pulmonary edema and the hemodynamic shift would appear to be an increase in resistance within the pulmonary veins and vessels, giving rise to an increase in capillary hydrostatic pressure and the increased probability of capillary permeability, with a further pooling of blood in the larger pulmonary vessels. This would result in a decreased supply of blood to the left heart, a decrease in systemic arterial pressure, and a decreased circulating blood volume, all consistent findings in the experimental preparations. These changes, in turn, could account for most of the ECG alterations, which are for the most part, a reflection of myocardial ischemia, the respiratory changes, and with decreased blood flow to the head, the ECG changes. The larger doses of venom obviously have a direct effect on the heart, since cardiac standstill is sometimes a finding. Whether or not the venom, even in small doses, has a direct effect on the coronary vessels cannot best be demonstrated by these particular preparations. There is no evidence that there is a primary effect on the respiratory center, although this may be so.

The changes on the venous side of the circulation are consistent with the proposed mechanism. Although systemic venous return is augmented in some animals for the first minute following the venom injection, it then decreases precipitously, as would be expected with a decrease in arterial pressure and a reduced circulating blood volume. Splenic constriction, evident in many of the dogs, increased capillary permeability, and hepatosplanchnic pooling probably also contributed to the venous deficit. The increase in the central hematocrit and portal venous pressure, which develops within 2 minutes, is consistent with the hemodynamic picture. At necropsy, animals have pulmonary congestion, engorgement of the great vessels of the chest and the heart, and congestion of the liver and viscera, although in the dog the spleen is usually small.

In a separate group of experiments, an attempt was made to pinpoint the sites of action. Venom was injected directly into various pulmonary vessels. However, the onset of vascular changes in these different vessels gave no further clue as to the initiating site(s) of action.

All evidence to date indicates that the site of action is the pulmonary vasculature and that the most probable mechanism is increased resistance in the postcapillary venules with pooling of blood in the capillaries and larger pulmonary arterial vessels. This possible mechanism is very similar to that described for the hypotensive crisis precipitated by rattlesnake venom (Russell *et al*., 1962). In view of the findings for the rattlesnake venom, and those recently described by Carlson *et al*. (1971) for *S. guttata* venom, it is apparent that the mechanism of action is not entirely a direct one but one that involves cholinergic mechanisms, autopharmacologic substances, and neurogenic reflexes.

The effects of the venom on the rat atria have been well described by Carlson *et al.* (1971). We will briefly summarize this work, since it complements the cardiovascular survey studies. The venom produces a primary, dose-related, muscarinic effect on the isolated rat atria. It also causes secondary beta adrenergic stimulation. The primary effect is probably due to the release of endogenous acetylcholine, which produces a typical negative inotropic and chronotropic effect on the atria. This is substantiated by studies with atria pretreated with atropine or hemicholium. Atropine inhibits the negative inotropic effect produced by the venom, indicating that muscarinic receptors are involved. Hemicholium inhibits acetylcholine synthesis, with subsequent depletion of endogenous stress. Atria pretreated with hemicholium show a normal response to exogenous acetylcholine, but do not respond to the venom. Thus, the venom does not appear to contain acetylcholine-like components, but causes the release of endogenous acetylcholine.

The secondary positive response to the venom can be inhibited by either a beta adrenergic blocking agent or by depleting endogenous norepinephrine with reserpine. The venom thus causes the release of both endogenous acetylcholine and catecholamines. The acetylcholine-releasing component and the lethal activity are associated with labile fractions, since stored venom which has lost its lethal capacity fails to produce a muscarinic response. However, stored venom does cause a mild secondary positive inotropic response.

It is interesting to note that the venom has a greater effect on atrial contraction than on atrial rate, a property also displayed by acetylcholine (Hollander and Webb, 1955) and interpreted to be due to a less dense accumulation of acetylcholinesterase at cholinergic receptors of contractile cells than at pacemaker cells. When atria are pretreated with an anticholinesterase agent the negative chronotropic effect of the venom is enhanced. This then is the first description of a primary muscarinic and a secondary beta adrenergic response by a piscine venom.

The methods and results for the remaining experiments noted in Table II can be summarized in brief, since the techniques used in these studies are common ones and ones applicable to investigations on venoms. The venom has little or no effect on a modified Bülbring nerve-muscle preparation or on the DEAM preparation (Parnas and Russell, 1967) of the crayfish. It appears to produce no deleterious changes in nerve conduction or neuromuscular transmission. The venom has a mild direct hemolytic effect *in vitro* but does not affect the blood clotting systems, nor does it produce significant hemolysis in the dog. No significant *in vitro* neutralization of the lethal activity was demonstrated when the venom was incubated with antivenin prepared against *Synanceja trachynis* venom. The stonefish antivenin did not produce precipitin bands against *S. guttata* venom or venom fractions, when studied by immunoelectrophoresis.

Although our study based on the protocol has not answered all our questions, we feel that the approach has been a very successful and gratifying one, and that the data obtained from the experiments have not only been significant but have fitted together in a manner that permits us to discuss the overall activities of a fish venom in a fairly composite way. The methods we have suggested are relatively simple and can be carried out in most laboratories. They lend themselves well to the study of marine venoms.

References

Austin, L., Gillis, R. G., and Youatt, G. (1965). Stonefish venom: Some biochemical and chemical observations. *Aust. J. Exp. Biol. Med. Sci.* **43**, 79.

Bottard, A. (1889). "Les poissons venimeux." Dion, Paris.

Carlisle, D. G. (1962). On the venom of the lesser weeverfish, *Trachinus vipera*. *J. Mar. Biol. Ass. U.K.* **42**, 155.

Carlson, R. W. (1971). Studies on the venom of the California scorpionfish *Scorpaena guttata* Girard. Unpublished Ph.D. dissertation, Univ. of Southern California.

Carlson, R. W., Schaffer, R. C., Jr., La Grange, R. G., Roberts, C. M., and Russell, F. E. (1971). Some pharmacological properties of the venom of the scorpionfish *Scorpaena guttata*. *Toxicon* **9**, 379.

Guyton, A. C. (1956). "Circulatory Physiology: Cardiac Output and its Regulation." Saunders, Philadelphia, Pennsylvania.

Halstead, B. W. (1965–70). "Poisonous and Venomous Marine Animals of the World," 3 vols. U.S. Govt. Printing Office, Washington, D.C.

Hollander, P. B., and Webb, J. L. (1955). Cellular membrane potentials and contractility of normal rat atrium and the effects of temperature, tension and stimulus frequency. *Circ. Res.* **3**, 604.

Parnas, I., and Russell, F. E. (1967). Effects of venoms on nerve, muscle and neuromuscular junction. *In* "Animal Toxins" (F. E. Russell and P. R. Saunders, eds.), pp. 401–415. Pergamon, Oxford.

Rand, W. M., Afifi, A. A., Palley, N., Sacks, S. J., and Weil, M. H. (1972). Cosine method for analysis of indicator dilution curves *J. Trauma* **12**, 708–714.

Russell, F. E. (1965). Marine toxins and venomous and poisonous marine animals. *In* "Advances in Marine Biology" (F. S. Russell, ed.), pp. 258–384. Academic Press, London.

Russell, F. E. (1968). Poisonous marine animals. *In* "The Safety of Foods" (H. D. Graham, ed.), pp. 68–81. Avi Publ., Westport, Connecticut.

Russell, F. E. (1969). Poisons and venoms. *In* "Fish Physiology" (W. S. Hoar and D. J. Randall, eds.), Vol. 3, pp. 401–449. Academic Press, New York.

Russell, F. E. (1972). Marine toxins. *In* "International Encyclopedia of Pharmacology" (G. Peters and G. Radouco-Thomas, eds.). Pergamon, Oxford. Section 71, Chapter 15, pages 3–114.

Russell, F. E., and Emery, J. A. (1960). Venom of the weevers *Trachinus draco* and *Trachinus vipera*. *Ann. N.Y. Acad. Sci.* **90**, 805.

Russell, F. E., and van Harreveld, A. (1954). Cardiovascular effects of the venom of the round stingray, *Urobatis halleri*. *Arch. Int. Physiol.* **62**, 322.

Russell, F. E., Fairchild, M. D., and Michaelson, J. (1958). Some properties of the venom of the stingray. *Med. Arts Sci.* **12**, 78.

Russell, F. E., Buess, F. W., and Strassberg, J. (1962). Cardiovascular response to *Crotalus* venom. *Toxicon* **1**, 15.

Saunders, P. R. (1960). Pharmacological and chemical studies of the venom of the stonefish (Genus *Synanceja*) and other scorpionfishes. *Ann. N.Y. Acad. Sci.* **90**, 798.

Schaeffer, R. C., Jr. (1970). The chemistry and physiopharmacology of *Scorpaena guttata* Girard venom. Unpublished M.S. thesis, Univ. of Southern California.

Schaeffer, R. C., Jr., Carlson, R. W., and Russell, F. E. (1971). Some chemical properties of the venom of the scorpionfish *Scorpaena guttata*. *Toxicon* **9**, 69.

Skeie, E. (1962). Weeverfish toxin: Extraction methods, toxicity determinations and stability examinations. *Acta Pathol. Microbiol. Scand.* **55**, 166.

Taylor, P. B. (1963). Venom and ecology of the California scorpionfish *Scorpaena guttata* Girard. Unpublished Ph.D. dissertation, Univ. of California, San Diego.

Van Harreveld, A., and Shadle, O. W. (1951). On hemodynamics. *Arch. Int. Physiol.* **59**, 165.

Weil, M. H., Mac Lean, L. D., Spink, W. W., and Visscher, M. B. (1956). Investigations on the role of the central nervous system in shock produced by endotoxin from gram-negative micro-organisms. *J. Lab. Clin. Med.* **48**, 661.

Weiner, S. (1959). Observations on the venom of the stonefish (*Synanceja trachynis*). *Med. J. Aust.* **1**, 620.

Chapter 8

Developmental Biology: Development in Marine Organisms

Ralph S. Quatrano

I. Introduction

During the development of a mature organism from a zygote, the processes of growth, differentiation, and morphogenesis give rise to different cell types characteristic of specific tissues and organs. Each process occurs in concert at the molecular, subcellular, cellular, tissue, and organ levels, and, all are under coordinated temporal, quantitative, and spatial control. A large body of evidence has accumulated indicating that the control of these developmental programs resides in the genome (c.f., Wilson, 1925; Waddington, 1966; Davidson, 1968; Gross, 1968). How specific sets of genes responsible for cellular differentiation are expressed and coordinated in space and time within developing cells and tissues, is a major unsolved problem of developmental biology. This central question was clearly expressed by E. B. Wilson in 1902, even before the chromosomal basis of heredity had been accepted:

> If chromatin inheres the sum total of hereditary forces, and if it be equally distributed at every cell division, how can its mode of action vary in different cells to cause diversity of structure, e.g. differentiation?

A. CONTROLS OF CELLULAR DIFFERENTIATION

Developmental biologists have focused their attention on defining the controls operative during *selective gene transcription* (synthesis of unique RNA molecules from a specific portion of the total genome), as well as how the products of transcription (RNA) are ultimately expressed via proteins as special structural and biochemical properties of the differentiated cell (c.f., Bonner *et al.*, 1968). This experimental approach is based on the concepts that in different cell types, at least one copy of every gene is present, and that there is no irreversible loss of genetic information during development.* The latter point has been well demonstrated in plants, where single cells (embryonic or mature; haploid or diploid) have given rise to flowering plants (Steward *et al.*, 1964; Vasil and Hildebrandt, 1965; Nitsch and Nitsch, 1969; Ohyama and Nitsch, 1972). Confirmation of the retention of total genetic capabilities in animal nuclei has been reported by Gurdon (1962). He demonstrated in amphibians that nuclei from differentiated cells (intestinal epithelium) can support normal embryonic development when transplanted into enucleated eggs. Accordingly then, cells of one tissue would differ from those of another by which sets of genes are transcribed and ultimately expressed.

Although selective transcription appears to be the prime mechanism of cellular differentiation, evidence has been presented in favor of an alternative, namely, *selective gene replication* (synthesis of unique DNA molecules from a specific portion of the total genome). The clearest, best documented case is the amplification of the ribosomal RNA genes (rDNA) in oocytes of amphibians (Gall, 1968; Brown and Dawid, 1968). The extra copies of these genes reside in nucleoli, and are used as templates to support the extra-

*Chromosome elimination or diminution is clearly illustrated in the development of certain cell types such as the germ-line stem cells of *Ascaris* (c.f. Wilson, 1925; Tobler *et al.*, 1972). For references and descriptions of similar examples found in other animals see Wilson (1925) and Davidson (1968), and plant roots see Berger and Witkus (1954).

ordinary demand for ribosomal RNA synthesis in immature oocytes. A similar example of rDNA amplification has been reported for the slime mold *Physarum* (Zellweger *et al*., 1972). It also appears that the genes for histones are present in multiple copies (Kedes and Birnsteil, 1971). In both of these cases, products of genes that are amplified or repeated are involved with the mechanism of gene expression; rRNA, a component of ribosomes, and histones, part of the chromosome (c.f., Bonner *et al*., 1968). A recent model of gene regulation proposed by Britten and Davidson (1969) requires that the repeated sequences found in the genome of higher organisms [which can be experimentally demonstrated by DNA annealing techniques (Britten and Kohne, 1968)] are basically regulatory in function. To further test this model, the location of the repeated and "single copy" or unique sequences in short fragments of sheared DNA is presently being investigated (e.g., Wu *et al*., 1972; Davidson and Hough, 1971).

Brown and Blackler (1972) have demonstrated that selective gene replication involves a chromosomal copy mechanism whereby a small region of the genome (0.2% in the case of rDNA) is selectively replicated. Another interesting mechanism of how selected regions of DNA may be amplified has been suggested by the experimental data of Crippa and Tocchini-Valentini (1971) and Ficq and Brachet (1971) using *Xenopus* oocytes. They implicated a role for the enzyme "reverse transcriptase," a RNA dependent-DNA polymerase (Baltimore, 1970; Scolnick *et al*., 1971). The amplified rDNA found in oocytes may be the result of a DNA polymerase acting upon an rRNA template rather than the chromosomal rDNA locus. The reverse transcriptase isolated from the avian myeloblastosis virus has synthesized *in vitro*, DNA from purified mRNA's for rabbit globin (Verma *et al*., 1972; Ross *et al*., 1972) and human globin (Kacian *et al*., 1972). Brown and Tocchini-Valentini (1972) recently demonstrated that a RNA–DNA complex (Mahdavi and Crippa, 1972) isolated from *Xenopus* serves as an *in vitro* intermediate for an enzyme system that synthesizes DNA from an RNA template.

Although selective gene replication appears to be an important mechanism, genes coding for specialized products characteristic of differentiated cells, do not appear to be amplified, but rather are present in one or a few copies. The accumulation of gene products in these cases occur by an increase in translation yield of a few, stable mRNA's (i.e., selective gene transcription), rather than by a reiteration of the gene itself (i.e., selective gene replication). Two genes that code for specific products of differentiated cells; namely, the hemoglobin gene (Bishop *et al*., 1972) and the silk fibroin gene (Suzuki *et al*., 1972) are not amplified. The presence of such "single copy" genes can be demonstrated in the genome of higher organisms by the technique of DNA annealing (Britten and Kohne, 1968). These sequences are probably analogous to the structural genes in the Jacob and Monod (1961) model of gene regulation, and the producer genes in the Britten and Davidson (1969) model.

If a control of molecular differentiation in cells resides at the level of transcription, i.e., once mRNA is formed, the specific gene product is expressed, one should be able to inject mRNA into a cell and have that recipient cell synthesize the specific product of the introduced RNA. Lane *et al.* (1971) and Marbaix and Lane (1972) have recently demonstrated that *Xenopus* oocytes were capable of synthesizing the α and β chains of hemoglobin when these cells were injected with a 9 S RNA solution from rabbit reticulocytes containing hemoglobin mRNA. Berns *et al.* (1972) have also demonstrated that *Xenopus* oocytes injected with 14 S RNA from calf-lens tissue can synthesize the α_2 chains of the α-crystallin lens cell protein.

Although it appears that selective gene transcription is the main mechanism of control for specialized products of differentiation at the level of the gene, additional controls of *gene expression* can be demonstrated at the level of: (1) packaging specific nuclear RNA messages (perhaps in the nucleolus) for transport with ribosomes to the cytoplasm for later use in protein synthesis (c. f., Harris, 1970), (2) translating RNA into specific polypeptide sequences (Ilan *et al.*, 1970; Harris, 1970), (3) localizing and activating specific polypeptides (Cabib and Farkas, 1971; Suelter, 1969), and (4) degrading or inactivating enzymes (c.f., Schimke and Doyle, 1970).

Within the last decade, most research aimed at elucidating how cells differentiate has centered around identifying and attempting to understand the control mechanisms at all of these levels which are responsible for selective gene expression. This has taken the experimental form of describing patterns of RNA and protein synthesis at different stages in the life cycles of various plants and animals. Much research of this type has been performed on marine animals during stages of oogenesis and early postfertilization development (Ursprung, 1967; Gross, 1968; Smith and Ecker, 1970). Results from these investigations support the existence of control mechanisms at some of the levels stated above. This experimental data can be summarized as follows:

(a) Protein synthesis increases soon after fertilization in sea urchins (Epel, 1967). These proteins appear to be necessary for subsequent growth and differentiation as judged by the ability of protein synthesis inhibitors to arrest development (Gross, 1968; Timourian and Watchmaker, 1970).

(b) The earliest detectable synthesis of RNA in sea urchins, after the third cleavage (Wilt, 1970), includes soluble or transfer RNA (tRNA) and messenger or heterogeneous RNA (mRNA). Ribosomal RNA synthesis cannot be detected until hatching (Emerson and Humphreys, 1970). Part of the mRNA becomes associated with an inactive form of polysome which becomes competent to synthesize proteins later in development (Infante and Nemer, 1967). The remainder of the mRNA turns over rapidly in the nucleus with only a small fraction reaching the cytoplasm (Aronson and Wilt, 1969). The fraction which enters the cytoplasm, becomes associated with active polysomes and is translated into proteins (Kedes and Gross, 1969). However, this RNA accounts for only a portion of the total proteins synthesized shortly after fertilization.

(c) Nuclear proteins are synthesized on these newly formed nuclear RNA templates (Nemer and Lindsay, 1969; Kedes and Gross, 1969). Specific RNA species (9–10 S), which are histone messengers, can be isolated from the class of active polysomes (Kedes and Gross, 1969). When this RNA fraction is used as a template in a cell-free protein synthesizing system, a product is formed which co-electrophoresis with authentic sea urchin histone (Kedes *et al.*, 1969; Moav and Nemer, 1971).

(d) Embryos raised in the presence of Actinomycin D (at levels that suppress most RNA synthesis) exhibit the initial increase in rate of protein synthesis (Gross and Cousineau, 1963), as well as changes in the soluble protein patterns when analyzed electrophoretically (Terman and Gross, 1965; Craig and Piatigorsky, 1971) and chromatographically (Ellis, 1966). This suggests that a stable RNA template present in the egg is responsible for the bulk of early protein synthesis after fertilization (Spirin, 1966). Metafora *et al.* (1971) have recently proposed that the increased rate of protein synthesis occurring after fertilization could be accounted for by the removal of an inhibitor protein from the maternal ribosomes. From their data, it appears that this translational inhibitor acts by preventing the binding of mRNA and aminoacyl-tRNA to the ribosome. Activation of proteases at fertilization (Lundblad, 1954), may be responsible for the removal of the protein inhibitor.

(e) From recent evidence, it appears that not all of the early postfertilization RNA synthesis occurs on a DNA template in the nucleus. The works of Dawid (1966) in amphibians and Piko *et al.* (1968) in sea urchins have demonstrated that up to 10 times more DNA is found in the cytoplasm than in the nucleus. Cytoplasmic RNA synthesis has been shown to take place on nonnuclear templates in sea urchins (c.f., Hartmann *et al.*, 1971; Craig and Piatigorsky, 1971). In 2-celled embryos of sea urchins, about 50% of the labeled RNA isolated from cells incubated in [^{3}H]-uridine is from mitochondria (Chamberlain and Metz, 1972). Ethidium bromide, an inhibitor of mitochondrial RNA synthesis in sea urchins, not only strongly inhibits cytoplasmic RNA synthesis, but also alters and inhibits various stages of postfertilization development (Craig and Piatigorsky, 1971). Hartmann *et al.* (1971) demonstrated that a significant portion of newly synthesized RNA in the cytoplasm of either early blastula or gastrula stage embryos is associated with a particulate fraction. This RNA is not released from the particles by RNAase or EDTA treatment, and bears some degree of nucleotide sequence homology to mitochondrial DNA.

From these results, two general conclusions arise:

1. Specific gene expression is regulated at the level of mRNA production (transcription) as evidenced by histone proteins, whereas the RNA templates for other proteins, e.g., "gastrula proteins," are synthesized in advance of their use in protein synthesis; i.e., translational level control.

2. Transcriptional control during early embryogenesis may occur not only at the nuclear template site, but also at a cytoplasmic one, presumably mitochondrial DNA.

B. Polarity as a Mechanism of Cellular Differentiation

Although the above investigations have contributed greatly to our understanding of control mechanisms in multicellular organisms, "this 'total embryo' approach does not measure biosynthesis occurring in a given cell line and, therefore, cannot in its present form give direct information on processes of cellular differentiation" (Ursprung, 1967). Also, the patterns of RNA and protein synthesis have not for the most part been correlated with specific morphological changes during postfertilization development, nor with the appearance of unique cytological and biochemical traits in cells undergoing differentiation. In attempting to understand the whole of cellular differentiation, the controls described must be integrated with less understood mechanisms dealing with the fate of gene products, i.e., the localization, activation, and degradation of RNA and protein with respect to specific cell differentiation and morphogenesis.

To approach these pertinent questions, one would ideally like to have a system whereby a single cell, free of surrounding cells, synchronously forms two cells which are different from each other in structure and function. In this simplest manifestation of cellular differentiation, two mechanisms, not necessarily mutually exclusive, may be involved. These are based on the time and manner in which the "determinative or differentiative factors" are acquired in each daughter cell. In one type (I in Fig. 1), these factors arise in the daughter cells at some time after cell division, presumably under control of microenvironment A. In some other microenvironment (C), these factors may give rise to those differentiative traits which distinguish the two cell types. In the other (II in Fig. 1), the determinative factors are already present and *localized* in the mother cell. Because of this polarity (an axis with different ends) at the time of cell division, these factors are separated into a daughter cell. In this case, the ultimate differentiation of daughter cell

A is preceded by the development of polarity in the mother cell. The establishment of polarity in the mother cell then, is the basis for the differentiation of the two cells resulting from cytokinesis. To understand the underlying mechanism of this type of differentiation, it is imperative not only to define the nature and time of appearance of the specific factors in the mother cell,

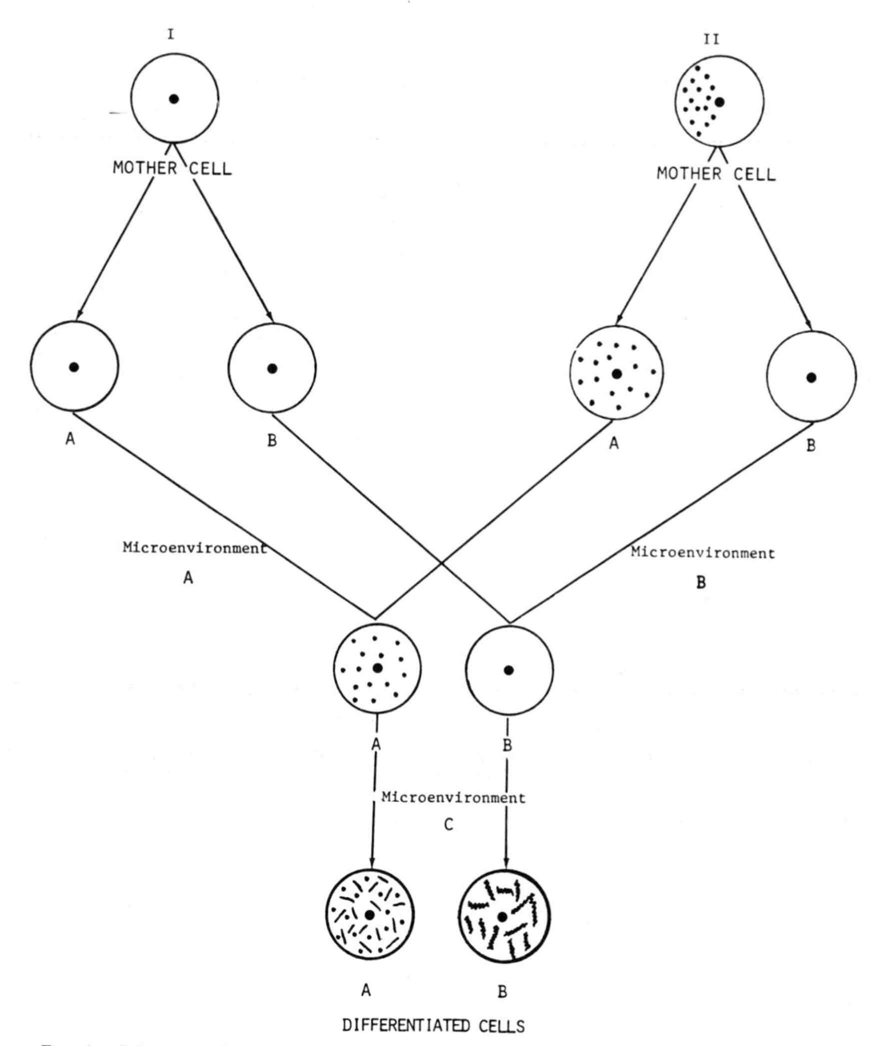

Fig. 1. Diagram of two general pathways (I, II) to demonstrate how specific traits are acquired during cellular differentiation. See text for discussion.

but also to determine how they become localized, stabilized, and in some cases, inactivated, until cell division partitions them into a daughter cell.

Cell polarity as described above coupled with unequal cell division is an important mechanism in the differentiation of many cell types in various plant groups; e.g., pollen grains, gonidial cells of *Volvox*, root hairs, and stomata of certain monocots, leaf cells of *Sphagnum* and sclereids in *Monstera* (c.f., Bünning, 1953; Sinnott, 1960; Stebbins, 1967; Starr, 1970). Also, since the formation of a multicellular blastula originates from a highly polar, regionalized egg cell in many animal groups (c.f., Davidson, 1968; Balinsky, 1970), as well as in certain egg cells of higher plants (Schulz and Jensen, 1968; Mogensen, 1972), problems of cell differentiation during early embryogenesis are also more closely related to this mechanism of polarity. After fertilization, each nucleus of the developing plant and animal embryo is segregated into a chemically and structurally unique cytoplasmic region which was formed during oogenesis. During the ensuing nucleocytoplasmic interactions, specific developmental patterns are regionally initiated, ultimately leading to tissue specialization and morphogenesis in various regions of the embryo. How the genome controls the specific localization of developmentally important macromolecules (i.e., determinative or differentiative factors) in the oocyte cytoplasm is not clear and experimentally difficult to approach. The flow sheet in Fig. 2 can serve as a framework to view the

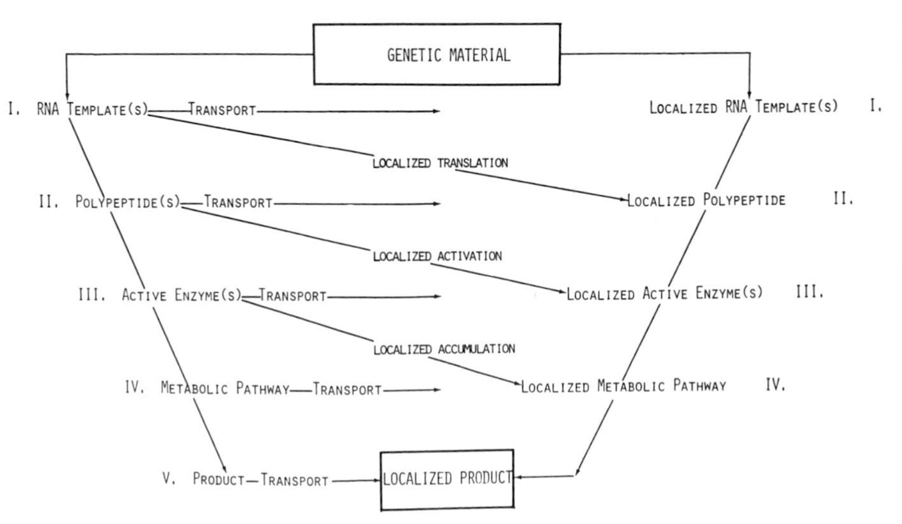

Fig. 2. Diagrammatic representation of the flow of genetic information and possible levels (I–V) where mechanisms (transport; localized translation, activation, accumulation) may be involved with the intracellular localization of specific macromolecules. See text for discussion.

problem of localization, its relation to gene products, and where controls may be imposed upon the ultimate polarity that is expressed.

The purpose of this review is to describe how two marine plants, the large unicellular green alga *Acetabularia* and the two-celled *Fucus* embryo, are being used as model systems to understand how polarity is established as a basic mechanism of cellular differentiation. Since the voluminous research effort on developmental phenomena in *Acetabularia* has been recently reviewed in two books (Puiseux-Dao, 1970; Brachet and Bonatto, 1970), most of this report will deal with the *Fucus* embryo and its potential as a system to investigate how gene products are linked to the establishment of a polar cell leading to cellular differentiation.

II. Life Cycles and Experimental Methods

A. FUCUS

The brown alga *Fucus* commonly grows in northern coastal areas attached to rocks in the intertidal regions. The swollen tips or receptacles of mature fronds possess small openings on the surface, each of which leads to a cavity (conceptacle) which contains either eggs or sperm in dioecious species, and both eggs and sperm in monoecious species (Fig. 3). The common species along the eastern North American coast, *Fucus vesiculosus* L., is dioecious, while the common west coast species, *Fucus distichus* Linnaeus subsp. *edentatus* (De la Pylaie) Powell,* is monoecious. In nature, the incoming tide washes the slightly desiccated receptacles causing them to expand and extrude the gametes into the open sea where fertilization occurs.

Good research material can be obtained throughout the year if receptacles are collected from the mid to low intertidal region as soon as possible after exposure by the outgoing tide, and kept in the dark (4°–10°C) in well-aerated plastic bags. As soon as possible after collecting, excess seawater should be removed from the receptacles to prevent premature shedding of gametes. These receptacles should be stored in a dark, well-aerated refrigerator (4°C) for use during the subsequent 2–3 weeks. Gamete discharge can be achieved at desired times during this period by washing the receptacles and placing them in the light (> 500 ft c.) at 12–15°C. Millipore-filtered seawater or a

*According to Smith (1969), the description and naming of the west coast species *Fucus distichus* by Powell (1957) is preferred; *F. distichus* Powell (1957) is identical to *F. furcatus* Gardner (1922) and *F. gardneri* Silva (1953). In this chapter, all references to the west coast species will be called *F. distichus* even though the original articles may have used *F. furcatus* and *F. gardneri*. A more geographical approach to the study of the Fucales is given by Nizamuddin (1970).

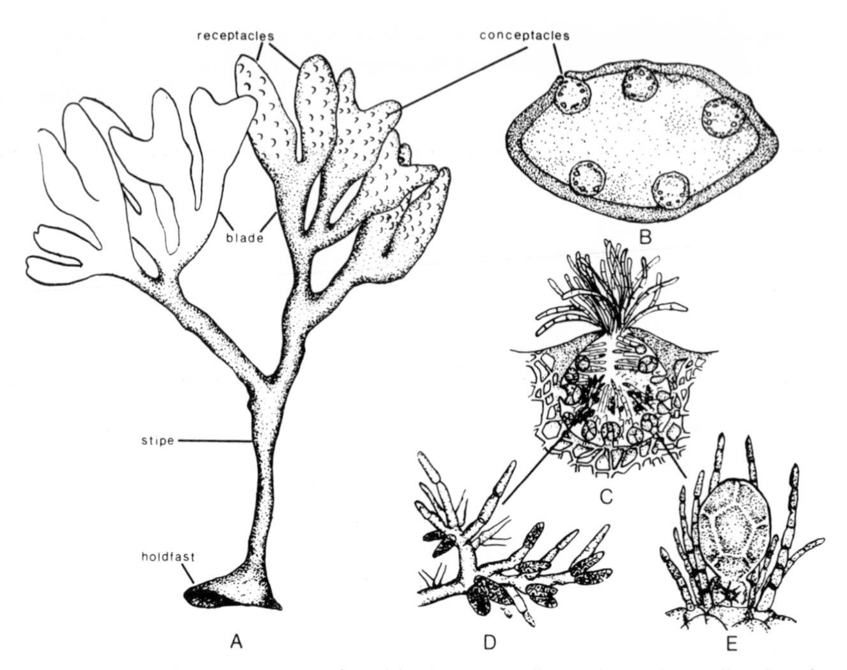

Fig. 3. Reproductive structures found in the mature *Fucus* plant. The swollen tips of the blades are called receptacles (A). When viewed in cross section (B), the receptacles are seen to contain numerous cavities called conceptacles. Each conceptacle (C) has an opening in which sterile filaments are located. The filaments are attached to the conceptacle wall as are the male reproductive structures (antheridia), which contain many sperm (D), and the female reproductive structures (oogonia), which contain several eggs (E). A full description of the cytological development of these structures is given by Farmer and Williams (1898) and Yamanouchi (1909). [Redrawn and modified from Robbins *et al.* (1964).]

variety of artificial seawater mixtures* stored at 4°C are routinely used as the development medium.

After washing in seawater, each receptacle of the dioecious *F. vesiculosus* is placed in a small petri dish and exposed to light at 15°C. After several hours, each dish may contain either several hundred to several thousand eggs (70–80 μm in diameter), or a sperm suspension. After passing the egg solution through a Nitex nylon mesh† (102 μm) to remove large debris and oogonia,

*In addition to commercially available seawater mixtures (e.g., *Instant Ocean*, Aquarium Systems, 1450 E. 28th St., Wickliffe, Ohio 44092), the medium of Müller (1962) and a simple salt solution (0.45 *M* NaCl, 0.03 *M* $MgCl_2$, 0.016 *M* $MgSO_4$, 0.01 *M* KCl, 0.009 *M* $CaCl_2$) are sufficient for gamete release, fertilization and embryo development, at least to the 3–4 cell stage.

†The Nitex nylon mesh can be purchased from Tobler, Ernst and Traber, Inc., 71 Murray St., New York, NY, and the plastic petri dishes can be obtained from Falcon Plastics, Oxnard, California.

eggs are collected by a brief low speed centrifugation (2 min, 100 *g*), resuspended, and then mixed with the sperm suspension. Thirty minutes later, sperm is removed by passing the sperm–egg suspension through another Nitex nylon mesh (35 or 53 μm) which permits sperm to pass but retains zygotes. The zygotes are washed several times with seawater containing either penicillin-streptomycin (25 μg/ml of each) or chloramphenicol (40 μg/ml), and pipetted into sterile plastic petri dishes* containing the same seawater–antibiotic mixture. Neither antibiotic mixture alters subsequent development and both are effective in eliminating bacterial contamination (Quatrano, 1968a,b; Peterson and Torrey, 1968). Time of fertilization ($\pm$15 minutes) is recorded 15 min after mixing the sperm–egg suspension. Four to six hours after fertilization, unfertilized eggs do not adhere to the dish and can be removed from the population by gently washing the culture with fresh seawater. In this way, the population within each dish consists entirely of fertilized eggs in a high degree of synchrony. More detailed procedures are given by Quatrano (1968b) and Torrey and Galun (1970).

In the monoecious species *F. distichus*, fertilization occurs during or shortly after shedding. To obtain synchronous populations of developing zygotes, shedding must be rapid and as synchronous as possible. This can be achieved by washing the receptacle with *cold* running tap water for 2–3 min and then drying between layers of paper toweling. By repeating this process several times, shedding will occur rapidly when receptacles are placed in seawater under light (500 ft c.) at 15°C. Receptacles are then removed 30 min after the first signs of shedding and zygotes are treated as described above for *F. vesiculosus*. Time of fertilization ($\pm$15 minutes) is recorded 15 min after the first signs of shedding. More details of these techniques for *F. distichus* are given by Pollack (1970) and Quatrano (1972). Large numbers of eggs of *F. distichus* can be obtained by incorporating sperm-inactivating agents such as 10^{-3} *M* ethylenediaminetetraacetic acid (EDTA), 10^{-3} *M* *m*-chlorocarbonylcyanide phenylhydrazone (CCCP), or 1% (w/v) chloral hydrate in the seawater (Kniep, 1907). Shedding gametes in the light at 2–4°C instead of 15°C prevents egg release from the oogonia and results in the discharge of intact oogonia which can be easily collected. Fertilization will not occur until the eggs are discharged from the oogonia (Pollack, 1970). Egg release can be easily induced by raising the temperature to 15°C.

Within minutes after fertilization, the diploid fertilized egg (zygote) initiates the development of a cell wall (Whitaker, 1940), similar to events in higher plants (Mogensen, 1972). The birefringent appearance of zygotes under a polarizing microscope at this time is due to the ordered deposition

*See dagger footnote p. 312.

of cellulose microfibrils and possibly the polyuronide alginic acid (composed of mannuronic and guluronic acids), which also possesses birefringent properties (Baardseth, 1966). The amount of an α-cellulose fraction (Corbett, 1963) increases linearly after fertilization until about 8–10 h after fertilization. Exogenously supplied [^{14}C]-glucose or Na_2[^{14}C]O_3 can be incorporated into this fraction (Quatrano, 1968a; unpublished observations). There is no increase in size of the zygote (Fig. 4A), similar to zygotes in angiosperms (Mogensen, 1972), or other morphogenetic events until 14 h after fertiliza-

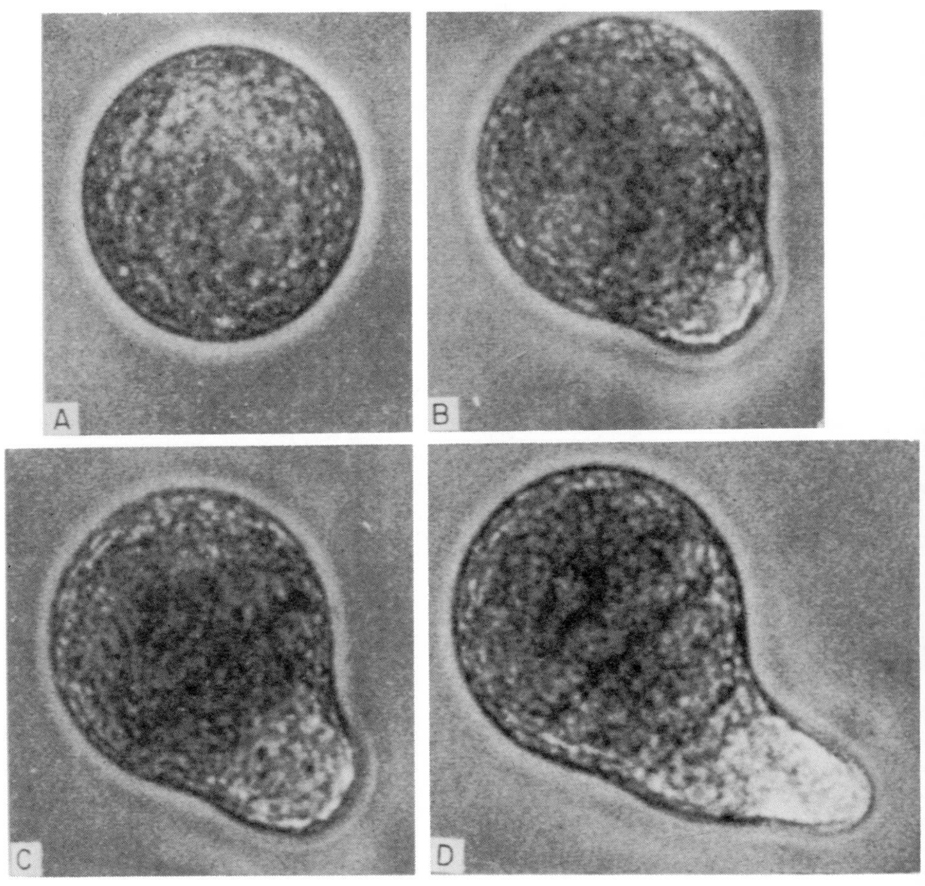

Fig. 4. Photomicrographs of zygote development in *Fucus*. The apolar zygote (A) forms a localized protuberance (rhizoid) at 14 hr after fertilization (B), which is subsequently partitioned from the rest of the cell by the first division at about 20 hr (C) and the second division at about 24 hr (D). (Photomicrographs courtesy of Dr. G. Benjamin Bouck.)

tion when a localized protuberance or rhizoid is formed (Fig. 4B). Since the cell wall is apparently laid down equally around the circumference of the eggs, rhizoid formation represents the first sign of morphological polarity. In the cytoplasmic region of the cell wall extension (rhizoid), one can also detect the first signs of intracellular polarity; the accumulation of RNA (Nakazawa, 1966), a sulfated α-1,2 linked fucose polymer, fucoidan (Mc-Cully, 1969; Quatrano, 1972) and certain organelles and inclusions (Quatrano, 1972). These localizations are eventually partitioned from the rest of the zygote by the first cell division which occurs at 20 h after fertilization (Fig. 4C,D). Sequestering of these components in only one of the first two cells of the embryo, the rhizoid cell, constitutes the major differences between these cells. All of these events occur in synchrony, free of surrounding cells and without an external energy source; i.e., in an aseptic, completely defined inorganic medium without the need of light. In addition, rhizoid formation and the accumulation of certain biochemical components occurs in a spherical cell, with a centrally located nucleus and a homogenous cytoplasm. It is an apolar cell. Without events associated with rhizoid formation there would be no differences in the cytoplasm into which the subsequent daughter nuclei could migrate. Hence, the *Fucus* system is not only amenable to experimental approaches aimed at identifying mechanisms of localization, but also how such localizations are maintained and related to cellular differentiation.

Despite the importance and attention this portion of the *Fucus* life cycle has received, there has been little work done on culturing embryos to adult size in the laboratory. Fulcher and McCully (1969) obtained growth of *F. vesiculosus* sporelings in an intertidal culture tank, and Burrows and Lodge (1953) grew offspring of the interspecific hybrid (*F. vesiculosus* × *F. serratus*) in laboratory culture. McLachlan *et al.* (1971) cultured four species of *Fucus* from zygotes to mature thalli and in one, *F. distichus*, viable gametes were obtained. However, the rate of growth of *F. distichus* in unialgal culture was slower than in nature and required 2 years to complete the life cycle. Recently McLachlan and Chen (1972) have reported the formation of adventitious embryos from rhizoid filaments. Obviously, much useful genetic, developmental, and biochemical knowledge could be gained by improving such techniques to obtain cell or tissue cultures as well as techniques to allow completion of the life cycle of *Fucus* in a short time period in the laboratory. Techniques of somatic cell genetics now being used in higher plants (e.g., Carlson *et al.*, 1972) might also be useful.

B. ACETABULARIA

The green alga *Acetabularia* is a single-celled marine organism containing one nucleus and capable of completing its life cycle in laboratory culture

(Fig. 5). Various details concerning the systematics, ecology, isolation, and sterile culturing techniques are given by Puiseux-Dao (1970). In the adult form, *Acetabularia* can achieve lengths of 5 cm and is composed of three portions; rhizoid, stalk, and cap. Following conjugation of the haploid isogametes, the zygote (50 μm) gives rise to a rhizoid, which contains the single large nucleus, and an elongating stalk which ultimately terminates its growth (3–4 months) by the differentiation of a species-specific cap (ca, 1 cm diameter) at the tip. At maturity the cap is subdivided into many rays, each of which is functionally a gametangium. At this time the large (150 μm) nucleus divides into thousands of secondary nuclei by mitosis. These nuclei migrate up the stalk into the rays of the cap where they form cysts. The cysts are set

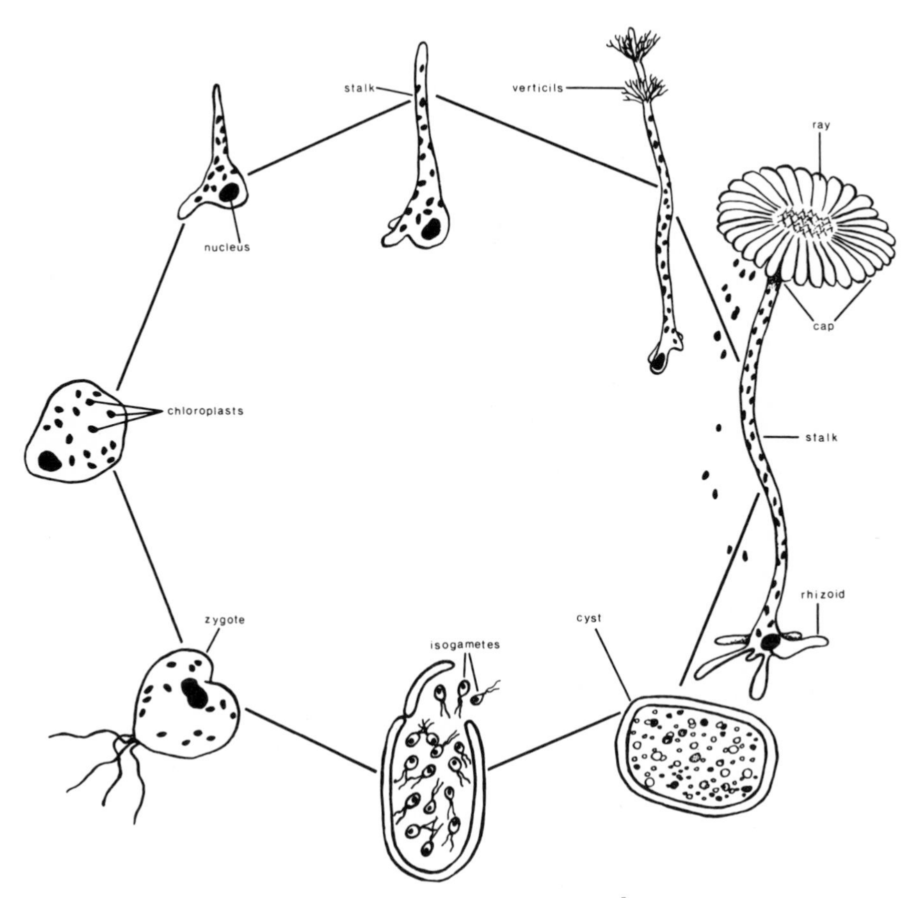

Fig. 5. Life cycle of the unicellular green alga *Acetabularia*. [Redrawn and modified from Brachet (1965).]

free when the cap degenerates and ultimately give rise, after many mitotic divisions and meiosis of each secondary nucleus, to numerous motile, haploid isogametes.

Classic experiments of nucleocytoplasmic interactions performed by Häemmerling in the 1940's and 1950's (c.f., 1953, 1963) were possible because of the durability and accessibility of the nucleus for removal as well as for transplantation into stalks of the same or different species. This early research demonstrated that anucleated fragments (stalk) could form caps months after the removal of the nucleus. This ability to undergo morphogenesis in the absence of the nucleus, however, was related to the amount of morphogenetic substances (MS) which were released from the nucleus prior to its removal. Häemmerling also demonstrated by interspecific nuclear transplants that the MS were species specific. For purposes of discussion in this review, the highly distinctive cap morphology offers another unique system to study the role of nuclear products in relation to a localized intracellular morphogenetic event (cap formation), similar to rhizoid formation in *Fucus* zygotes.

III. Experimental Basis of Polar Development

From one of the earliest studies of polarity (Mirbel, 1835), numerous marine algae have been used to investigate, or have the potential to elucidate, events responsible for the establishment of a visibly polar cell: isolated, multinucleate internodal cells of *Bryopsis*, *Dasycladus*, and *Caulerpa* (Bloch, 1943); the apical cells of *Griffithsia* (Waaland and Cleland, 1972) and *Cladophora* (Bloch, 1943), *Chara* and *Nitella* (Sandan, 1955), and *Enteromorpha* (Müller-Stoll, 1952); and, eggs, zygotes, and embryos of the brown algae, Fucaceae (Whitaker, 1940; Nakazawa, 1960; 1962). In *Fucus*, the most widely studied genus, eggs and young zygotes are visibly apolar as evidenced by morphological, ultrastructural, and cytochemical criteria (Quatrano, 1968a; 1972). Since the polar axis in zygotes can be oriented by a variety of imposed gradients (Table I), and bipolar embryos induced by polarized light (Jaffe, 1958a), the polar axis which is finally realized is initially labile and "arises in some more epigenetic manner than through the directed rotation of some preformed asymmetric structure" (Jaffe, 1958b).

A. Requirements for Polar Development in Fucus Embryos

1. *Fertilization*

Although a wide variety of external gradients can determine the polar axis during early zygote development (Table I), none appear to be *required*. For example, if zygotes of *F. vesiculosus* were tumbled in darkness in a paraffin-lined tube containing sand (to prevent attachment), rhizoids were

TABLE 1
Vectors Effective in Orienting Rhizoids of *Fucus* Zygotes[a]

	Vector	pH	Response half-maximal at	Rhizoid
1	Light (unpolarized)	8	—	Dark
1a		6	—	Dark
1b	Light (polarized)	8	—	E vector
2	Sperm	8	—	Entry point
3	Shape of egg made prolate	8	—	Long axis
4	Heat gradient	8	0.4°C	Hot end
5	Centrifugal force	8	1 min at 800 *g* (Cb)	Centrifugal
5a		6	—	Centripetal
6	Osmotic *P* gradient	8	5×10^{-2} *M*	High water
7	Voltage gradient	8	25 mV	+
		—	10 mV	−
8	K^+ gradient	8	2×10^{-2} *M*	High K^+
9	pH gradient	8	0.2 pH unit	High H^+
10	Dinitrophenol gradient	8	2×10^{-5} *M*	High DNP
11	Indoleacetic acid gradient	8	$<5 \times 10^{-5}$ *M*	High IAA
12	Flow	8	10 μm/sec	Upstream
12a		6	0.1 μm/sec	Downstream
13	Diffusion barrier	8.5	—	Away
13a		6	—	Toward
14	Another egg	8	1 egg diameter	Away
14a		6	4 egg diameters	Toward
15	Various thalli	8	2–10 mm	Toward

[a] From Jaffe (1968).

still initiated in this gradient-free environment (Quatrano, 1968a). Similar results were obtained in other fucaceous eggs (Whitaker, 1940; Nakazawa, 1956; Sussex, 1967). Since Overton (1913) reported in *F. vesiculosus* the formation of polar embryos from eggs parthenogenically activated with butyric or acetic acids, sperm entry also appears not to be required for polar development. Knapp (1931) reported in *Cystoseira*, a related brown alga, that the site of sperm entry corresponded to the site of polar growth. It may be that the site of sperm entrance into the egg is another localized event which can orient an already established but labile polar axis. It would appear from these data, that the site of sperm entry is not the initial determinant of the site of rhizoid formation, but rather some other event occurring prior to fertilization.

On the basis of these observations, one can predict that a parthenogenically activated egg, if kept in a gradient-free environment should develop a rhizoid and give rise to a polar embryo. This experiment, formulated by Whitaker (1940), has been attempted in our lab without success. Recent experiments done in the Marine Biology Laboratory at Woods Hole, indicated that unfertilized and unactivated eggs possessed the same number of polar embryos as populations of eggs activated with butyric or acetic acids (less than 1%). Despite precautions against unwanted fertilization (separation of plants, careful washing of receptacles, media containing 10^{-3} M EDTA), sperm contamination represents the most obvious explanation of our results. Although no quantitative data was given, Overton's observation may have been on a very small number of eggs that were already fertilized. Until a clear case of parthenogenic activation is reported, we will assume that sperm entry is required, and the site of entry is the subsequent site of polar development in a gradient-free environment. It is interesting to note that if 1 hour old zygotes of *F. vesiculosus* are placed in distilled water, the cell contents will be extruded from a *localized* break in the cell wall. A polar plasmolysis is also evident prior to rhizoid formation (Reed and Whitaker, 1941). Experiments in which sperm entry is localized (use of capillary tubes or fine metal screens) should lead to cytological and biochemical information concerning a possible localized alteration in the cell wall and/or cell membrane that is related to subsequent rhizoid formation.

2. *Biochemical Events*

It is clear that *at least* RNA and protein syntheses from the zygote genome are needed for the *expression* of polarity (Nakazawa and Takamura, 1967; Quatrano, 1968b). By applying inhibitors of RNA synthesis (actinomycin D) and protein synthesis (cycloheximide, puromycin) to zygotes at various times after fertilization, specific periods of RNA (1–5 hr) and protein (9–13 hr) syntheses were delineated that are necessary for the final expression of the polarity. Experiments as to the specificity and the time necessary for these inhibitors to prevent macromolecular synthesis, have been reported (Quatrano, 1968a,b). An increase in the respiration (Whitaker, 1931) and protein synthesis (Peterson and Torrey, 1968) rates soon after fertilization have not been studied with respect to their possible role in the manufacture and deposition of the cell wall and subsequent polar growth.

Water uptake is also required by zygotes for expression of cell polarity. Rhizoid formation can be prevented by a nonspecific osmotic inhibition caused by sucrose (0.7 *M* in seawater) and other sugars and sugar alcohols such as mannitol (Torrey and Galun, 1970). The internal pressure potential, which increases by 180–300 milliosmoles between 2 and 8 hr after fertilization, can be entirely accounted for by an increase in $[K^+]$ and $[Cl^-]$ during this time (Allen *et al.*, 1972). This accumulation of salts is presumably necessary for the water uptake. The magnitude of $[K^+]$ increase during the first 24 hr after fertilization (greater than 2.5-fold) may also have a role in control of enzyme activation and/or protein synthesis (Evans and Sorger, 1966; Suelter, 1969; Meeker, 1970).

B. COMPONENT PROCESSES OF POLAR DEVELOPMENT

Processes involved in polar cell formation and embryonic cell differentiation in *Fucus* can be divided into those events responsible for: (1) the irreversible fixation of a polar axis determined originally by sperm entry and possibly modified by some external environmental gradient, (2) the localization of material needed for cell wall expansion (rhizoid formation) and rhizoid specific macromolecules and particles, and (3) the partitioning of this material into the new rhizoid cell by cell division.

1. *Fixation of a Polar Axis*

Various gradients are perceived by zygotes at different times after fertilization, light being the last environmental cue to cause a shift in the polar axis (Fig. 6). Several hours before an observable polar growth, the zygote is no longer able to respond to any of these environmental gradients. The polarity then present becomes irreversibly fixed as the permanent rhizoid–thallus axis.

Unfortunately, polar axis fixation and the localized accumulation of macromolecules induced by light cannot be separated in time from each other (Fig. 7) or from cell division, even in synchronously developing cultures of zygotes (Murphy *et al.*, 1970; Quatrano, 1972). Some success was achieved by Torrey and Galun (1970) when they demonstrated in *F. vesicu-*

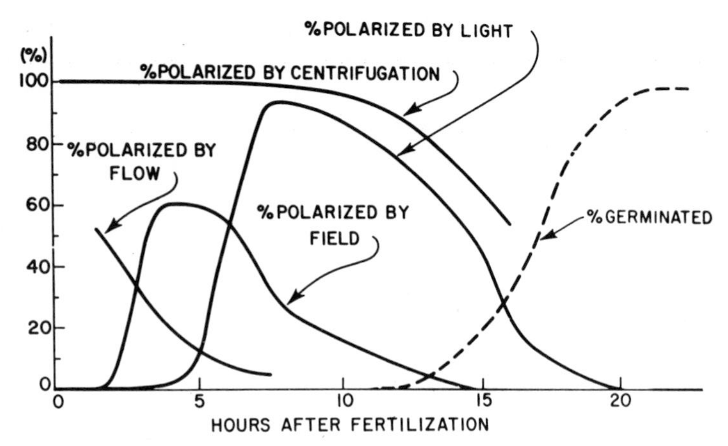

Fig. 6. Susceptibility of *Fucus* zygotes to various types of fields or gradients (see Table I) at different times after fertilization. [From Jaffe (1968).]

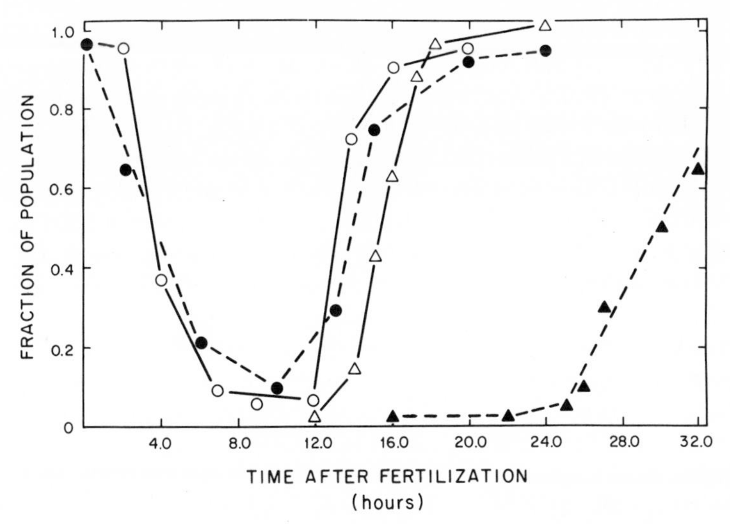

Fig. 7. Time course of sensitivity to polarity inducing light and the initiation of rhizoid outgrowth in *F. distichus*, with (---------) and without (———) cycloheximide treatment (1.0 μg/ml from 7 to 15 hr after fertilization). Each point represents the average of more than 200 zygotes that were exposed to a 2 hour pulse of unilateral light beginning 60 min before and ending 60 min after the designated time. Circles (0) represent fraction of population lacking photosensitivity, while triangles (△) represent fraction of the population possessing rhizoids. Notice how treatment with cycloheximide (-------) effectively separated the fixation of a stable polar axis (●) from the formation of the rhizoid (▲). Similar results were obtained with sucrose. [From Quatrano (1973).]

losus that zygotes grown in a seawater–sucrose medium and unilaterally illuminated, formed spherical, multicellular embryos. When the apolar embryos were removed from the hypertonic medium, rhizoids were formed only from the shaded hemisphere. Since unilateral light was applied from only one direction and continuously during this period, it is not known if a stable polar axis was present in the zygote and passed on to selected cells of the spherical embryo, or, if the polarity was irreversibly fixed in embryonic cells of the shaded hemisphere after immersion in normal seawater just before or concurrent with the emergence of the rhizoids. In either case, however, only a portion of the cell population would possess an internal fixed polar axis (hence, biochemical characteristics of the fixation processes may be masked), and neither event is separated from cell division.

The aim of a recent study in our lab was to separate the fixation of a stable polar axis from polar cell formation (rhizoid initiation) and cell division in an entire population of synchronously developing zygotes. In a previous study with *F. distichus* (Quatrano, 1972) it was demonstrated that: (1) zygotes are maximally photopolarizable between 7 and 11 hr after fertilization, (2) the site of rhizoid formation is irreversibly fixed or determined at 12–16 hr, and (3) the actual formation of the rhizoid occurs at 14–18 hr. Our approach was to treat zygotes with reversible inhibitors of rhizoid formation and determine what effect they had on the fixation of a stable polarity. The procedure was to administer unilateral light to dark-grown cultures from 7 to 9 hr (LT I). Half of this light oriented population of zygotes was treated with a reversible inhibitor of rhizoid formation; either cycloheximide (Quatrano, 1968b), sucrose (Torrey and Galun, 1970), or cytochalasin B (Quatrano, 1973; Nelson and Jaffe, 1973). When more than 25% of the zygotes in the other half of the population (untreated controls) formed rhizoids, the inhibitor was washed out of the treated cultures (which at this time had not formed rhizoids) and replaced with fresh seawater. A second light pulse (LT II), 90° from LT I, was then given to both treated and control cultures from 15 to 17 hr to determine if the polar axis was irreversibly fixed. At 48 hr the precentage of rhizoids oriented by LT I and LT II was determined. If the reversible inhibitor of rhizoid formation did not block the fixation of the first light induced polarity, all rhizoids should be oriented with respect to LT I and not LT II. If the second light treatment did influence the site of rhizoid formation, the inhibitor would appear to prevent events associated with the fixation of a stable polar axis as well. In the latter case, then, the two events would not be separated from each other in time.

Cycloheximide and sucrose delayed the appearance of a polar cell (rhizoid formation) without inhibiting the irreversible determination (fixation) of a polar axis (Fig. 7), whereas cytochalasin B was shown to delay the fixation of a light induced polar axis (Table II). Orientation of polar growth by unilateral light was blocked by cytochalasin B if it were present during the light pulse (Quatrano, 1973; Nelson and Jaffe, 1973). All these reversible inhibitors had no effect on cell division. Disruption of the mitotic apparatus and prevention of cell division by colchicine had no effect on rhizoid formation or on the photopolarization of the developmental axis.

Use of these various inhibitors clearly delimits three separate events in the differentiation of the first two cells of the *Fucus* embryo: (1) the fixation of a stable polar axis, (2) the accumulation of material in the rhizoid, and (3) cell division which partitions these specific localizations into the two cells. Whether the structural localizations in the rhizoidal hemisphere of the zygote (Quatrano, 1972) and the localized deposition of fucoidan in the cell wall (Moon and Forman, 1972)—both of which occur 2 hr before rhizoid visualization—represent the actual fixation of a stable axis and/or the initial steps of rhizoid formation (see Section III,B,2) may now be analyzed using these inhibitors to separate these events in an entire population of single cells. The biochemical basis of "axis fixation" can now be approached.

TABLE II
Effects of Different Inhibitors on Photopolarization in *Fucus*[a]

Treatment[b]	Time (hr) Inhibitor Added	Experiment	Oriented LT I[c]	Oriented LT II[c]	No. Cells per Embryo
None	—	A	90.5	9.5	4–6
		B	92.1	7.9	4–6
Cycloheximide	9–15	A	85.5	14.5	3–5
		B	88.0	12.0	3–5
Sucrose	9–15	A	88.9	11.1	3–5
		B	88.2	11.8	3–5
Colchicine	7–17	A	88.5	11.5	1
		B	89.8	10.2	1
Cytochalasin B	9–15	A	19.2	80.8	2–4
		B	25.0	75.0	2–4

[a] From Quatrano (1973).

[b] Techniques and methods used for synchronization and photopolarization were described elsewhere (Quatrano, 1972). All treatments were subject to unilateral light from 7–9 hr (LT I) and again 15–17 hr from a different direction (LT II). The number of rhizoids oriented by each light treatment was determined at 48 hr. Each experiment represents a population of at least 200 zygotes.

[c] Percent of the total population of zygotes that formed rhizoids in the shaded quadrant when exposed to unilateral LT I or LT II.

2. *Localization of Components for Differentiation*

In analyzing what becomes localized in the zygote which is essential for the differentiation of the rhizoid cell, we can first ask, what structurally and biochemically makes the rhizoid cell different from the thallus cell? The differences include:

i. *Cell shape and growth kinetics.* The rhizoid cell is a rapidly elongating cell which increases in length presumably by tip growth. It is actively synthesizing *new* cell wall components (Moon and Forman, 1972; Nakazawa *et al*., 1969), and Golgi derived vesicles can be seen depositing material into the new cell wall (Bouck and Quatrano, unpublished observations). The thallus cell does not change in cross-sectional diameter (axis of cell plate), nor does there appear to be any qualitative or quantitative alteration in its cell wall.

ii. *Staining properties and organelle composition.* When stained with toluidine blue (TBO), metachromatically colored (pink) material is localized only in the rhizoid cell, whereas the orthochromatic color (blue) is concentrated in the thallus cell (Quatrano, unpublished observation). The polyanion in the rhizoid cell mostly responsible for this differential stain is the sulfated polysaccharide fucoidan. The secretion of this compound from the rhizoid tip, detected cytochemically and autoradiographically, is apparently involved with adhesion of the embryo to the substratum (Fulcher and McCully, 1971). Nakazawa (1966) and Nakazawa and Takamura (1967) observed a localized accumulation of RNA (Unna stain with RNase controls) in this region of rhizoid elongation. Few if any chloroplasts and osmiophilic granules are found in the rhizoid (c.f., Jaffe, 1968), but are concentrated in the thallus cell. Golgi bodies and mitochondria, however, are localized predominantly in the rhizoid cell. At the time of rhizoid initiation, then, one can observe in the region of the cell wall protuberance characteristic organelles, sulfated polysaccharides and RNA.

Similarly, in the region of cap differentiation in *Acetabularia*, localization of organelles (Werz, 1965), acidic polysaccharides (Werz, 1960a) basic proteins (Werz, 1960b) and RNA (Werz, 1959; Olszewska *et al.*, 1961) have been reported. Also, the sugar composition of the cap cell wall is different from those sugars found in the stalk cell wall (Zetsch *et al.*, 1970). Localizations like these were also reported in the stalk tip of the red alga *Polysiphonia* by Werz (1961), as well as the accumulation of [^{32}P] and [^{35}S] into macromolecules in the rhizoid-forming region of the unicellular green alga *Cladophora* (Schoser, 1956). All of these studies with algae indicate the intracellular localization of proteins, sulfated polysaccharides and nucleic acids are related to a subsequent morphologenetic event at that site.

a. *Organelles.* We approached the problem of when these subcellular localizations were first visible in *Fucus* zygotes by taking advantage of the fact that unilateral white light can cause a specific area of the cytoplasm to become the subsequent site of the localization. By unilaterally illuminating zygotes from 8–10 hr, all of the rhizoids formed 6–8 hr later would be on the shaded side. If these zygotes are embedded in agar during the light treatment, one can orient these agar blocks containing the zygotes in plastic, so that the subsequent site of localization can be observed in sections viewed under the light and electron microscope (Quatrano, 1972). Now one could ask, is there any characteristic subcellular localizations of rhizoid-specific components in the region of subsequent rhizoid formation after a polar axis has been imposed? If so, when do they appear in relation to rhizoid formation?

Light-oriented (8–10 hr) zygotes of *F. vesiculosus* or *F. distichus* fixed for cytological observations between 8 and 12 hr after fertilization gave no indication of a localized accumulation of organelles or inclusions (Quatrano, 1972). From 12 hr on, one could demonstrate in zygotes that did not possess a rhizoid, an intracellular accumulation of osmiophilic bodies and some undefined extracellular material in the area of future rhizoid growth. However, if a true median plane section is viewed, the perinuclear region is highly polarized, with fingerlike projections radiating toward the site of rhizoid initiation. In such sections, the nucleolus was situated at the opposite side of the nucleus from the localization and often was in the process of vacuolation and disintegration. The perinuclear localization is heavily concentrated with mitochondria, ribosomes, osmiophilic bodies, densely fibrillar vesicles and what appears to be Golgi derived vesicles filled with finely fibrillar material. Only mitochondria and Golgi bodies are subsequently accumulated in the rhizoid cell. Numerous membranous extensions from the nuclear envelope were observed on the rhizoid side of the nucleus. In all cases these observations occurred only at a time when zygotes did not possess a protuberance or rhizoid, but had a stable, fixed polar axis. In addition, these organelle localizations occurred only in the shaded hemisphere of the imposed axis of light (Quatrano, 1972).

One can also observe the localization of subcellular particles just before and during rhizoid formation in *Fucus*, by incubating zygotes for several hours in a seawater solution containing the vital dyes Neutral Red (c.f., Mahlberg, 1972) or Rhodamine B (Quatrano, unpublished observations). Similar localizations were observed in the tip of an elongating root hair cell (Schumacher, 1936), in rhizoid tips of the fern *Polypodium* (Smith, 1972), in the apical regions of each cell in *Cladophora*, and at specific sites in vacuoles from a number of plants (Goldacre, 1952). Kühn (1955) has observed the accumulation of chloroplasts and "plasma" into specific cell regions of the marine alga *Bryopsis* during the establishment of polarity.

b. *Cell Wall Alterations.* Since the cell wall of plants is rigid and fulfills important structural roles (Albersheim, 1965), many important morphogenetic phenomena of plant development are associated with cell wall metabolism. Cleland (1971) has summarized the extensive evidence supporting two views that have emerged concerning the biochemical mechanism of cell wall elongation in higher plants: (1) alteration in amount or pattern of cell wall synthesis and deposition, and (2) action of polysaccharide hydrolases. The process of cell elongation might first involve the cleavage of a cross-link in the wall (polysaccharide or protein) followed by a small amount of turgor driven viscoelastic extension (Cleland, 1971). Since this viscoelastic extension is partly "reversible," this unit of elongation must be "finalized" by subsequent cleavage and repetition of viscoelastic extension, deposition (synthesis) of new cell wall, or some other biochemical modification of the wall.

Since these conclusions are based on an analysis of the elongation of complex tissues, evidence for the role of hydrolases in *cell* elongation in plants is indirect. However, much evidence is available from lower plants and specialized cells of higher plants that link hydrolase activity to localized cell wall alterations. Hormone induced cellulase activity has been found to correlate with branch formation in the water mold *Achlya* (Thomas and Mullins, 1967). Comparative studies with a wild type and morphological mutant of *Neurospora crassa* have shown important correlations between branching patterns and wall bound glucanases (Mahadevan and Mahadkar, 1970). Altered colony morphology (i.e., increased branching) of *Schizophyllum commune* induced by cellobiose has been related to elevated levels of cell wall hydrolyzing enzyme(s), as well as altered ratios of the major cell wall components as a result of growth on cellobiose (Niederpruem and Wessels, 1969). The formation of new buds in yeast, *Saccharomyces cerevisiae*, is also thought to be caused by enzymes acting on cell wall constituents. Recently, Matile *et al.* (1971) have isolated and characterized a type of vesicle, presumably derived from endoplasmic reticulum, which is found only in budding cells of yeast. The vesicles are secreted locally into the cell wall of growing buds, and contain exo- and endo-β-1, 3-glucanases. Similar vesicles functioning in pollen tube formation (Van Der Woude *et al.*, 1971) have also been described.

In the following two sections, rhizoid formation in *Fucus* and cap formation in *Acetabularia* will be discussed as two single-celled marine systems to investigate the role of cell wall enzymes in localized cellular morphogenesis.

i. *Rhizoid formation in Fucus*. Differentiation of a normal-shaped rhizoid cell in *Fucus* embryos, depends upon a localized extension and growth of the cell wall. There are several important characteristics of the *Fucus* system that make it amenable to study further the biochemistry of cell wall extension and growth, as well as the role of localized cell wall alteration in cellular morphogenesis:

1. The unfertilized egg does not possess a cell wall, but after fertilization, wall deposition occurs and increases at a linear rate until just before rhizoid formation. Not only can one follow the buildup of the chemical and physical properties of the wall, but various components can be tagged and followed through subsequent morphogenetic stages. Cell walls devoid of cytoplasm but possessing structural integrity can be isolated in large quantities (Moon and Forman, 1973; Quatrano, unpublished observations).

2. The cell wall extension is confined to a specific site in the wall which can be predicted with certainty 6–8 hr *before* extension occurs. Therefore, events associated directly with the mobilization of products and initiation of wall extension that occur at the site of wall growth, can be followed by cytochemical, ultrastructural, and autoradiographic techniques. This, then, will serve as essential correlative evidence with whole cell biochemical changes that can be monitored by standard enzymatic and associated techniques.

Preliminary evidence from our laboratory (Quatrano *et al.*, 1973) indicates that zygotes do possess β-glucosidase activity and suggests that it is related to the localized cell elongation. Approximately 80–90% of the β-glucosidase activity, as measured by release of *p*-nitrophenyl from the synthetic substrate *p*-nitrophenyl-β-*D*-glucopyranoside, is found in a particular fraction. This particle sediments with mitochondria in a sucrose gradient (Fig. 8), and exhibits properties similar to the glucanase particles found in budding yeast cells (Matile *et al.*, 1971), and elongating pollen tubes (Van Der Woude *et al.*, 1971). Whether the particles found in *Fucus* homogenates are identical to those vesicles which accumulate at the site of cell wall extension is not yet known. However, complete release of the enzyme from the particle can be achieved by treatment with the detergent Triton X-100, which suggests the enzyme is membrane bound.

A sixteenfold purification of the β-glucosidase has been obtained by ammonium sulfate fractionation (30–60% fraction) of the Triton X solubilized enzyme. This partially purified enzyme can release reducing sugars from β-1,3 and β-1,4 synthetic substrates as well as from cell wall "ghosts" and two polysaccharide fractions isolated from *Fucus*. The enzyme is competitively inhibited *in vitro* by the glucosidase inhibitor $\Delta^{1,5}$-gluconolactone (50% inhibition at 60 n*M*). At 4mM the inhibitor also prevents rhizoid formation in synchronized cultures. Future experiments are designed to determine if the vesicles and enzyme are localized in the region of cell wall extension using the fluorescent-antibody technique (Goldman, 1968), as well as to determine the genetic controls operative in localizing this structure and active enzyme (c.f., Fig. 2).

ii. *Cap formation in Acetabularia.* Cap differentiation in *Acetabularia* depends upon "morphogenetic substances" (RNA) released from the

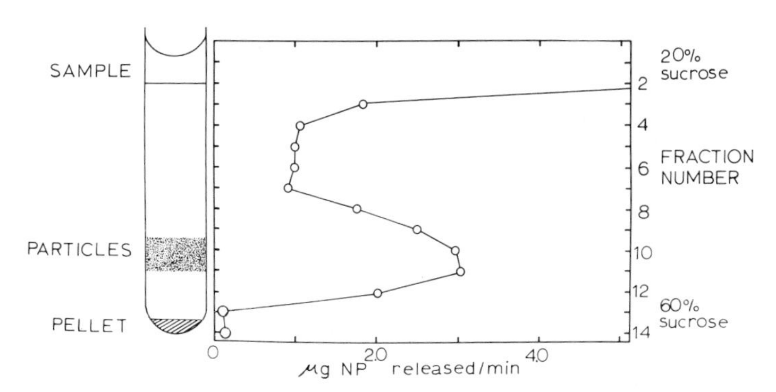

Fig. 8. Distribution of β-glucosidase activity [μg nitrophenyl (NP) released/min] in a sucrose gradient (20–60%) showing the association of enzyme activity with a white particulate layer which sediments at the same location as mitochondria. A 39,000 *g* pellet from a homogenate of 30-hr embryos, was resuspended in buffer (pH 6.8) containing EDTA (1mM) and sucrose (10%), layered on the gradient, and centrifuged at 40,000 *g* for 60 min.

nucleus, but does not require the presence of the nucleus for expression. Anucleated stalks can form the species-specific cap (Hämmerling, 1963) as well as regulate the levels of certain enzymes (Spencer and Harris, 1964). In view of this, long-lived or stable mRNA has been implicated in the genetic control of this localized differentiation (Brachet *et al.*, 1964). One feature of the new cap structure is that the cell wall contains rhamnose and large quantities of glucose and galactose compared to that present in the cell walls of stalks and rhizoids (Zetsch *et al.*, 1970). These investigators have focused attention on enzymes which may play a role in this localized alteration in the cell wall. One such enzyme, UDPG-pyrophosphorylase (UDPG-PYRO) participates in the synthesis of sugar nucleotides, intermediates in the incorporation of sugars into polysaccharides (Hassid, 1969; Loewus, 1971).

A close temporal correlation was observed by Zetsch *et al.* (1970) between a 3- to 4-fold increase in the enzyme UDPG-PYRO and the appearance of the cap. Puromycin and cycloheximide (protein synthesis inhibitors) blocked cap formation and the increase in the UDPG-PYRO in both anucleated and nucleated cells. However, the RNA synthesis inhibitor actinomycin D inhibited enzyme activity only in nucleated cells. RNA synthesis from the chloroplast genome appears not to be involved since rifampicin, at concentrations that block chloroplast RNA synthesis in anucleated cells, has no effect on UDPG-PYRO, cap formation nor nuclear RNA synthesis (Brändle and Zetsch, 1971). This evidence, although based entirely on inhibitor experiments, is consistent with the interpretation that the nuclear genes code for this enzyme and a stable nuclear template RNA is present in stalks which is activated for translation at the time of cap initiation.

To determine if a similar spatial correlation exists between the phosphorylase and site of cap initiation, Zetsch *et al.* (1970) found that the activity of this enzyme is *not* limited to the apical region of the stalk. However, a clear apico–basal distribution does exist. If normal cells are sectioned into three pieces containing identical protein concentrations, the apical region has roughly twice as much activity as the stalk and four times the activity of the rhizoid section. This apico–basal distribution of enzyme is maintained and amplified fourfold during cap formation. Differences in activity that may be caused by the polar distribution of enzyme inhibitors, activators, and localized enzyme destruction have been ruled out (Zetsch *et al.*, 1970). It appears then that the polar distribution of activity is most probably the result of different amounts of enzyme.

If the three sections obtained from capless cells are incubated separately for 31 days, only the apical section forms caps commencing at 16 days. Each of the other sections exhibits normal increases in length. However, the increase in UDPG-PYRO activity beginning at 16 days and doubling by 31 days, occurs only in apical segments. No significant increase is observed after 16 days in the stalk or rhizoid portions. This data coupled with the inhibitor studies discussed above led Zetsch to conclude that "the polar distribution of the enzyme is the result of a preferential synthesis of the enzyme in the apical region of the stalk. The rate of enzyme synthesis is higher within the apical piece than in the middle and basal segments, and this is particularly evident at the time of onset of cap formation" (Zetsch *et al.*, 1970).

Although a similar apico–basal distribution of RNA has been reported (Hammerling, 1963), this most probably represents total RNA and not template or specific mRNA's. In terms of the framework outlined earlier (Fig. 2, Section II,A), the data suggests that the localization of UDPG-PYRO is either due to more template specific RNA in the apical regions (Level I in Fig. 2), a greater rate of translation of equally distributed templates (Level II) or transport and accumulation of randomly synthesized enzyme (Level III). Local activation of a previously synthesized enzyme appears not to be the mechanism operative in the localized distribution of UDPG-PYRO activity.

c. *Sulfated Polysaccharides.* In analyzing polarity as a mechanism of differentiation, the main advantage of *Fucus* compared to other plant and animal embryonic systems is that the pattern of structural and biochemical localizations occurs in a previously homogeneous egg cytoplasm. Thus, in a synchronously developing unicellular system like *Fucus*, one can follow the cytological and biochemical development of a polar accumulation of specific macromolecules such as the sulfated polysaccharide of brown algae, fucoidan (Fig. 9). The possible role of sulfate and sulfated mucopolysaccharides, such as heparin, in polar development and gene expression has

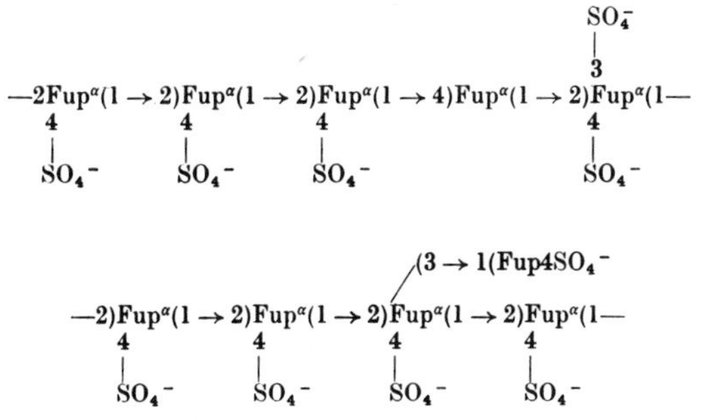

SO_4^- (on C-3 of the last residue)

—2Fup$^{\alpha}$(1 → 2)Fup$^{\alpha}$(1 → 2)Fup$^{\alpha}$(1 → 4)Fup$^{\alpha}$(1 → 2)Fup$^{\alpha}$(1—

(each residue: 4 — SO_4^-, except the fourth)

(3 → 1(Fup4SO_4^-

—2)Fup$^{\alpha}$(1 → 2)Fup$^{\alpha}$(1 → 2)Fup$^{\alpha}$(1 → 2)Fup$^{\alpha}$(1—

(each residue: 4 — SO_4^-)

Fup = L-fucopyranose

Fig. 9. Diagrammatic representation of fucoidan, the α-1,2 linked fucan sulfate, found in brown algae. Evidence from molecular fine structure studies indicate that most if not all of the ester-linked sulfate in on carbon-4. However, whether only α-1,2 linkages are present with some sulfate on carbon-3 (top figure), or if some other linkages are present which would give rise to branch points (bottom figure), is not known. From Percival and McDowell (1967).

been discussed with respect to sea urchin (Runnström *et al.*, 1964; Kinoshita, 1971) and *Acetabularia* (DeCarli and Brachet, 1968) development. The mechanism by which the sulfated polysaccharide fucoidan becomes localized in the rhizoid cell of the two-celled *Fucus* embryo is the main subject of this section.

There is no cytochemical localization of sulfated polysaccharides in eggs or young zygotes of *Fucus* (McCully, 1970). By using the stain TBO cytochemically and $Na_2[^{35}S]O_4$ autoradiographically, McCully (1969, 1970) first detected the appearance of a sulfated polymer at the time of rhizoid formation. The macromolecule is initially observed around the rhizoid-half of the nucleus radiating toward the site of rhizoid initiation, and eventually becomes localized in the region of the cell wall protuberance. This same pattern and timing has been described for certain organelles and inclusions (Quatrano, 1972). During rhizoid elongation and after the first cell division, most if not all of the sulfated polysaccharide is cytochemically detected only in the rhizoid cell (Fulcher and McCully, 1971; Quatrano, 1968a), including the extracellular region of the rhizoid tip where it is presumably involved with the adhesion of the two celled embryo to the substratum.

i. *Sulfation of fucoidan in* Fucus. In line with the framework cited earlier (Fig. 2; Section II,A), several possibilities exist that might account for the localization of the TBO-stained material (fucoidan) in the rhizoidal region. Since TBO is staining the negatively charged sulfate groups in fucoidan under the conditions employed, the appearance of localized stain may be the result of the unmasking of sulfate groups or the biochemical attachment of sulfate to a previously synthesized fucan, as well as the *de novo* synthesis of the entire molecule. Whether one of these possibilities occurs throughout the cell at random with the resulting product transported to the site of rhizoid formation, or whether a specific localized region is the site of *de novo* synthesis, sulfation, or unmasking of fucoidan is the critical question. These possibilities can be visualized as follows:

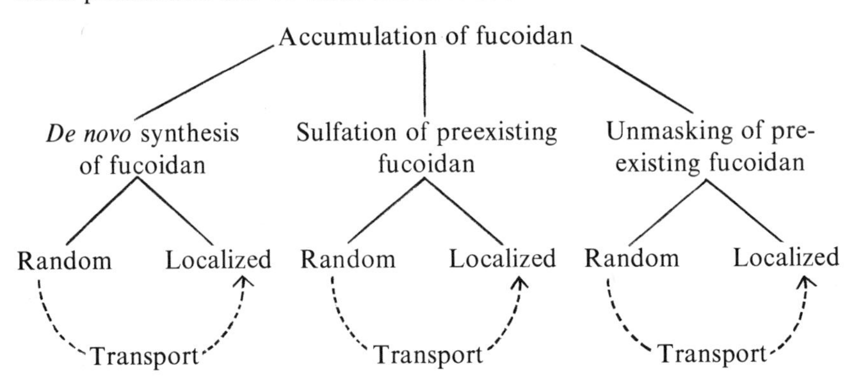

Sulfate incorporation into a polymer can be detected cytochemically and autoradiographically during rhizoid initiation, i.e., 14–18 hr after fertilization (McCully, 1969, 1970). To determine the exact time of sulfate accumulation, and if the acceptor was the sulfated polysaccharide fucoidan, $Na_2[^{35}S]O_4$ was added for 60-min periods at various times after fertilization. A fucoidan fraction was then assayed for radioactivity. It appeared that incorporation of sulfate into fucoidan did not occur until 10–11 hr after fertilization, and reached a maximum rate at 18–19 hr, before exhibiting a rate decrease (Fig. 10). When the radioactivity in fucoidan is expressed as a fraction of the sulfate incorporated into the soluble cytoplasm, the same labeling pattern is observed. Hence, the pattern of sulfate accumulation into fucoidan does not appear to be a result of changes in the sulfate pool or permeability of embryos to sulfate (Quatrano and Crayton, 1973). Sulfation is dependent upon protein synthesis as judged by the ability of

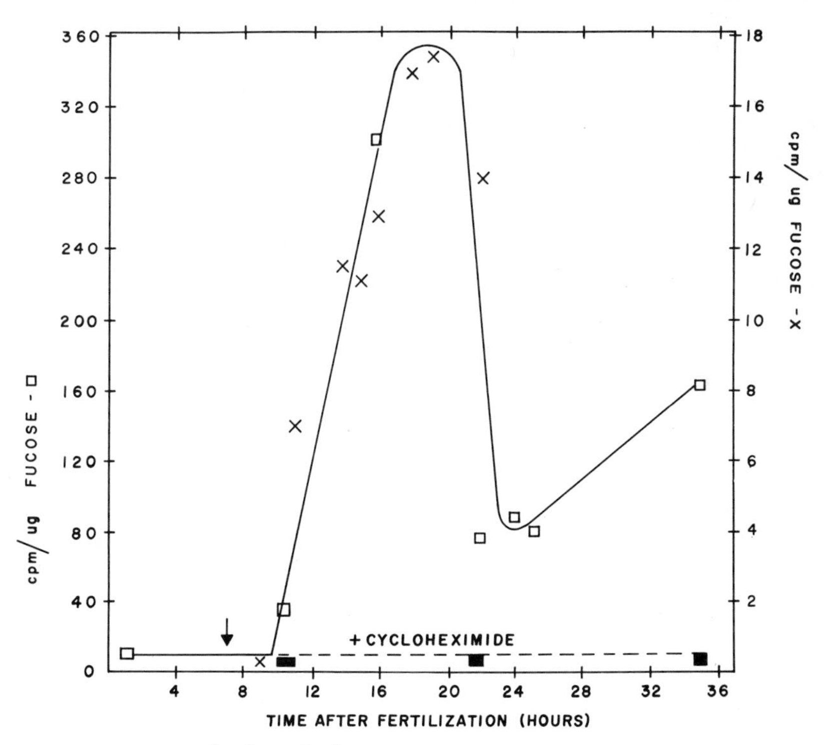

Fig. 10. Patterns of $[^{35}S]$ ($Na_2[^{35}S]O_4$ – 2 μCi/ml) accumulation into the fucoidan fraction of *F. distichus* when zygotes were pulsed for 60 min at different times after fertilization. The specific activity of fucoidan in the two experiments is expressed as cpm/μg fucose in the fucoidan fraction. [From Quatrano and Crayton (1973).]

cycloheximide to prevent sulfate accumulation into fucoidan (Fig. 10). This lack of sulfate incorporation is not due to inhibition of sulfate uptake or by an altered permeability due to the presence of cycloheximide, since the level of sulfate in the soluble pool is comparable to controls. Protein synthesis is inhibited by 94% in zygotes treated in this manner with cycloheximide, and no rhizoid formation or cell division is observed (Quatrano, 1968b). The timing of sulfate accumulation in fucoidan is consistent with the autoradiographic and cytochemical data of McCully (McCully, 1966, 1968, 1969, 1970; Fulcher and McCully, 1971), and directly demonstrates the incorporation of sulfate into fucoidan. Our data also suggest that sulfation is probably not the biochemical equivalent of polar axis "fixation," since cycloheximide prevented sulfation but not axis fixation (see Section III,B,1; Fig. 7).

Kelly (1955) has demonstrated that basic proteins such as protamine sulfate and histones can drastically suppress the metachromsia of TBO solutions of various mucopolysaccharides such as heparin and chrondroitin sulfate. Trypsinization of such complexes can restore the full metachromasia of the TBO-mucopolysaccharide. *In vitro* and *in vivo* treatments with proteases did not alter the staining properties of fucoidan in early or late zygote development (Quatrano and Crayton, 1973). The initiation of fucoidan sulfation then, is most likely controlled either at the level of the *de novo* synthesis of the entire molecule, or at sulfation of a previously synthesized fucan.

Bidwell and Ghosh (1963) demonstrated by incubating sterile frond tips of *F. vesiculosus* in $Na_2[^{35}S]O_4$ and $NaH[^{14}C]O_3$ that the rate of sulfate incorporation was considerably greater (about 20–600 times, depending on the duration of the experiment) than the low rate of CO_2 incorporation into fucoidan. They concluded that the sulfate component of fucoidan is rapidly exchanged without complete breakdown and resynthesis of the polysaccharide; or, at least without the concurrent synthesis of the carbon moieties of the fucose chain. Results of our study using 4 hr pulses of radioactive CO_2, fucose or mannose, also demonstrated a low rate of turnover in the fucose component of the ethanol insoluble fucoidan fraction at the time of fucoidan sulfation. Also, the amount of fucoidan/zygote remained constant during the first 24 hr after fertilization. These data and others (Quatrano and Crayton, 1973), strongly support the alternative that a fucan, sulfated at trace levels, is present in eggs and young zygotes, but is not metabolically active. Therefore, at 10 hr after fertilization, there does not appear to be *de novo* synthesis of the entire molecule nor the unmasking of the sulfate groups on fucoidan, but rather the sulfation of a previously existing fucan.

Concerning the intracellular site of fucoidan sulfation, it has been shown autoradiographically (McCully, 1969, 1970) and cytochemically (McCully, 1970; Quatrano, 1972), that the sulfation is initially associated with the accumulation of organelles in the rhizoidal hemisphere and ultimately localized entirely in the rhizoid cell (c.f., Quatrano, 1972). However, the length of time zygotes and embryos were exposed to $Na_2[^{35}S]O_4$ (McCully, 1969, 1970) was not short enough in duration to determine if the localization of sulfation was real or apparent; e.g., sulfation could have been at random in the cell and the sulfated product localized by some other mechanism. Preliminary results using short-pulses of $Na_2[^{35}S]O_4$ followed by different chase periods indicated a localization of $[^{35}S]$ in the wall of the emerging rhizoid. Consistent with this, is the finding by Moon and Forman (1972) that local regions of isolated cell walls of *Fucus* zygotes were stained metachromatically with TBO. Further experiments are needed to convincingly demonstrate the localized sulfation in cell walls. No information is now available concerning the *cytoplasmic* localization of the sulfation of free fucoidan or particles containing the sulfated polysaccharide. It is interesting to note that Godman and Lane (1964) and Lane *et al.* (1964) reported mucopolysaccharide sulfation associated with the vesicular component of the Golgi apparatus. In addition to the fact that Golgi bodies and mitochondria are the primary organelles in the rhizoid cell (Jaffe, 1968; Bouck and Quatrano, 1972), the vesicles which become localized around the nucleus at the time of rhizoid site determination (Quatrano, 1972) and ultimately sequestered in the rhizoid cell, appear to be Golgi-derived (Bouck and Quatrano, 1972). It would be interesting to see if in fact the site of fucoidan sulfation is the Golgi complex, and more specifically, if the Golgi and associated vesicles localized at the site of rhizoid formation are those reponsible for the increased sulfation at 12 hr after fertilization.

ii. *Mechanism of transport.* Lund (1921) demonstrated in animals that the *direction* in which differentiation occurs can be determined by an imposed electrical field of similar strength to the bioelectric field. He also demonstrated that if *Fucus* zygotes are grown in an electric field, rhizoids will all form on the positive side (Lund, 1923). Went (1932) proposed that polar electrical differences in plants might determine the direction of movement of charged molecules such as hormones. Rose (1970a,b) has also shown in his work with *Tubularia* regeneration that not only does an electrical polarity determine the polarity of morphogenesis by controlling the movement of charged repressors, but also that a bioelectric field of sufficient strength is *necessary* for regeneration to occur.

On the basis of an electric current he detected passing through zygotes of *Fucus*, Jaffe (1966) proposed that the mechanism involved in the redistribution and spatial orientation of polymers and cellular components at the time of rhizoid initiation, may be one involving a cellular electrophoresis (Jaffe, 1968, 1970). The sign of the voltage which develops indicates that a flow of positive ions (current) enters the rhizoid site and ultimately exits from the thallus end of the cytoplasm. If the entering cations are immobilized locally by an anionic gel in the cytoplasm, local binding would initiate a fixed charged gradient and thus a field. The intracellular current thus developed may serve to localize negatively charged components toward the growth point, i.e. the rhizoid (Jaffe, 1968, 1970). Not only might this localization of positively-charged ions initiate the current, but it may also serve to activate enzymes locally (c.f., Fig. 2; Section II,A) that would amplify the localization in terms of enhancing or inhibiting various biochemical events (c.f., Allen *et al.*, 1972 and Section III,A,2).

We have demonstrated that the sulfation of fucoidan occurring at 10 hr after fertilization results in the macromolecule having a sufficient negative charge to move towards a positive electric pole *in vitro* (Quatrano and Crayton, 1973). Cytochemical and autoradiographic data (see Section III,B,2) are consistent with this mechanism accounting for the observed *in vivo* localization of fucoidan in *F. distichus* (Fulcher and McCully, 1971; Moon and Forman, 1972; Quatrano, 1972). However, the biochemical sulfation of fucoidan as well as the observed structural localizations at the rhizoid site occur 2 hr before rhizoid initiation, the time at which the current is detected in *F. distichus* (Fig. 11). However, the current in *P. fastiagata* is detected 2 hr before rhizoid initiation, but no cytological or biochemical date is available in *Pelvetia*. In any case the timing is close enough to suggest that the negatively charged fucoidan may be electrophoretically accumulated at the rhizoid site. On the other hand, could a localized enzymatic sulfation of fucoidan serve as the anionic gel to immobilize the incoming cations at the rhizoid pole, thereby setting up the fixed charged gradient? In either instance, biochemical mechanisms under genetic control that regulate macromolecular or particulate charge, might be of general importance is elucidating the basis of cytoplasmic localizations.

IV. Conclusions and Summary

Developmental biologists have elucidated specific sites of genetic control for cellular differentiation from the initial transcription of a gene to its ultimate expression. Recent studies demonstrating the multiplicity of such control sites were summarized. These included controls at the level of the gene itself (e.g., selective gene transcription and selective gene replication), stability of template RNA and its transport to the site of translation, availability of components for template translation into protein, and, assembly, activation, and degradation of the polypeptide. During sea urchin embryogenesis, these levels of control involved with cellular differentiation were

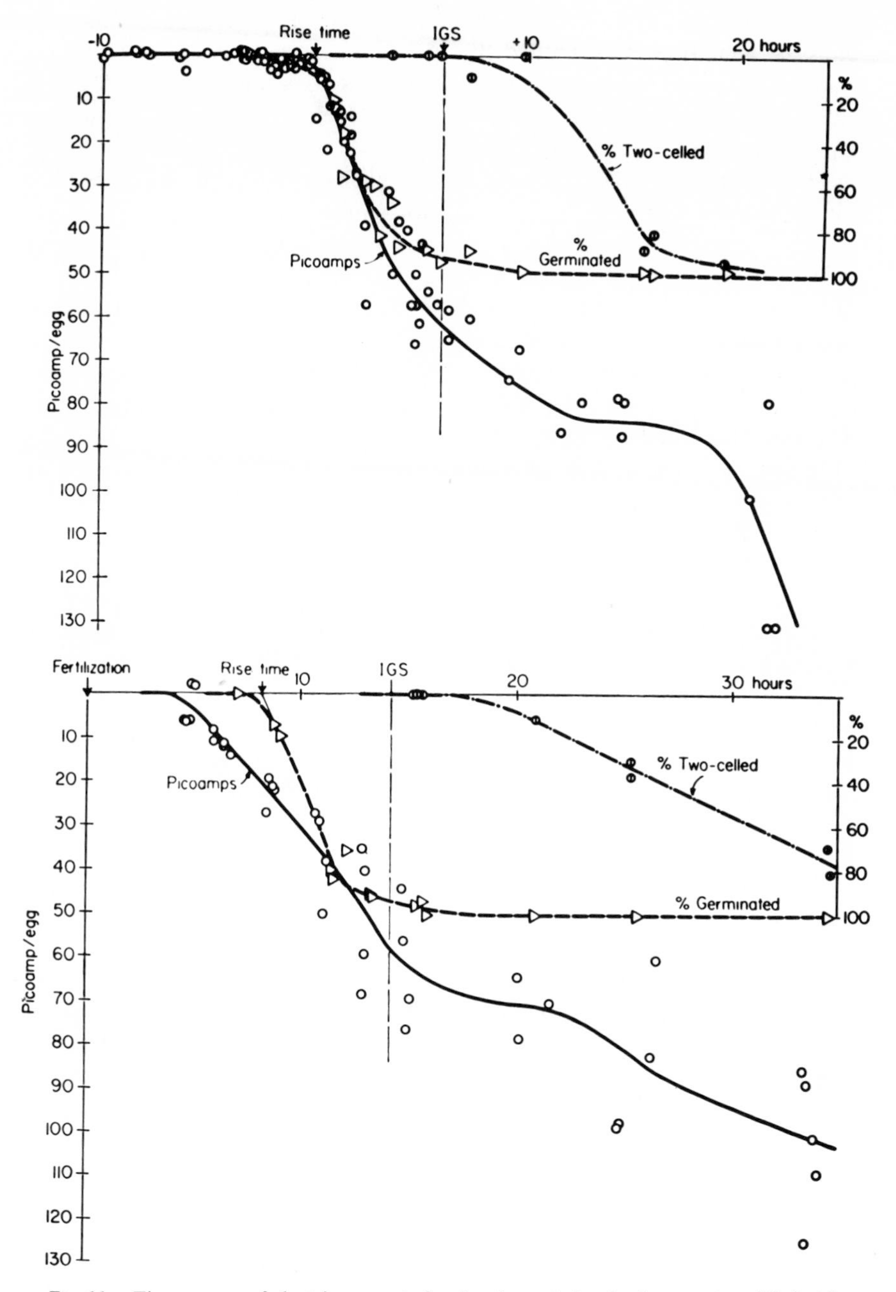

Fig. 11. Time course of electric currents flowing through developing zygotes of *F. distichus* (top) and *Pelvetia fastigiata* (bottom) when measured in series according to Jaffe (1966). [From Jaffe (1968).]

shown not only to be operative on nuclear gene products, but cytoplasmic ones as well. It was concluded that coordination between events of gene transcription and expression at two different locations within the cell is essential for proper implementation of a developmental program during embryogenesis.

The experimental basis of these control mechanisms involved with cellular differentiation has been built from biochemical and biophysical analyses of DNA and the products of transcription, as well as use of radioactive tracers to analyze patterns of RNA and protein synthesis. It was pointed out that, for the most part, these results were obtained from multicellular systems, and the appearance of unique cytological and biochemical traits within individual cells cannot be specifically associated with the macromolecular synthetic events measured in the whole multicellular system. To understand cellular differentiation, not only must more information be obtained from *single cells*, but the above mentioned controls must be integrated with less understood mechanisms, such as how macromolecules and subcellular structures are localized in a single cell.

Establishment of a polar or regionalized cell is an important step in the differentiation of many plant cell types, as well as normal development of both plant and animal embryos. Relatively little is known of the levels of genomic control which are operative in specifying, initiating and maintaining these intracellular localizations. A diagram was presented in this review which depicted possible sites of control involved with localization of specific products (Fig. 2). Rhizoid formation in zygotes of the brown alga *Fucus*, and cap formation in the unicellular green alga *Acetabularia* were shown to be model systems to explore how gene products and control mechanisms of gene expressions are linked to the establishment of a polar cell which leads to cellular differentiation. Their life cycles were briefly described, while a more detailed coverage was given to those portions of the life cycle used for experimental purposes. In addition, methods were given for the collection, storage and handling of these cellular systems.

Recent experiments have shown that cellular differentiation in the two-celled *Fucus* embryo can be separated into three distinct events which are separable in time from each other in synchronously developing cultures; (1) fixation or stabilization of a predetermined polar axis, (2) localization of organelles and the sulfated polysaccharide fucoidan in the region of cell wall expansion (rhizoid), and (3) cell division which partitions the localizations into one of the two cells thereby giving rise to the cellular differentiation. Various factors have been defined which are required for a polar zygote to form in *Fucus*, and numerous gradients have been determined to orient the direction of the polarity. Experiments with specific inhibitors indicate that at specific times after fertilization, RNA and protein synthesis are required for rhizoid formation. Technical difficulties have prevented determination of the specific types of RNA that are required to be synthesized, but purified RNA has now been isolated from zygotes (Quatrano and Melouk, unpublished observations), and analysis of this RNA and its distribution into the two-celled embryo can now be undertaken. Stabilization of the polar axis, however, is not dependent upon postfertilization protein synthesis or on the presence of microtubules in the form of the spindle apparatus. However, Cytochalasin B, a reversible inhibitor of rhizoid formation, can reversibly delay axis fixation as well, and interfere with the directionality of the polarity imposed by unilateral light. Use of such inhibitors as "handles" will now allow investigators to correlate biochemical events and possible control mechanisms with each of the major events involved in cellular differentiation. Isolation of a series of genetic mutants, devoid in the ability to undergo one or more of these events responsible for cellular differentiation, would be the preferred way to approach these problems. Several algal systems (e.g., *Volvox*) have the potential for such investigations, and a start has been made in a number of labs, including our own, to utilize these unique systems.

Concerning the mechanism of localization, recent work has centered around a description of what is chemically and structurally localized and subsequently sequestered into one of the two cells. Fucoidan, found only in the rhizoid cell of the two-celled *Fucus* embryo, is present in the egg and young zygote with little, if any, sulfate attached. Radioactive tracer experiments are consistent with the interpretation that fucoidan is metabolically inactive, and that 10 hr after fertilization a chemical sulfation of the previously synthesized fucan occurs. There does not appear to be the *de novo* synthesis of the entire macromolecule nor the unmasking of sulfate groups on the fucan. The alternatives now being investigated are, whether the sulfation which occurs is localized in the zygote at 10 hr, or if sulfation is at random and the sulfated polymer subsequently accumulated in the rhizoid by some other mechanism. Determination of the intracellular site of sulfation, possibly the Golgi apparatus, and the biochemical pathway involved with sulfation, possibly through the intermediates adenosine phosphosulfate and phosphoadenosine phosphosulfate, will allow one to determine the mechanism of sulfation, its control, and possible role in cell polarity. Similar types of experiments are being performed in the *Fucus* system to investigate the possible localization of enzymes involved with the local extension of the cell wall that occurs during rhizoid formation.

More progress has been made using the *Acetabularia* system in elucidating sites of genetic control for the localization of enzymes related to an intracellular morphogenetic event. Recent results were discussed which demonstrated a close temporal and spatial correlation between the activity and distribution of the enzyme UDPG-pyrophosphorylase, and the local appearance of the species-specific cap of *Acetabularia*. A localized activation of a previously synthesized enzyme does not appear to be the mechanism that can account for the distribution of enzyme activity in different regions of the cell. Rather, the data suggests that the localization of the UDPG-pyrophosporylase is either due to a larger amount of specific template RNA in the apical regions of the stalk, a greater rate of translation of equally distributed templates, or transport and accumulation of randomly synthesized enzyme.

Basic to an understanding of localization is the elucidation of mechanisms of directional transport of particles and macromolecules within the cells. Significant and elegant biophysical experimentation by Jaffe has demonstrated in the *Fucus* system that an intracellular electric current (driven by membrane permeability changes) may be the causal or amplification event in the subsequent distribution of charged compounds and particles into the two cells of the embryo. The intriguing possibility exists that the genetic control of localization in *Fucus* may involve an enzymatic mechanism(s) which alters molecular and particle charge (e.g., sulfation of the polysaccharide fucoidan). Distribution of such charged components into specific localizations within the cell would then be driven by a voltage gradient of a specific direction within the cell.

It is hoped that this review demonstrated how two marine plants have contributed to our understanding of cellular differentiation. Further investigations with *Fucus* and *Acetabularia* along the lines outlined in the review will continue to provide important information to all of developmental biology. The potential these and other marine algae offer to developmental biologists to approach important questions concerning cellular differentiation is great.

Acknowledgments

I would like to thank Dr. Ian Sussex for his critical reading of the manuscript, my wife for her help with the illustrations, and Mrs. Linda Fletcher for typing this chapter. Investigations from the author's laboratory reported in this review were supported by grants from the National Science Foundation (GB 14835) and the Public Health Service (GM 19247). The literature review was completed in January, 1973.

References

Albersheim, P. (1965). Biogenesis of the cell wall. *In* "Plant Biochemistry" (J. Bonner and J. E. Varner, eds.), pp. 298–321. Academic Press, New York.

Allen, D. A., Jacobsen, L., Joaquin, J., and Jaffe, L. F. (1972). Ionic concentrations in developing *Pelvetia* eggs. *Develop. Biol.* **27**, 538–545.

Aronson, A., and Wilt, F. (1969). Properties of nuclear RNA in sea urchin embryos. *Proc. Nat. Acad. Sci.* U.S. **62**, 186–193.

Baardseth, E. (1966). Localization and structure of alginate gels. *In Proc. Int. Seaweed Symp., 5th* p. 19. Pergamon, Oxford.

Balinsky, B. I. (1970). "An Introduction to Embryology," 3rd ed. Saunders, Philadelphia, Pennsylvania.

Baltimore, D. (1970). Viral RNA-dependent DNA polymerase. *Nature (London)* **226**, 1209–1211.

Berger, C. A., and Witkus, E. R. (1954). The cytology of *Xanthisma texanum* D.C. I. Differences in the chromosome number of root and shoot. *Bull. Torrey Bot. Club* **81**, 489–491.

Berns, A. J. M., van Kraaikamp, M., Bloemendal, H., and Lane, C. D. (1972). Calf crystallin synthesis in frog cells: The translation of lens-cell 14s RNA in oocytes. *Proc. Nat. Acad. Sci.* U.S. **69**, 1606–1609.

Bidwell, R. G. S., and Ghosh, N. R. (1963). Photosynthesis and metabolism in marine algae. VI. The uptake and incorporation of S^{35} sulphate in *Fucus vesiculosus. Can. J. Botan.* **41**, 209–220.

Bishop, J. O., Pemberton, R., and Baglioni, C. (1972). Reiteration frequency of haemoglobin genes in the duck. *Nature New Biol.* **235**, 231–234.

Bloch, R. (1943). Polarity in plants. *Botan. Rev.* **9**, 261–310.

Bonner, J., Dahmus, M. E., Fambrough, D., Huang, R. C., Marushige, K., and Tuan, D. Y. H. (1968). The biology of isolated chromatin. *Science* **159**, 47–56.

Bouck, G. B. and Quatrano, R. S. (1972). Cytochemical and fine structural analysis in *Fucus*. I. The egg and developing zygote (In preparation).

Brachet, J. L. A. (1965). *Acetabularia. Endeavor* **24**, 155–161.

Brachet, J. and Bonotto, S. (1970). "Biology of Acetabularia." Academic Press, New York.

Brachet, J., Denis, H., and Vitry, F. de. (1964). The effects of actinomycin D and puromycin on morphogenesis in amphibian eggs and *Acetabularia mediterranea. Develop. Biol.* **9**, 398–434.

Brändle, E., and Zetsch, K. (1971). Die Wirkung von Rifampicin auf die RNA-und Proteinsynthese sowie die Morphogenese und den Chlorophyllgehalt kernhaltigen und kernloser *Acetabularia*-Zellen. *Planta* **99**, 46–55.

Britten, R. J., and Davidson, E. H. (1969). Gene regulation for higher cells: A theory. *Science* **165**, 349–357.

Britten, R. J., and Kohne, D. E. (1968). Repeated sequences in DNA. *Science* **161**, 529–540.

Brown, D. D. and Blackler, A. W. (1972). Gene amplification proceeds by a chromosome copy mechanism. *J. Mol. Biol.* **63**, 75–83.

Brown, D. D., and Dawid, I. B. (1968). Specific gene amplification in oocytes. *Science* **160**, 272–280.

Brown, R. D., and Tocchini-Valentini, G. P. (1972). On the role of RNA in gene amplification. *Proc. Nat. Acad. Sci.* U.S. **69**, 1746–1748.

Bünning, E. (1953). "Entwicklungs-und Bewegungs physiologie Der Pflanze." Springer-Verlag, Berlin.

Burrows, E. M., and Lodge, S. M. (1953). Culture of *Fucus* hybrids. *Nature* (*London*) **172**, 1009–1010.

Cabib, E., and Farkas, V. (1971). The control of morphogenesis: An enzymatic mechanism for the initiation of septum formation in yeast. *Proc. Nat. Acad. Sci.* U.S. **68**, 2052–2056.

Carlson, P. S., Smith, H. H., and Dearing, R. D. (1972). Parasexual interspecific plant hybridization. *Proc. Nat. Acad. Sci.* U.S. **69**, 2292–2294.

Chamberlain, J. P., and Metz, C. B. (1972). Mitochondrial RNA synthesis in sea urchin embryos. *J. Mol. Biol.* **64**, 593–607.

Cleland, R. (1971). Cell wall extension. *Ann. Rev. Plant Physiol.* **22**, 197–222.
Corbett, W. M. (1963). Determination of the alpha-cellulose content of cotton and wood cellulose. *In* "Methods of Carbohydrate Chemistry" (R. Whistler, ed.), Vol. III, p. 27. Academic Press, New York.
Craig, S. P., and Piatigorsky, J. (1971). Protein synthesis and development in the absence of cytoplasmic RNA synthesis in nonnucleate egg fragments and embryos of sea urchins: Effect of ethidium bromide. *Develop. Biol.* **24**, 214–232.
Crippa, M., and Tocchini-Valentini, G. P. (1971). Synthesis of amplified DNA that codes for ribosomal RNA. *Proc. Nat. Acad. Sci.* U.S. **68**, 2769–2773.
Davidson, E. H. (1968). "Gene Activity in Early Development." Academic Press, New York.
Davidson, E. H., and Hough, B. R. (1971). Genetic information in oocyte RNA. *J. Mol. Biol.* **56**, 491–506.
Dawid, I. B. (1966). Evidence for the mitochondrial origin of frog egg cytoplasmic DNA. *Proc. Nat. Acad. Sci.* U.S. **56**, 269– 276.
De Carli, H., and Brachet, J. (1968). Morphologie animale: Action de l'héparine sur la Morphogénèse. *Bull. Cl. Sci. Acad. Roy. Belg.* **54**, 1158–1164.
Ellis, C. H. (1966). The genetic control of sea urchin development: A chromatographic study of protein synthesis in the *Arbacia punctulata* embryo. *J. Exp. Zool.* **163**, 1–22.
Emerson, C. P., Jr., and Humphreys, T. (1970). Regulation of DNA-like RNA and the apparent activation of ribosomal RNA synthesis in sea urchin embryos: Quantitative measurements of newly synthesized RNA, *Develop. Biol.* **23**, 86–112.
Epel, D. (1967). Protein synthesis in sea urchin eggs: A "late" response to fertilization. *Proc. Nat. Acad. Sci.* U.S. **57**, 899–906.
Evans, H. J., and Sorger, G. J . (1966). Role of mineral elements with emphasis on the univalent cations. *Ann. Rev. Plant. Physiol.* **17**, 47–76.
Farmer, J. B., and Williams, J. L. (1898). Contributions to our knowledge of the Fucaceae: Their life-history and cytology. *Phil. Trans. Roy. Soc. London B* **190**, 623–645.
Ficq, A., Brachet, J. (1971). RNA-dependent DNA polymerase: Possible role in the amplification of ribosomal RNA in Xenopus oocytes. *Proc. Nat. Acad. Sci.* U.S. **68**, 2774–2776.
Fulcher, R. G., and McCully, M. E. (1969). Laboratory culture of the intertidal brown alga *Fucus vesiculosus. Can. J. Bot.* **47**, 219–222.
Fulcher, R. G., and McCully, M. E. (1971). Histological studies on the genus *Fucus*. V. An autoradiographic and electron microscope study of the early stages of regeneration. *Can. J. Bot.* **49**, 161–165.
Gall, J. G. (1968). Differential synthesis of the genes for ribosomal RNA during amphibian oogenesis. *Proc. Nat. Acad. Sci.* U.S. **60**, 553–560.
Gardner, N. L. (1922). The genus *Fucus* on the Pacific coast of North America. *Univ. Calif. Publ. Bot.* **10**, 1–180.
Godman, G. C., and Lane, N. (1964). On the site of sulfation in the chondrocyte. *J. Cell Biol.* **21**, 353–366.
Goldacre, R. J. (1952). The folding and unfolding of protein molecules as a basis of osmotic work. *Int. Rev. Cytol.* **1**, 135–164.
Goldman, M. (1968). "Fluorescent Antibody Methods." Academic Press, New York.
Gross, P. R. (1968). Biochemistry of differentiation. *Ann. Rev. Biochem.* **37**, 631–660.
Gross, P. R., and Cousineau, G. (1963). Effects of actinomycin D on macromolecular synthesis and early development in sea urchin eggs. *Biochem. Biophys. Res. Commun.* **10**, 321–326.
Gurdon, J. B. (1962). Adult frogs derived from the nuclei of single somatic cells. *Develop. Biol.* **4**, 256–273.

Hämmerling, J. (1953). Nucleocytoplasmic relationships in the development of *Acetabularia*. *Int. Rev. Cytol.* **2**, 475–498.

Hämmerling, J. (1963). Nucleocytoplasmic interactions in *Acetabularia* and other cells. *Ann. Rev. Plant. Physiol.* **14**, 65–92.

Harris, H. (1970). "Nucleus and Cytoplasm," 2nd ed. Oxford Univ. Press, London and New York.

Hartmann, J. F., Ziegler, M. M., and Comb, D. G. (1971). Sea urchin embryogenesis: I. RNA synthesis by cytoplasmic and nuclear genes during development. *Develop. Biol.* **25**, 209–231.

Hassid, W. Z. (1969). Biosynthesis of oligosaccharides and polysaccharides in plants. *Science* **165**, 137–144.

Ilan, J., Ilan, J., and Patel, N. (1970). Mechanism of gene expression in *Tenebrio molitor*. *J. Biol. Chem.* **245**, 1275–1281.

Infante, A. A., and Nemer, M. (1967). Accumulation of newly synthesized RNA templates in a unique class of polyribosomes during embryogenesis. *Proc. Nat. Acad. Sci.* U.S. **58**, 681–688.

Jacob, F., and Monod, J. (1961). Genetic regulatory mechanisms in the synthesis of proteins. *J. Mol. Biol.* **3**, 318–356.

Jaffe, L. F. (1958a). Tropistic responses of zygotes of the Fucaceae to polorized light. Exp. Cell Res. **15**, 282–299.

Jaffe, L. F. (1958b). Morphogenesis in lower plants. *Ann. Rev. Plant. Physiol.* **9**, 359–384.

Jaffe, L. F. (1966). Electrical currents through the developing *Fucus* egg. *Proc. Nat. Acad. Sci.* U.S. **56**, 1102–1109.

Jaffe, L. F. (1968). Localization in the developing *Fucus* egg and the general role of localizing currents. *Advan. Morphogenesis* **7**, 295–328.

Jaffe, L. F. (1970). On the centripetal course of development, the *Fucus* egg, and self-electrophoresis. *Develop. Biol. Suppl.* **3**, 83–111.

Kacian, D. L., Spiegelman, S., Bank, A., Tarada, M., Metafora, S., Dow, L., and Marks, P. A. (1972). *In vitro* synthesis of DNA components of human genes for globins. *Nature New Biol.* **235**, 167–169.

Kedes, L. H., and Birnstiel, M. L. (1971). Reiteration and clustering of DNA sequences complementary to histone messenger RNA. *Nature New Biol.* **230**, 165–169.

Kedes, L. H., and Gross, P. R. (1969). Identification in cleaving embryos of three RNA species serving as templates for the synthesis of nuclear proteins. *Nature* (*London*) **223**, 1335–1339.

Kedes, L. H., Gross, P. R., Cognetti, G., and Hunter, A. L. (1969). Synthesis of nuclear and chromosomal proteins on light polyribosomes during cleavage in sea urchin embryo. *J. Mol. Biol.* **45**, 337–351.

Kelly, J. W. (1955). Suppression of metachromasy by basic proteins. *Arch. Biochem. Biophys.* **55**, 130–137.

Kinoshita, S. (1971). Heparin as a possible initiator of genomic RNA synthesis in early development of sea urchin embryos. *Exp. Cell. Res.* **64**, 403–411.

Knapp, E. (1931). Entwicklungsphysiologische Untersuchungen an Fucaceen-Eiern. *Planta* **14**, 731–751.

Kniep, H. (1907). Beiträge zur Keimungs-Physiologie und Biologie von *Fucus*. *Jahrb. Wiss. Bot.* **44**, 635–724.

Kühn, A. (1955). "Vorlesungen über Entwicklungsphysiologie." Springer-Verlag, Berlin.

Lane, C. D., Marbaix, G., and Gurdon, J. B. (1971). Rabbit haemoglobin synthesis in frog cells; The translation of reticulocyte 9S RNA in frog oocytes. *J. Mol. Biol.* **61**, 73–91.

Lane, N., Caro, L., Oterovilardebó, L. R., and Godman, G. C. (1964). On the site of sulfation in colonic goblet cells. *J. Cell Biol.* **21**, 339–351.

Loewus, F. (1971). Carbohydrate interconversions. *Ann Rev. Plant. Physiol.* **22**, 337–364.

Lund, E. J. (1921). Experimental control of organic polarity by the electric current. I. Effects of electric current on regenerating internodes of *Obelia commisuralis*. *J. Exp. Zool.* **34**, 471–493.

Lund, E. J. (1923). Electrical control of organic polarity in the egg of *Fucus*. *Bot. Gaz.* **76**, 288–301.

Lundblad, G. (1954). Proteolytic activity in sea urchin gametes. 4. Further investigations of the proteolytic enzymes of the egg. *Ark. Kemi* **7**, 127–157.

Mahadeŭan, P. R., and Mahadkar, U. R. (1970). Role of enzymes in growth and morphology of *Neurospora crassa*: Cell-wall-bound enzymes and their possible role in branching. *J. Bacteriol.* **101**, 941–947.

Mahdavi, Y., and Crippa, M. (1972). An RNA-DNA complex intermediate in ribosomal gene amplification. *Proc. Nat. Acad. Sci.* U.S. **69**, 1749–1752.

Mahlberg, P. (1972). Localization of neutral red in lysosome structures in hair cells of *Tradescantia virginiana Can. J. Bot.* **50**, 857–859.

Marbaix, G., and Lane, C. D. (1972). Rabbit haemoglobin synthesis in frog cells. II. Further characterization of the products of translation of reticulocyte 9S RNA. *J. Mol. Biol.* **67**, 517–524.

Matile, P., Cortat, M., Wiemken, A., and Frey-Wyssling, A. (1971). Isolation of glucanase-containing particles from budding *Saccharomyces cerevisiae*. *Proc. Nat. Acad. Sci.* **68**, 636–640.

McCully, M. E. (1966). Histological studies on the genus *Fucus*. I. Light microscopy of the mature vegetative plant. *Protoplasma* **62**, 287-305.

McCully, M. E. (1968). Histological studies on the Genus *Fucus*. II. Histology of the reproductive tissues. *Protoplasma* **66**, 205–230.

McCully, M. E. (1969). The synthesis and secretion of polysaccharides during early development of the *Fucus* embryo. *Int. Bot. Congr. 11th, Seattle.*

McCully, M. E. (1970). The histological localization of the structural polysaccharides of seaweeds. *Ann. N.Y. Acad. Sci.* **175**, 702–711.

McLachlan, J., and Chen, L. C-M. (1972). Formation on adventive embryos from rhizoidal filaments in sporelings of four species of *Fucus* (Phaeophyceae). *Can. J. Bot.* **50**, 1841–1844.

McLachlan, J., Chen, L. C-M., and Edelstein, T. (1971). The culture of four species of *Fucus* under laboratory conditions. *Can. J. Bot.* **49**, 1463–1469.

Meeker, G. L. (1970). Intracellular potassium requirement for protein synthesis and mitotic apparatus formation in sea urchin eggs. *Exp. Cell. Res.* **63**, 165–170.

Metafora, S., Felicetti, L., and Gambino, R. (1971). The mechanism of protein synthesis activation after fertilization of sea urchin eggs. *Proc. Nat. Acad. Sci.* **68**, 600–604.

Mirbel, C. F. (1835). Remarques sur la nature et l'origine des couches corticales et du liber des arbres dicotylédons. *Ann. Sci. Nat. III (Bot.)* 143–148.

Moav, B., and Nemer, M. (1971). Histone synthesis. Assignment to a special class of polyribosomes in sea urchin embryos. Biochemistry **10**, 881–888.

Mogensen, H. L. (1972). Fine structure and composition of the egg apparatus before and after fertilization in *Quercus gambelii*: The functional ovule. *Amer. J. Bot.* **59**, 391–941.

Moon, A., and Forman, M. (1972). Cell wall synthesis and morphogenesis of *Fucus furcatus* embryos. *Amer. J. Bot.* **59**, 654.

Moon, A., and Forman, M. (1973). A procedure for the isolation of cell walls of *Fucus furcatus* embryos. *Protoplasma* **75**, 461–464.

Müller, D. (1962). Über jahres-und lunarperiodische Erscheinuhgen bei einigen Braunalgen. *Bot. Mar.* **4**, 140–155.
Müller-Stoll, W. R. (1952). Über regeneration und polarität bei *Enteromorpha. Flora* **139**, 148–180.
Murphy, T. M., Kahn, A., and Lang, A. (1970). Stabilization of the polarity axis in the zygotes of some Fucaceae. *Planta* **90**, 97–108.
Nakazawa, S. (1956). Development mechanics of Fucaceous algae. I. The preexistent polarity in *Coccophora* eggs. *Sci. Rep. Tohoku Univ., 4th Ser. Biol.* **22**, 175–179.
Nakazawa, S. (1960). Nature of the protoplasmic polarity. *Protoplasma* **52**, 274–294.
Nakazawa, S. (1962). Polarity. *In* "Physiology and Biochemistry of Algae" (R. A. Lewin, ed), pp. 653–661. Academic Press, New York.
Nakazawa, S. (1966). Regional concentration of cytoplasmic RNA in *Fucus* eggs in relation to polarity. *Naturwissenschaften* **53**, 138–139.
Nakazawa, S., and Takamura, K. (1967). An analysis of rhizoid differentiation in *Fucus* eggs. *Cytologia* **32**, 408–415.
Nakazawa, S., Takamura, K., and Abe, M. (1969). Rhizoid differentiation in *Fucus* eggs labeled with calcofluor white and birefringence of cell wall. *Bot. Mag.* (*Tokyo*) **82**, 41–44.
Nelson, D. R., and Jaffe, L. F. (1973). Cells without cytoplasmic movements respond to cytochalasin. *Develop. Biol.* **30**, 206–208.
Nemer, M., and Lindsay, D. T. (1969). Evidence that the S-polysomes of early sea urchin embryos may be responsible for the synthesis of chromosomal histones. *Biochem. Biophys. Res. Commun.* **35**, 156–160.
Niederpruem, D. J., and Wessels, J. G. H. (1969). Cytodifferentiation and morphogenesis in *Schizophyllum commune. J. Bacteriol.* **94**, 1594–1602.
Nitsch, J. P., and Nitsch, C. (1969). Haploid plants from pollen grains. *Science* **163**, 85–87.
Nizamuddin, M. (1970). Phytogeography of the Fucales and their seasonal growth. *Bot. Mar.* **13**, 131–139.
Ohyama, K., and Nitsch, J. P. (1972). Flowering haploid plants obtained from protoplasts of tobacco leaves. *Plant. Cell Physiol.* **13**, 229–236.
Olszweska,, M., Vitry, F. de, and Brachet, J. (1961). Incorporation de la DL-methionine-^{35}S dan les fragments nucléés et anucléés d'*Acetabularia mediterranean. Exp. Cell Res.* **22**, 370–380.
Overton, J. B. (1913). Artificial parthenogenesis in *Fucus. Science* **37**, 841–844.
Percival, E., and McDowell, R. H. (1967). "Chemistry and Enzymology of Marine Algal Polysaccharides." Academic Press, New York.
Peterson, D. M., and Torrey, J. G. (1968). Amino acid incorporation in developing *Fucus* embryos. *Plant. Physiol.* **43**, 941–947.
Piko, L. D., Blair, G., Tyler, A., and Vinograd, J. (1968). Cytoplasmic DNA in the unfertilized sea urchin egg; Physical properties of circular mitochondrial DNA and the occurrence of catenated forms. *Proc. Nat. Acid. Sci.* U.S. **59**, 838–845.
Pollack, E. G. (1970). Fertilization in *Fucus*. Planta **92**, 85–99.
Powell, H. T. (1957). Studies in the genus *Fucus* L. I. *Fucus distichus* L. emend. Powell. *Mar. Biol. Ass. U.K. J.* **36**, 407–432.
Puiseux-Dao, S. (1970). "Acetabularia and Cell Biology." Springer-Verlag, New York.
Quatrano, R. S. (1968a). Biochemical and fine structural studies on rhizoid formation in *Fucus* zygotes. *Ph.D.* dissertation, Yale Univ. New Haven, Connecticut.
Quatrano, R. S. (1968b). Rhizoid formation in *Fucus* zygotes: Dependence on protein and ribonucleic and syntheses. *Science* **162**, 468–470.

Quatrano, R. S. (1972). An ultrastructural study of the determined site of rhizoid formation in *Fucus* zygotes. *Exp. Cell Res.* **70**, 1–12.

Quatrano, R. S. (1973). Separation of processes associated with differentiation of two-celled *Fucus* embryos. *Develop. Biol.* **30**, 209–213.

Quatrano, R. S., and Crayton, M. A. (1973). Sulfation of fucoidan in *Fucus* embryos. I. Possible role in localization. *Develop. Biol.* **30**, 29–41.

Quatrano, R. S., Stevens, P. A., and Melouk, H. (1973). B-glucosidase in *Fucus* embryos and its relationship to rhizoid formation (In preparation).

Reed, E. A., and Whitaker, D. M. (1941). Polarized plasmolysis of *Fucus* eggs with particular reference to ultraviolet light. *J. Cell. Comp. Physiol.* **18**, 329–338.

Robbins, W. W., Weier, T. E., and Stocking, C. R. (1964). "Botany," 3rd ed. Wiley, New York.

Rose, S. M. (1970a). Differentiation during regeneration caused by migration of repressors in bioelectric fields. *Amer. Zool.* **10**, 91–99.

Rose, S. M. (1970b). Regeneration: Key to Understanding Normal and Abnormal Growth and Development. Meredith Corp., New York.

Ross, J., Aviv, H., Scolnick, E., and Leder, P. (1972). *In vitro* synthesis of DNA complementary to purified rabbit globin m–RNA. *Proc. Nat. Acad. Sci.* U.S. **69**, 264–268.

Runnström, J., Horstadius, S. S., Immers, J., and Fudge-Mastrangelo, M. (1964). An analysis of the role of sulfate in the embryonic differentiation of the sea urchin. *Rev. Swiss. Zool.* **71**, 21–54.

Sandan, T. (1955). Physiological studies on growth and morphogenesis of the isolated plant cell cultivated *in vitro*. I. General feature on the growth and morphogenesis of the internodal cell of Characeae. *Bot. Mag.* (*Tokyo*) **68**, 274–280.

Schimke, R. T., and Doyle, D. (1970). Control of enzyme levels in animal tissue. *Ann. Rev. Biochem.* **39**, 929–976.

Schoser, G. (1956). Uber die Regeneration bei den Cladophoraceen, *Protoplasma* **47**, 103–134.

Schulz, S. R., and Jensen, W. A. (1968). *Capsella* embryogenesis: the egg, zygote, and young embryo. *Amer. J. Bot.* **55**, 807–819.

Schumacher, W. (1936). Untersuchungen Über die Wanderung des Fluoresceins in den Haaren von *Cucurbita Pepo*. *Jahr. Wiss. Bot.* **82**, 507–533.

Scolnick, E. M., Aaronson, S., Todaro, G., and Parks, W. (1971). RNA dependent DNA polymerase activity in mammalian cells. *Nature* (*London*) **229**, 318–321.

Silva, P. C. (1953). The identity of certain Fuci of Esper. *Wasmann J. Biol.* **11** (2), 221–232.

Sinnott, E. W. (1960). "Plant Morphogenesis." McGraw-Hill, New York.

Smith, D. L. (1972). Staining and osmotic properties of young gametophytes of *Polypodium vulgare* L. and their bearing on rhizoid function. *Protoplasma* **74**, 465–479.

Smith, G. M. (1969). "Marine Algae of the Monterey Peninsula, California," 2nd ed. Stanford Univ. Press, Stanford, California.

Smith, L. D., and Ecker, R. E. (1970). Regulatory processes in the maturation and early cleavage of amphibian eggs. *Develop. Biol.* **5**, 1–38.

Spencer, T., and Harris, H. (1964). Regulation of enzyme synthesis in an enucleate cell. *Biochem. J.* **91**, 282–286.

Spirin, A. S. (1966). On "masked" forms of messenger RNA in early embryogenesis and in other differentiating systems. *Current Topics Develop. Biol.* **1**, 1–38.

Starr, R. C. (1970). Control of differentation in *Volvox*. *Develop Biol. Suppl.* **4**, 59–100.

Stebbins, G. L. (1967). Gene action, mitotic frequency and morphogenesis in higher plants. *Develop. Biol. Suppl.* **1**, 113–135.

Steward, F. C., Mapes, M. O., Kent, A. E., and Holsten, R. D. (1964). Growth and development of cultured plant cells. *Science* **143**, 20–27.

Suelter, C. H. (1969). Enzymes activated by monovalent cations. *Science* **168**, 789–794.

Sussex, I. M. (1967). Polar growth of *Hormosira banksii* zygotes in shake culture. *Amer. J. Bot.* **54**, 535–510.

Suzuki, Y., Gage, L. P., and Brown, D. D. (1972). The genes for silk fibroin in *Bombyx Mori. J. Mol. Biol.* **70**, 637–649.

Terman, S. H., and Gross, P. R. (1965). Translational-level control of protein synthesis during early development. *Biochem. Biophys. Res. Commun.* **21**, 595–600.

Thomas, D. S., and Mullins, J. T. (1967). Role of enzymatic wall-softening in plant morphogenesis: Hormonal induction in *Achlya. Science* **156**, 84–85.

Timourian, H., and Watchmaker, G. (1970). Protein synthesis in sea urchin eggs. II. Changes in amino acid uptake and incorporation at fertilization. *Develop. Biol.* **23**, 478–491.

Tobler, H., Smith, K. D., and Ursprung, H. (1972). Molecular aspects of chromatin elimination in *Ascaris lumbricoides. Develop. Biol.* **27**, 190–203.

Torrey, J. G., and Galun, E. (1970). Apolar embryos of *Fucus* resulting from osmotic and chemical treatment. *Amer. J. Bot.* **57**, 111–119.

Ursprung, H. (1967). Developmental genetics. *Ann. Rev. Genet.* **1**, 139–162.

Van Der Woude, W. J., Morre, D. J ., and Bracker, C. E. (1971). Isolation and characterization of secretory vesicles in germinated pollen of *Lilium longiflorum. J. Cell Sci.* **8**, 331–351.

Vasil, V., and Hildebrandt, A. C. (1965). Differentiation of tobacco plants from single, isolated cells in microcultures. *Science* **180**, 889–892.

Verma, I. M., Temple, G. F., Fan, H., and Baltimore, D. (1972). *In vitro* synthesis of DNA complementary to rabbit reticulocyte 10S RNA. *Nature New Biol.* **235**, 163–167.

Waaland, S. D., and Cleland, R. (1972). Development in the red alga, *Griffithsia pacifica*: control by internal and external factors. *Planta* **105**, 196–204.

Waddington, C. H. (1966). "Principles of Development and Differentiation." Macmillan, New York.

Went, F. W. (1932). Eine Botanische Polaritatstheorie. *Jahr. Wiss. Bot.* **76**, 528–554.

Werz, G. (1959). Über Polare, Plasmaunterschiede bei *Acetabularia. Planta* **53**, 502–521.

Werz, G. (1960a). Über Strukturierungen der Wuchszone von *Acetabularia mediterranea. Planta* **55**, 38–56.

Werz, G. (1960b). Anreicherung von Ribonucleinsaüre in der Wuchszone von *Acetabularia mediterranea. Planta* **55**, 22–37.

Werz, G. (1961). Über Anreicherungen von Ribonukleinsäure, speziellem Protein und anionischem Polysaccharid in jungen Zellen von *Polysiphonia* spec. *Naturwissenschaften* **48**, 221–222.

Werz, G. (1965). Determination and realization of morphogenesis in *Acetabularia. Brookhaven Symp. Biol.* **18**, 185–203.

Whitaker, D. M. (1931). On the rate of oxygen consumption by fertilized and unfertilized eggs. I. *Fucus vesiculosus. J. Gen. Physiol.* **15**, 167–182.

Whitaker, D. M. (1940). Physical factors of growth. Growth Suppl. 75–90.

Wilson, E. B. (1902). "The Cell in Development and Inheritance," 2nd ed. Macmillan, New York.

Wilson, E. B. (1925). "The Cell in Development and Inheritance," 3rd ed. Macmillan, New York.

Wilt, F. H. (1970). The acceleration of ribonucleic acid synthesis in cleaving sea urchin embryos. *Develop. Biol.* **23**, 444–455.

Wu, J-R., Hurn, J., and Bonner, J. (1972). Size and distribution of the repetitive segments of the *Drosophila genome. J. Mol. Biol.* **64**, 211–219.

Yamanouchi, S. (1909). Mitosis in *Fucus. Botan. Gaz.* **47**, 173–197.

Zellweger, A., Ryser, U., and Braun, R. (1972). Ribosomal genes of *Physarum*: Their isolation and replication in the mitotic cycle. *J. Mol. Biol.* **64**, 681–691.

Zetsche, K., Grieninger, G. E., and Anders, J. (1970). Regulation of enzyme activity during morphogenesis of nucleate and anucleate cells of *Acetabularia. In* "Biology of Acetabularia" (J. Brachet and S. Bonotto, eds.), pp. 87–110. Academic Press, New York.

Author Index

Numbers in italics refer to the pages on which the complete references are listed.

D

E

H

I

J

Q

R

T

Subject Index

D

T

U

V

W

X

Y

Z